Energy Services Fundamentals and Financing

Energy Services Fundamentals and Financing

Edited by

David Borge-Diez
Department of Electric, System and Automatic Engineering, University of León, León, Spain

Enrique Rosales-Asensio
Department of Electrical Engineering, University of Las Palmas de Gran Canaria, Campus de Tafira s/n, 35017 Las Palmas de Gran Canaria, Canary Islands, Spain

Academic Press is an imprint of Elsevier
125 London Wall, London EC2Y 5AS, United Kingdom
525 B Street, Suite 1650, San Diego, CA 92101, United States
50 Hampshire Street, 5th Floor, Cambridge, MA 02139, United States
The Boulevard, Langford Lane, Kidlington, Oxford OX5 1GB, United Kingdom

Copyright © 2021 Elsevier Inc. All rights reserved.

No part of this publication may be reproduced or transmitted in any form or by any means, electronic or mechanical, including photocopying, recording, or any information storage and retrieval system, without permission in writing from the publisher. Details on how to seek permission, further information about the Publisher's permissions policies and our arrangements with organizations such as the Copyright Clearance Center and the Copyright Licensing Agency, can be found at our website: www.elsevier.com/permissions.

This book and the individual contributions contained in it are protected under copyright by the Publisher (other than as may be noted herein).

Notices
Knowledge and best practice in this field are constantly changing. As new research and experience broaden our understanding, changes in research methods, professional practices, or medical treatment may become necessary.

Practitioners and researchers must always rely on their own experience and knowledge in evaluating and using any information, methods, compounds, or experiments described herein. In using such information or methods they should be mindful of their own safety and the safety of others, including parties for whom they have a professional responsibility.

To the fullest extent of the law, neither the Publisher nor the authors, contributors, or editors, assume any liability for any injury and/or damage to persons or property as a matter of products liability, negligence or otherwise, or from any use or operation of any methods, products, instructions, or ideas contained in the material herein.

British Library Cataloguing-in-Publication Data
A catalogue record for this book is available from the British Library

Library of Congress Cataloging-in-Publication Data
A catalog record for this book is available from the Library of Congress

ISBN: 978-0-12-820592-1

For Information on all Academic Press publications
visit our website at https://www.elsevier.com/books-and-journals

Publisher: Brian Romer
Acquisitions Editor: Peter Adamson
Editorial Project Manager: Andrea Akeh
Production Project Manager: Sojan P. Pazhayattil
Cover Designer: Christian J. Bilbow

Typeset by MPS Limited, Chennai, India

Contents

List of contributors — xi

Part 1 Energy services

1. **Energy services: concepts, applications and historical background** — 3
   Abdeen Mustafa Omer
   1.1 Introduction — 3
   1.2 Energy and population growth — 5
   1.3 Energy saving in buildings — 7
   1.4 Energy use in agriculture — 9
   1.5 Renewable energy technologies — 12
   1.6 Energy and sustainable development — 25
   1.7 Global warming — 28
   1.8 Recommendations — 30
   1.9 Conclusion — 31
   References — 31

Part 2 Energy financing schemas

2. **The promotion of renewable energy communities in the European Union** — 37
   Theodoros G. Iliopoulos
   2.1 Overview — 37
   2.2 The link between the provision of energy services and the increase of energy efficiency — 38
   2.3 The efficiency gains stemming from distributed generation of energy production — 39
   2.4 The concept of renewable energy community — 41
   2.5 The promotion of renewable energy communities in EU law — 44
   2.6 The promotion of renewable energy communities in the draft National Energy and Climate Plans — 47
   2.7 Conclusion — 50
   References — 51

3. **Financial schemes for energy efficiency projects: lessons learnt from in-country demonstrations** — 55
Charikleia Karakosta, Aikaterini Papapostolou, George Vasileiou and John Psarras
  3.1 Introduction — 55
  3.2 The proposed methodology — 58
  3.3 Innovative financing schemes — 59
  3.4 Case study countries — 62
  3.5 Key actors identification — 67
  3.6 Knowledge transfer — 68
  3.7 Conclusions — 72
  References — 73

Part 3 Energy systems in buildings

4. **Energy in buildings and districts** — 81
Jacopo Vivian
  4.1 Introduction — 81
  4.2 Thermal comfort — 84
  4.3 User behavior — 88
  4.4 Weather conditions under climate change and growing urbanization — 91
  4.5 Envelope and materials — 94
  4.6 From passive to nearly zero-energy building design — 96
  4.7 Smart buildings and home automation — 99
  4.8 From smart buildings to smart districts and cities — 100
  4.9 Concluding discussion — 103
  References — 104

5. **Renewable energy integration as an alternative to the traditional ground-source heat pump system** — 109
Cristina Sáez Blázquez, David Borge-Diez, Ignacio Martín Nieto, Arturo Farfán Martín and Diego González-Aguilera
  Nomenclature — 109
  5.1 Introduction — 110
  5.2 Methodology — 111
  5.3 Technical calculation — 116
  5.4 Economic and environmental analysis — 122
  5.5 Discussion — 123
  5.6 Conclusions — 128
  Acknowledgments — 128
  References — 129

6.	**Energy-saving strategies on university campus buildings: Covenant University as case study**	**131**
	Sunday O. Oyedepo, Emmanuel G. Anifowose, Elizabeth O. Obembe and Shoaib Khanmohamadi	
	6.1 Introduction	131
	6.2 Materials and methods	136
	6.3 Results and discussions	139
	6.4 Conclusion	152
	References	153
7.	**Energy conversion systems and Energy storage systems**	**155**
	Jian Zhang, Heejin Cho and Pedro J. Mago	
	7.1 Introduction	155
	7.2 Energy systems in buildings	156
	7.3 Conclusion	175
	References	175
8.	**Energy systems in buildings**	**181**
	Getu Hailu	
	8.1 Introduction	181
	8.2 Energy-efficient building envelopes	182
	8.3 Renewable energy sources for building energy application	183
	8.4 Solar thermal energy storage	191
	8.5 Wind energy	196
	8.6 Heat pumps	199
	8.7 Biomass	204
	8.8 Summary	205
	References	205

Part 4   Energy efficiency in industrial sector

9.	**Energy efficiency and renewable energy sources for industrial sector**	**213**
	Kamil Kaygusuz	
	9.1 Introduction	213
	9.2 Global energy trends	215
	9.3 Energy consumption and emissions in industry	216
	9.4 Energy efficiency in industry for climate change mitigation	221
	9.5 Energy efficiency and renewable sources in industry	224
	9.6 Case study in Turkey	229
	9.7 Policy options	233
	9.8 Conclusions	236
	Acknowledgment	237
	References	237

10 Energy efficiency in tourism sector: eco-innovation measures and energy   **239**
Sánchez-Ollero José-Luis, Sánchez-Cubo Francisco, Sánchez-Rivas García Javier and Pablo-Romero-Gil-Delgado María
10.1 Introduction   239
10.2 State of the arts   240
10.3 Methods and data   242
10.4 Results and discussion   244
10.5 Conclusions   245
References   246

Part 5   Energy services markets: development and status quo

11 Energy service markets: status quo and development   **251**
Marc Ringel
11.1 Introduction   251
11.2 The European framework for energy services   253
11.3 The German energy service market   259
11.4 Developments of segments of the service market   264
11.5 Market development   268
11.6 Conclusions: lessons learned from the German case   271
References   271

12 Worldwide trends in energy market research   **277**
Esther Salmerón-Manzano, Alfredo Alcayde and Francisco Manzano-Agugliaro
12.1 Introduction   277
12.2 Data   278
12.3 Results   278
References   290

13 Which aspects may prevent the development of energy service companies? The impact of barriers and country-specific conditions in different regions   **293**
Marina Yesica Recalde
13.1 Introduction   293
13.2 Which are the problems confronted by energy efficiency actions and policy instruments?   296
13.3 Which are the most relevant barriers confronted by energy service companies in different regions?   300

13.4	Removing barriers and promoting energy service companies	306
13.5	Lessons learned and conclusions	308
Acknowledgments		312
References		312
Further reading		315

Index   **317**

List of Contributors

Alfredo Alcayde Department of Engineering, ceiA3, University of Almeria, Almeria, Spain

Emmanuel G. Anifowose Department of Mechanical Engineering, Covenant University, Ota, Nigeria

David Borge-Diez Department of Electric, System and Automatic Engineering, University of León, León, Spain

Heejin Cho Department of Mechanical Engineering, Mississippi State University, Mississippi State, MS, United States

Sánchez-Cubo Francisco Department of Applied Economics, Faculty of Tourism, University of Malaga, Malaga, Spain

Diego González-Aguilera Department of Cartographic and Land Engineering, Higher Polytechnic School of Ávila, University of Salamanca, Ávila, Spain

Getu Hailu Department of Mechanical Engineering, University of Alaska Anchorage, Anchorage, AK, United States

Theodoros G. Iliopoulos Hasselt University, Hasselt, Belgium

Sánchez-Rivas García Javier Department of Economic Analysis and Political Economy, Faculty of Economics and Business, University of Sevilla, Sevilla, Spain

Sánchez-Ollero José-Luis Department of Applied Economics, Faculty of Tourism, University of Malaga, Malaga, Spain

Charikleia Karakosta Energy Policy Unit (EPU-NTUA), Decision Support Systems Laboratory, School of Electrical and Computer Engineering, National Technical University of Athens, Athens, Greece

Kamil Kaygusuz Department of Chemistry, Karadeniz Technical University, Trabzon, Turkey

Shoaib Khanmohamadi Department of Mechanical Engineering, Kermanshah University of Technology, Kermanshah, Iran

Pedro J. Mago Department of Mechanical and Aerospace Engineering, West Virginia University, Morgantown, WV, United States

Francisco Manzano-Agugliaro Department of Engineering, ceiA3, University of Almeria, Almeria, Spain

Pablo-Romero-Gil-Delgado María Department of Economic Analysis and Political Economy, Faculty of Economics and Business, University of Sevilla, Sevilla, Spain

Arturo Farfán Martín Department of Cartographic and Land Engineering, Higher Polytechnic School of Ávila, University of Salamanca, Ávila, Spain

Ignacio Martín Nieto Department of Cartographic and Land Engineering, Higher Polytechnic School of Ávila, University of Salamanca, Ávila, Spain

Elizabeth O. Obembe Department of Mechanical Engineering, Covenant University, Ota, Nigeria

Abdeen Mustafa Omer Energy Research Institute, Nottingham, United Kingdom

Sunday O. Oyedepo Department of Mechanical Engineering, Covenant University, Ota, Nigeria

Aikaterini Papapostolou Energy Policy Unit (EPU-NTUA), Decision Support Systems Laboratory, School of Electrical and Computer Engineering, National Technical University of Athens, Athens, Greece

John Psarras Energy Policy Unit (EPU-NTUA), Decision Support Systems Laboratory, School of Electrical and Computer Engineering, National Technical University of Athens, Athens, Greece

Marina Yesica Recalde National Scientific and Technical Research Council (CONICET) / Environment and Development Program, Bariloche Foundation (BF), Argentina

Marc Ringel Faculty of Economics and Law, Energy Economics, Nuertingen Geislingen University, Geislingen, Germany

Cristina Sáez Blázquez Department of Cartographic and Land Engineering, Higher Polytechnic School of Ávila, University of Salamanca, Ávila, Spain

Esther Salmerón-Manzano Faculty of Law, International University of La Rioja, Logroño, Spain

George Vasileiou Energy Policy Unit (EPU-NTUA), Decision Support Systems Laboratory, School of Electrical and Computer Engineering, National Technical University of Athens, Athens, Greece

Jacopo Vivian University of Padua, Padua, Italy

Jian Zhang Department of Mechanical Engineering, University of Wisconsin Green Bay, Green Bay, WI, United States

Part 1

Energy services

Energy services: concepts, applications and historical background

Abdeen Mustafa Omer
Energy Research Institute, Nottingham, United Kingdom

Chapter Outline

1.1 Introduction 3
1.2 Energy and population growth 5
1.3 Energy saving in buildings 7
1.4 Energy use in agriculture 9
1.5 Renewable energy technologies 12
    1.5.1 Solar energy 12
    1.5.2 Efficient bioenergy use 14
    1.5.3 Combined heat and power 19
    1.5.4 Hydrogen production 20
    1.5.5 Hydropower generation 21
    1.5.6 Wind energy 22
1.6 Energy and sustainable development 25
1.7 Global warming 28
1.8 Recommendations 30
1.9 Conclusion 31
References 31

1.1 Introduction

Over millions of years ago plants covered the earth, converting the energy of sunlight into living tissue, some of which was buried in the depths of the earth to produce deposits of coal, oil, and natural gas. During the past few decades has found many valuable uses for these complex chemical substances, manufacturing from them plastics, textiles, fertilizers, and the various end products of the petrochemical industry. Each decade sees increasing uses for these products. Coal, oil, and gas are nonrenewable natural resources, which will certainly be of great value to future generations, as they are to ours. The rapid depletion of nonrenewable fossil resources need not continue, since it is now or soon will be technically and economically feasible to supply all of man's need from the most abundant energy source of all, the sun. The sunlight is not only inexhaustible but also the only energy source, which is completely nonpolluting (United Nations, 2001).

Energy Services Fundamentals and Financing. DOI: https://doi.org/10.1016/B978-0-12-820592-1.00001-4
© 2021 Elsevier Inc. All rights reserved

Industry's use of fossil fuels has been blamed for our warming climate. When coal, gas, and oil are burnt, they release harmful gases, which trap heat in the atmosphere and cause global warming. However, there has been an ongoing debate on this subject, as scientists have struggled to distinguish between changes that are human induced and those that could be put down to natural climate variability. Industrialized countries have the highest emission levels and must shoulder the greatest responsibility for global warming. However, action must also be taken by developing countries to avoid future increases in emission levels as their economies develop and population grows. Human activities that emit carbon dioxide (CO_2), the most significant contributor to potential climate change, occur primarily from fossil fuel production. Consequently, efforts to control CO_2 emissions could have serious negative consequences for economic growth, employment, investment, trade, and the standard of living of individuals everywhere. Scientifically, it is difficult to predict the relationship between global temperature and greenhouse gas (GHG) concentrations. The climate system contains many processes that will change if global warming occurs. Critical processes include heat transfer by winds and currents; the hydrological cycle involving evaporation, precipitation, runoff, and groundwater; and the formation of clouds, snow, and ice, all of which display enormous natural variability. The equipment and infrastructure for energy supply and use are designed with long lifetimes, and the premature turnover of capital stock involves significant costs. Economic benefits occur if capital stock is replaced with more efficient equipment in step with its normal replacement cycle. Likewise, if opportunities to reduce future emissions are taken in a timely manner, they should be less costly. Such flexible approaches would allow society to take account of evolving scientific and technological knowledge and to gain experience in designing policies to address climate change (Rees, 1999).

The World Summit (WS) on sustainable development in Johannesburg committed itself to "encourage and promote the development of renewable energy sources to accelerate the shift toward sustainable consumption and production." The WS aimed at breaking the link between resource use and productivity. It is about:

- trying to ensure economic growth does not cause environmental pollution,
- improving resource efficiency,
- examining the whole life cycle of a product,
- enabling consumers to receive more information on products and services,
- examining how taxes, voluntary agreements, subsidies, regulation, and information campaigns can best stimulate innovation and investment to provide cleaner technology.

The energy conservation scenarios include rational use of energy policies in all economy sectors and use of combined heat and power (CHP) systems, which are able to add to energy savings from the autonomous power plants. Electricity from renewable energy sources is by definition the environmental green product. Hence, a renewable energy certificate system is an essential basis for all policy systems, independent of the renewable energy support scheme. It is, therefore, important that all parties involved support the renewable energy certificate system in place. Existing renewable energy technologies (RETs) could play a significant mitigating role, but the economic and political climate will have to change first. Climate change is real. It is happening now, and GHGs produced by human activities are significantly contributing to it. The predicted global temperature increase between 1.5°C and 4.5°C could lead to potentially catastrophic environmental impacts. These include sea level rise, increased frequency of extreme weather events, floods, droughts, disease migration from various places, and possible stalling of the Gulf Stream. This has led scientists to argue that climate change issues are not the ones that politicians can afford to ignore, and policy makers tend to agree (Bos et al., 1994). However, reaching international agreements on climate change policies is not a trivial task.

Renewable energy is the term used to describe a wide range of naturally occurring, replenishing energy sources. The use of renewable energy sources and the rational use of energy are the fundamental inputs for any responsible energy policy. The energy sector is encountering difficulties because increased production and consumption levels entail higher levels of pollution and eventually climate change, with possibly disastrous consequences. Moreover, it is important to secure energy at an acceptable cost in order to avoid negative impacts on economic growth. On the technological side, renewables have an obvious role to play. In general, there is no problem in terms of the technical potential of renewables to deliver energy. Moreover, there are very good opportunities for RETs to play an important role in reducing the emissions of GHGs into the atmosphere, certainly far more than that have been exploited so far. However, there are still some technical issues to address in order to cope with the intermittency of some renewables, particularly wind and solar. Yet, the biggest problem with relying on renewables to deliver the necessary cuts in GHG emissions is more to do with politics and policy issues than with technical ones (Bos et al., 1994). The single most important step that governments could take to promote and increase the use of renewables is to improve access for renewables to the energy market. This access to the market would need to be under favorable conditions and, possibly, under favorable economic rates as well. One move that could help, or at least justify, better market access would be to acknowledge that there are environmental costs associated with other energy supply options, and that these costs are not currently internalized within the market price of electricity or fuels. This could make a significant difference, particularly if appropriate subsidies were applied to renewable energy in recognition of the environmental benefits it offers. Similarly, cutting energy consumption through end-use efficiency is absolutely essential. This suggests that issues of end-use consumption of energy will have to come into the discussion in the foreseeable future.

1.2 Energy and population growth

Throughout the world, urban areas have increased in size during recent decades. About 50% of the world's population and approximately 7.6% in more developed countries are urban dwellers. Even though there is evidence to suggest that in many "advanced" industrialized countries there has been a reversal in the rural-to-urban shift of populations, virtually all population growth expected between 2000 and 2030 will be concentrated in urban areas of the world. With an expected annual growth of 1.8%, the world's urban population will double in 38 years (United Nations, 2001).

With increasing urbanization in the world, cities are growing in number, population, and complexity. At present, 2% of the world's land surface is covered by cities, yet the people living there consume 75% of the resources (Rees, 1999). Indeed, the ecological footprint of cities is many times larger than the areas they physically occupy. Economic and social imperatives often dictate that cities must become more concentrated, making it necessary to increase the density to accommodate the people, to reduce the cost of public services, and to achieve required social cohesiveness. The reality of modern urbanization inevitably leads to higher densities than in traditional settlements, and this trend is particularly notable in developing countries.

The world population is rising rapidly, notably in the developing countries. Historical trends suggest that increased annual energy use per capita is a good surrogate for the standard of living factors, which promote a decrease in population growth rate. If these trends continue, the stabilization of the world's population will require the increased use of all sources of energy, particularly as cheap oil and gas are depleted. The improved efficiency of energy use and renewable energy sources will therefore be essential in stabilizing population, while providing a decent standard of living all over the world (Bos et al., 1994). Moreover, energy is the vital input for economic and social development of any country. With an increase in industrial and agricultural activities, the demand for energy is also rising. It is a well-accepted fact that commercial energy use has to be minimized. This is because of the environmental effects and the availability problems. The focus has now shifted to noncommercial energy resources, which are renewable in nature. This is found to have less environmental effects, and also the availability is guaranteed. Even though the ideal situation will be to enthuse people to use renewable energy resources, there are many practical difficulties that need to be tackled. The people groups who are using the noncommercial energy resources, such as urban communities, are now becoming more demanding and wish to have commercial energy resources that are made available for their use. This is attributed to the increased awareness, improved literacy level, and changing culture (Bos et al., 1994). The quality of life practiced by people is usually represented as being proportional to the per capita energy use of that particular country. It is not surprising that people want to improve their quality of life. Consequently, it is expected that the demand for commercial energy resources will increase at a greater rate in the years to come (Bos et al., 1994). Because of this emerging situation, the policy makers are left with two options: either concentrate on renewable energy resources and have them as substitutes for commercial energy resources or have a dual approach in which renewable energy resources will contribute to meet a significant portion of the demand, whereas the conventional commercial energy resources would be used with caution whenever necessary. Even though the first option is the ideal one, the second approach will be more appropriate for a smooth transition (Bos et al., 1994). Worldwide, renewable energy contributes as much as 20% of the global energy supplies (Duchin, 1995). Over two-thirds of this comes from biomass use, mostly in developing countries, some of it is unsustainable. Yet, the potential for energy from sustainable technologies is huge.

The RETs have the benefit of being environmentally benign when developed in a sensitive and appropriate way with the full involvement of local communities. In addition, they are diverse, secure, locally based, and abundant. In spite of the enormous potential and the multiple benefits, the contribution from renewable energy still lags behind the ambitious claims for it due to the initially high development costs, concerns about local impacts, lack of research funding, and poor institutional and economic arrangements (Duchin, 1995).

An approach is needed to integrate renewable energies in a way to meet high building performance. However, because renewable energy sources are stochastic and geographically diffuse, their ability to match demand is determined by adoption of one of the following two approaches (Energy Use in Offices, 2000): the utilization of a capture area greater than that occupied by the community to be supplied or the reduction of the community's energy demands to a level commensurate with the locally available renewable resources.

1.3 Energy saving in buildings

The prospects for development in power engineering are, at present, closely related to ecological problems. Power engineering has harmful effects on the environment, as it discharges toxic gases into atmosphere and also oil-contaminated and saline waters into rivers, while polluting the soil with ash and slag and having adverse effects on living things on account of electromagnetic fields, and so on. There is thus an urgent need for new approaches to provide an ecologically safe strategy. Substantial economic and ecological effects for thermal power projects can be achieved by improvement, upgrading the efficiency of the existing equipment, reduction of electricity loss, saving of fuel, and optimization of its operating conditions and service life.

Improving access for rural and urban low-income areas in developing countries through energy efficiency and renewable energies is important. Sustainable energy is a prerequisite for development. Energy-based living standards in developing countries, however, are clearly below standards in developed countries. Low levels of access to affordable and environmentally sound energy in both rural and urban low-income areas are therefore a predominant issue in developing countries. In recent years, many programs for development aid or technical assistance have been focusing on improving access to sustainable energy, many of them with impressive results. Apart from success stories; however, experience also shows that positive appraisals of many projects evaporate after completion and vanishing of the implementation expert team. Altogether, the diffusion of sustainable technologies such as energy efficiency and renewable energies for cooking, heating, lighting, electrical appliances, and building insulation in developing countries has been slow. Energy efficiency and renewable energy programs could be more sustainable, and pilot studies could be more effective and pulse releasing if the entire policy and implementation process were considered and redesigned from the outset. New financing and implementation processes are needed that allow reallocating financial resources and thus enabling countries themselves to achieve a sustainable energy infrastructure. The links between the energy policy framework, financing, and implementation of renewable energy and energy-efficient projects have to be strengthened, and capacity building efforts are required.

The admission of daylight into buildings alone does not guarantee that the design will be energy efficient in terms of lighting. In fact, the design for increased daylight can often raise concerns relating to visual comfort (glare) and thermal comfort (increased solar gain in the summer and heat losses in the winter from larger apertures). Such issues will clearly need to be addressed in the design of the window openings, blinds, shading devices, heating system, and so on. In order for a building to benefit from daylight energy terms, it is a prerequisite that lights are switched off when sufficient daylight is available. The nature of the switching regime—manual or automated, centralized or local, switched, stepped, or dimmed—will determine the energy performance. Simple techniques can be implemented to increase the probability that lights are switched off (Givoni, 1998).

These include:

- making switches conspicuous,
- loading switches appropriately in relation to the lights,
- switching banks of lights independently,
- switching banks of lights parallel to the main window wall.

There are also a number of methods that help reduce the lighting energy use, which, in turn, relate to the type of occupancy pattern of the building (Givoni, 1998). The light switching options include:

- centralized timed off (or stepped)/manual on
- photoelectric off (or stepped)/manual on
- photoelectric on (or stepped), photoelectric dimming
- occupant sensor (stepped) on/off (movement or noise sensor)

Likewise, energy savings from the avoidance of air-conditioning can be very substantial. While day-lighting strategies need to be integrated with artificial lighting systems in order to become beneficial in terms of energy use, reductions in overall energy consumption levels by employment of a sustained program of energy consumption strategies and measures would have considerable benefits within the buildings sector. The perception is often given however is that rigorous energy conservation as an end in itself imposes a style on building design resulting in a restricted aesthetic solution. It would perhaps be better to support a climate-sensitive design approach, which encompassed some elements of the pure conservation strategy together with strategies that work with the local ambient conditions making use of energy technology systems, such as solar energy, where feasible. In practice, low-energy environments are achieved through a combination of measures that include:

- the application of environmental regulations and policy
- the application of environmental science and best practice

- mathematical modeling and simulation
- environmental design and engineering
- construction and commissioning
- management and modifications of environments in use.

While the overriding intention of passive solar energy design is to achieve a reduction in purchased energy consumption, the attainment of significant savings is in doubt. The nonrealization of potential energy benefits is mainly due to the neglect of the consideration of postoccupancy user and management behavior by energy scientists and designers alike. Buildings consume energy mainly for cooling, heating, and lighting. The energy consumption was based on the assumption that the building operates within ASHRAE-thermal comfort zone during the cooling and heating periods (ASHRAE, 1993). Most of the buildings incorporate energy-efficient passive cooling, solar control, photovoltaic (PV), lighting and day lighting, and integrated energy systems. It is well known that thermal mass with night ventilation can reduce the maximum indoor temperature in buildings in summer (Kammerud et al., 1984). Hence, comfort temperatures may be achieved by proper application of passive cooling systems. However, energy can also be saved if an air-conditioning unit is used (Shaviv, 1989). The reason for this is that in summer, heavy external walls delay the heat transfer from the outside into the inside spaces. Moreover, if the building has a lot of internal mass the increase in the air temperature is slow. This is because the penetrating heat raises the air temperature as well as the temperature of the heavy thermal mass. The result is slow heating of the building in summer as the maximal inside temperature is reached only during the late hours when the outside air temperature is already low. The heat flowing from the inside heavy walls can be removed with good ventilation in the evening and night. The capacity to store energy also helps in winter, since energy can be stored in walls from one sunny winter day to the next cloudy one.

1.4 Energy use in agriculture

The land area required to provide all our energy is a small fraction of the land area required to produce our food, and the land best suited for collecting solar energy (rooftops and deserts) is the land least suited for other purposes. The economical utilization of solar energy in all its varied forms—PV, direct solar thermal, renewable fuels, ocean thermal, and wind—can offer the world the technology, then can conserve valuable nonrenewable fossil resources for future generations to enjoy, and all can live in a world of abundant energy without pollution. Energy in agriculture is important in terms of crop production and agro-processing for value adding. Human, animal, and mechanical energy are extensively used for crop production in agriculture. Energy requirements in agriculture are divided into two groups being direct and indirect.

Direct energy is required to perform various tasks related to crop production processes such as land preparation, irrigation, interculture, threshing, harvesting, and transportation of agricultural inputs and farm produce. It is seen that direct energy is directly used at farms and on fields. Indirect energy, on the other hand, consists of the energy used in the manufacture, packing, and transport of fertilizers, pesticides, and farm machinery. As the name implies, indirect energy is not directly used on the farm. Major items for indirect energy are fertilizers, seeds, machinery production, and pesticides (Table 1.1).

Table 1.1 Energy equivalent of inputs and outputs.

Input	Unit	Equivalent energy (MJ)
1. *Human labor*	h	2.3
2. *Animal labor*		
Horse	h	10.10
Mule	h	4.04
Donkey	h	4.04
Cattle	h	5.05
Water buffalo	h	7.58
3. *Electricity*	kWh	11.93
4. *Diesel*	L	56.31
5. *Chemical fertilizers*		
Nitrogen	kg	64.4
P_2O_5	kg	11.96
K_2O	kg	6.7
6. *Seed*		
Cereals and pulses	kg	25
Oil seed	kg	3.6
Tuber	kg	14.7
Output		
7. *Major products*		
Cereal and pulses	kg	14.7
Sugar beet	kg	5.04
Tobacco	kg	0.8
Cotton	kg	11.8
Oil seed	kg	25
Fruits	kg	1.9
Vegetables	kg	0.8
Water melon	kg	1.9
Onion	kg	1.6
Potatoes	kg	3.6
Olive	kg	11.8
Tea	kg	0.8
8. *By-products*		
Husk	kg	13.8
Straw	kg	12.5
Cob	kg	18.0
Seed cotton	kg	25.0

Calculating energy inputs in agricultural production is more difficult in comparison to the industry sector due to the high number of factors affecting agricultural production. However, considerable studies have been conducted in different countries on energy use in agriculture (Singh, 2000; CAEEDAC, 2000; Yaldiz et al., 1993; Dutt, 1982; Baruah, 1995; Thakur and Mistra, 1993).

Energy use in the agricultural sector depends on the size of the population engaged in agriculture, the amount of arable land, and the level of mechanization. To calculate the energy used in agricultural production or repair of machinery, the following formula is used:

$$\mathrm{ME} = \frac{(G \times E)}{(T \times C_a)} \tag{1.1}$$

where ME is the machine energy (MJ/ha), G is the weight of tractor (kg), E is the constant that is taken 158.3 MJ/kg for tractor, T is the economic life of tractor (h), and C_a is the effective field capacity (ha/h).

For calculation of C_a, the following equation is used:

$$C_a = \frac{(S \times W \times E_f)}{10} \tag{1.2}$$

where C_a is the effective field capacity (ha/h), W is the working width (m), S is the working speed (km/h), and E_f is the field efficiency (%).

Agricultural greenhouses have a very poor efficiency of thermal conversion of the received solar energy. This is particularly evident in Europe, where, in a cycle of 24 hours, and in winter period, the following constraints are observed:

- During the day to maintain through ventilation, an inside temperature at a level lower than the excessive temperatures, harmful for the growth and the development of the cultures.
- At night to assure, by a supply of heating energy, an optional temperature higher than the crucial level of the culture.

This low thermal efficiency is due to the fact that, in a classic greenhouse, the only usable thermal support is the greenhouse soil, which has a weak thermal inertia. Storage of most of the daily excess energy, in order to reuse it during the night where the temperature is low, is therefore impossible. Among the other climatic factors that contribute in the development of greenhouse cultivation, the inside air temperature, in contact with the aerial part of the plant, constitutes a dominant representative factor.

The impact of heating on the increase of the inside air temperature is very important, because a significant increase of agronomic efficiency in the experimental greenhouse.

Explanations for the use of inefficient agricultural–environmental polices include the high cost of information required to measure benefits on a site-specific basis, information asymmetries between government agencies, and farm decision-makers that result in high implementation costs, distribution effects, and political considerations (Wu and Boggess, 1999). The aim of agri-environment schemes is achieved by:

- sustaining the beauty and diversity of the landscape,
- improving and extend wildlife habitats,
- conserving archeological sites and historic features,
- improving opportunities for countryside enjoyment,
- restoring neglected land or features,
- creating new habitats and landscapes.

1.5 Renewable energy technologies

Sustainable energy is the energy that, in its production or consumption, has minimal negative impacts on human health and on the healthy functioning of vital ecological systems, including the global environment. It is an accepted fact that renewable energy is a sustainable form of energy, which has attracted more attention during recent years. A great amount of potential renewable energy, environmental interest, as well as economic consideration of fossil fuel consumption and high emphasis of sustainable development for the future will be needed. Nearly one-fifth of all global power is generated by renewable energy sources, according to a new book published by the OECD/IEA (OECD/IEA, 2004). Renewables for power generation: status and prospects claim that renewables are the second largest power source after coal (39%) and ahead of nuclear (17%), natural gas (17%), and oil (8%). From 1973 to 2000 renewables grew at 9.3% a year, and the authors predict that this will increase 10.4% a year to 2010. Wind power grew fastest at 52% and will multiply by seven times to 2010, overtaking biopower. Reducing GHGs by production of environmental technology (wind, solar, fuel cells, etc.). The challenge is to match leadership in GHG reduction and production of renewable energy with developing a major research and manufacturing capacity in environmental technologies.

More than 50% of world's area is classified as arid, representing the rural and desert part, which lack electricity and water networks. The inhabitants of such areas obtain water from borehole wells by means of water pumps, which are driven by diesel engines. The diesel motors are associated with maintenance problems, high running cost, and environmental pollution. Alternative methods are pumping by PV or wind systems. Renewable sources of energy are regional and site specific. It has to be integrated in the regional development plans.

1.5.1 Solar energy

The availability of data on solar radiation is a critical problem. Even in developed countries, very few weather stations have been recording the detailed solar radiation data for a period of time long enough to have statistical significance. Solar radiation arriving on earth is the most fundamental renewable energy source in nature. It powers the biosystem, the ocean, and atmospheric current system and affects the global climate. Reliable radiation information is needed to provide input data in modeling solar energy devices, and a good database is required in the work of energy planners, engineers, and agricultural scientists. In general, it is not easy to design solar energy conversion systems when they have to be installed in remote locations. First, in most cases, solar radiation measurements are not available for these sites. Second, the radiation nature of solar radiation makes the computation of the size of such systems difficult. While solar energy data are recognized as very important, their acquisition is by no means straightforward. The measurement of solar radiation requires the use of costly equipment such as pyrheliometers and pyranometers. Consequently, adequate facilities are often not available in developing countries to mount viable monitoring programs. This is partly due to the equipment cost and also due to the cost of technical manpower. Several attempts have, however, been made to estimate solar radiation through the use of meteorological and other physical parameters in order to avoid the use of expensive network of measuring instruments (Duffie and Beckman, 1980; Sivkov, 1964a,b; Barabaro et al., 1978).

Two of the most essential natural resources for all life on the earth and for man's survival are sunlight and water. Sunlight is the driving force behind many of the RETs.

The worldwide potential for utilizing this resource, both directly by means of the solar technologies and indirectly by means of biofuels, wind, and hydrotechnologies, is vast. During the last decade, interest has been refocused on renewable energy sources due to the increasing prices and foreseeable exhaustion of presently used commercial energy sources. The most promising solar energy technology is related to thermal systems: industrial solar water heaters, solar cookers, solar dryers for peanut crops, solar stills, solar-driven cold stores to store fruits and vegetables, solar collectors, solar water desalination, solar ovens, and solar commercial bakers. Solar PV system includes solar PV for lighting, solar refrigeration to store vaccines for human and animal use, solar PV for water pumping, solar PV for battery chargers, solar PV for communication network, microwave, receiver stations, radio systems in airports, VHF and beacon radio systems in airports, and educational solar TV posts in villages. Solar pumps are the most cost-effective system for low power requirement (up to 5 kW) in remote places. Applications include domestic and livestock drinking water supplies for which the demand is constant throughout the year and irrigation. The suitability of solar pumping for irrigation is uncertain because the demand may vary greatly with seasons. Solar systems may be able to provide trickle irrigation for fruit farming, but not usually the large volumes of water needed for wheat growing.

The hydraulic energy required to deliver a volume of water is given by the formula:

$$E_w = \rho_w g V H \quad (1.3)$$

where E_w is the required hydraulic energy (kWh/day), ρ_w is the water density, g is the gravitational acceleration (m/s), V is the required volume of water (m^3/day$^-$), and H is the head of water (m).

The solar array power required is given by:

$$P_{sa} = \frac{E_w}{E_{sr}\eta F} \qquad (1.4)$$

where P_{sa} is the solar array power (kW$_p$), E_{sr} is the average daily solar radiation (kWh/m^2/day), F is the array mismatch factor, and η is the daily subsystem efficiency.

Substituting Eq. (1.1) in Eq. (1.2), the following equation is obtained for the amount of water that can be pumped:

$$V = \frac{P_{sa} E_{sr} \eta F}{\rho_w g H} \qquad (1.5)$$

PV consists of 32 modules; $P_{sa} = 1.6$ kW$_p$, $F = 0.85$, $\eta = 40\%$.

A further increase of the PV depends on the ability to improve the durability, performance, and the local manufacturing capabilities of the PV. Moreover, the availability of credit schemes (e.g., solar funds) would increase the annual savings of oil and foreign currency and would further improve the security of energy supply, and further employment could be created.

1.5.2 Efficient bioenergy use

The data required to perform the trade-off analysis simulation can be classified according to the divisions given in Table 1.2. The overall system or individual plants and the existing situation or future development are shown in Table 1.2.

The effective economic utilizations of these resources are shown in Table 1.3, but their use is hindered by many problems such as those related to harvesting,

Table 1.2 Classifications of data requirements.

	Plant data	**System data**
Existing data	Size Life Cost (fixed and var. O&M) Forced outage Maintenance Efficiency Fuel Emissions	Peak load Load shape Capital costs Fuel costs Depreciation Rate of return Taxes
Future data	All of above, plus Capital costs Construction trajectory Date in service	System lead growth Fuel price growth Fuel import limits Inflation

Table 1.3 Effective biomass resource utilization.

Subject	Tools	Constraints
Utilization and land clearance for agriculture expansion	• Stumpage fees • Control • Extension • Conversion • Technology	• Policy • Fuelwood planning • Lack of extension • Institutional
Utilization of agricultural residues	• Briquetting • Carbonization • Carbonization and briquetting • Fermentation • Gasification	• Capital • Pricing • Policy and legislation • Social acceptability

Table 1.4 Agricultural residues routes for development.

Source	Process	Product	End use
Agricultural residues	Direct	Combustion	Rural poor Urban household Industrial use
	Processing	Briquettes	Industrial use Limited household use
	Processing	Carbonization (small scale)	Rural household (self-sufficiency)
	Carbonization	Briquettes Carbonized	Urban fuel Energy services
	Fermentation	Biogas	Household
Agricultural and animal residues	Direct	Combustion	Industry (Save or less efficiency as wood)
	Briquettes	Direct combustion	(Similar end-use devices or improved)
	Carbonization	Carbonized	Use
	Carbonization	Briquettes	Briquettes use
	Fermentation	Biogas	Use

collection, and transportation, besides the photo-sanitary control regulations. Biomass energy is experiencing a surge in interest stemming from a combination of factors, for example, greater recognition of its current role and future potential contribution as a modern fuel, global environmental benefits, its development and entrepreneurial opportunities, and so on. Possible routes of biomass energy development are shown in Table 1.4.

1. Biomass energy for domestic needs
    a. Population increase
    b. Urbanization
    c. Agricultural expansion
    d. Fuelwood crisis
    e. Ecological crisis
    f. Fuelwood plantations
    g. Community forestry
    h. Improved stoves
    i. Agroforestry
    j. Improved charcoal production
    k. Residue utilization
2. Biomass energy for petroleum substitution
    a. Oil price increase
    b. Balance of payment problems
    c. Economic crisis
    d. Fuelwood plantations
    e. Residue utilization
    f. Wood-based heat and electricity
    g. Liquid fuels from biomass
    h. Producer gas technology
3. Biomass energy for development
    a. Electrification
    b. Irrigation and water supply
    c. Economic and social development
    d. Fuelwood plantations
    e. Community forestry
    f. Agroforestry
    g. Briquettes
    h. Producer gas technology

The use of biomass through direct combustion has long been, and still is, the most common mode of biomass utilization as shown in Tables 1.2–1.4. Examples for dry (thermochemical) conversion processes are charcoal making from wood (slow pyrolysis), gasification of forest and agricultural residues (fast pyrolysis—this is still in demonstration phase), and of course, direct combustion in stoves, furnaces, and so on. Wet processes require substantial amount of water to be mixed with the biomass. Biomass technologies include briquetting, improved stoves, biogas, improved charcoal, carbonization, and gasification.

1.5.2.1 Briquette processes

Charcoal stoves are very familiar to African society. As for the stove technology, the present charcoal stove can be used and can be improved upon for better efficiency. This energy term will be of particular interest to both urban and rural households and to all the income groups due to the simplicity, convenience, and lower air-pollution characteristics. However, the market price of the fuel together with that of its end-use technology may not enhance its early high market penetration especially in the urban low-income and rural households.

Briquetting is the formation of a char (an energy-dense solid fuel source) from otherwise wasted agricultural and forestry residues. One of the disadvantages of wood fuel is that it is bulky with a low energy density and is therefore required to transport. Briquette formation allows for a more energy-dense fuel to be delivered, thus reducing the transportation cost and making the resource more competitive. It also adds some uniformity, which makes the fuel more compatible with systems that are sensitive to the specific fuel input.

1.5.2.2 Improved cook stoves

Traditional wood stoves can be classified into four types: three stone, metal cylindrical shaped, metal tripod, and clay type. Another area in which rural energy availability could be secured, where woody fuels have become scarce, is by the improvements of traditional cookers and ovens to raise the efficiency of fuel saving and by planting fast growing trees to provide a constant fuel supply. The rural development is essential and economically important since it will eventually lead to better standards of living, people's settlement, and self-sufficient in the following:

- food and water supplies,
- better services in education and health care,
- good communication modes.

1.5.2.3 Biogas technology

Biogas technology can not only provide fuel, but is also important for comprehensive utilization of biomass forestry, animal husbandry, fishery, agricultural economy, protecting the environment, realizing agricultural recycling, as well as improving the sanitary conditions, in rural areas. The introduction of biogas technology on wide scale has implications for macro planning such as the allocation of government investment and effects on the balance of payments. Factors that determine the rate of acceptance of biogas plants, such as credit facilities and technical backup services, are likely to have to be planned as part of general macro policy, as do the allocation of research and development funds (Hall and Scrase, 1998).

1.5.2.4 Improved forest and tree management

Dry cell batteries are a practical but expensive form of mobile fuel that is used by rural people when moving around at night and for powering radios and other small appliances. The high cost of dry cell batteries is financially constraining for rural households, but their popularity gives a good indication of how valuable a versatile fuel such as electricity is in rural area. Dry cell batteries can constitute an environmental hazard unless they are recycled in a proper fashion (Table 1.5). Direct burning of fuelwood and crop residues constitutes the main usage of biomass, as is the case with many developing countries. However, the direct burning of biomass in an

Table 1.5 Energy carrier and energy services in rural areas.

Energy carrier	Energy end use
Fuelwood	Cooking
	Water heating
	Building materials
	Animal fodder preparation
Kerosene	Lighting
	Ignition fires
Dry cell batteries	Lighting
	Small appliances
Animal power	Transport
	Land preparation for farming
	Food preparation (threshing)
Human power	Transport
	Land preparation for farming
	Food preparation (threshing)

Table 1.6 Biomass residues and current use.

Type of residue	Current use
Wood industry waste	Residues available
Vegetable crop residues	Animal feed
Food processing residue	Energy needs
Sorghum, millet, wheat residues	Fodder, building materials
Groundnut shells	Fodder, brick making, direct fining oil mills
Cotton stalks	Domestic fuel considerable amounts available for short period
Sugar, bagasse, molasses	Fodder, energy need, ethanol production (surplus available)
Manure	Fertilizer, brick making, plastering

inefficient manner causes economic loss and adversely affects human health. In order to address the problem of inefficiency, research centers around the world have investigated the viability of converting the resource to a more useful form, namely solid briquettes and fuel gas. Biomass resources play a significant role in energy supply in all developing countries. Biomass resources should be divided into residues or dedicated resources, the latter including firewood, and charcoal can also be produced from forest residues (Table 1.6).

1.5.2.5 Gasification application

Gasification is based on the formation of a fuel gas (mostly CO and H_2) by partially oxidizing raw solid fuel at high temperatures in the presence of steam or air. The technology can use wood chips, groundnut shells, sugarcane bagasse, and other similar fuels to generate capacities from 3 to 100 kW. Three types of gasifier designs have been developed to make use of the diversity of fuel inputs and to meet the requirements of the product gas output (degree of cleanliness, composition, heating value, etc.) (Hall and Scrase, 1998).

1.5.3 Combined heat and power

Denmark has broadly seen three scales of the CHP which were largely implemented in the following chronological order (Pernille, 2004):

- large-scale CHP in cities (> 50 MWe)
- small (5 kWe−5 MWe) and medium-scale CHP (5−50 MWe)
- industrial and small-scale CHP

Most of the heat is produced by large-scale CHP plants (gas-fired combined cycle plants using natural gas, biomass, waste, or biogas) as shown in Table 1.7. The domestic hot water (DHW) is energy efficient because of the way the heat is produced, and the required temperature level is an important factor. Buildings can be heated to temperature of 21°C, and DHW can be supplied with a temperature of 55°C using energy sources that are most efficient when producing low temperature levels (<95°C) for the DHW. Most of these heat sources are CO_2 neutral or emit low levels. Only a few of these sources are available to small individual systems at a reasonable cost, whereas DHW schemes can have access to most of the heat sources and at a low cost because of the plant's size and location. Low-temperature

Table 1.7 Sources of renewable energy.

Energy source	Technology	Size
Solar energy	Domestic solar water heaters	Small
	Solar water heating for large demands	Medium−large
	PV roofs: grid connected systems generating electric energy	Medium−large
Wind energy	Wind turbines (grid connected)	Medium−large
Hydraulic energy	Hydro plants in derivation schemes	Medium−small
	Hydro plants in existing water distribution networks	Medium−small
Biomass	High-efficiency wood boilers	Small
	CHP plants fed by agricultural wastes or energy crops	Medium
Animal manure	CHP plants fed by biogas	Small
CHP	High-efficiency lighting	Wide
	High-efficiency electric	Wide
	Householders appliances	Wide
	High-efficiency boilers	Small−medium
	Plants coupled with refrigerating absorption machines	Medium−large

Table 1.8 Final energy projections including biomass (Mtoe) (D'Apote, 1998).

Region	Biomass	Conventional energy	Total	Share of biomass (%)
1995				
Africa	205	136	341	60
China	206	649	855	24
East Asia	106	316	422	25
Latin America	73	342	416	18
South Asia	235	188	423	56
Total developing countries	825	1632	2456	34
Other non-OECD countries	24	1037	1061	1
Total non-OECD countries	849	2669	3518	24
OECD countries	81	3044	3125	3
World	930	5713	6643	14
2025				
Africa	371	266	631	59
China	224	1524	1748	13
East Asia	118	813	931	13
Latin America	81	706	787	10
South Asia	276	523	799	35
Total developing countries	1071	3825	4896	22
Other non-OECD countries	26	1669	1695	1
Total non-OECD countries	1097	5494	6591	17
OECD countries	96	3872	3968	2
World	1193	9365	10,558	11

DHW, with return temperatures of around 30°C−40°C can utilize the following heat sources:

- efficient use of the CHP by extracting heat at low calorific value (CV),
- efficient use of biomass or gas boilers by condensing heat in economizers (Table 1.8),
- efficient utilization of geothermal energy.

1.5.4 Hydrogen production

Hydrogen is now beginning to be accepted as a useful form for storing energy for reuse on, or for export off, the grid. Clean electrical power harvested from wind and wave power projects can be used to produce hydrogen by electrolysis of water—splitting this into its constituent parts of hydrogen and oxygen. Electrolyzers split water molecules into its constituent parts: hydrogen and oxygen. These are collected as gases: hydrogen at the cathode and oxygen at the anode. The process is quite simple. Direct current is applied to the electrodes to initiate the electrolysis process. The reaction that occurs is:

At the anode:

$$4OH^- \Rightarrow O_2 + 2H_2O + 4\acute{e}$$

At the cathode:

$$4H_2O + 4\acute{e} \Rightarrow 2H_2 + 4OH^-$$

The overall reaction is:

$$2H_2O \Rightarrow 2H_2 + O_2$$

Production of hydrogen is an elegant environmental solution. Hydrogen is the most abundant element on the planet; it cannot be destroyed (unlike hydrocarbons); it simply changes state—water to hydrogen and back to water—during consumption. In its production and consumption, there is no CO or CO_2 production and depending upon methods of consumption, the production of oxides of nitrogen can be avoided too. The transition will be very messy, and will take many technological paths—converting fossil fuels and methanol to hydrogen, building hybrid engines, and so on—but the future will be hydrogen fuel cells. Hydrogen is already produced in huge volumes and used in a variety of industries. Current worldwide production is around 500 billion Nm^3 per year (David, 2002).

Most of the hydrogen produced today is consumed on-site, such as at oil refineries, and it is not sold on the market. From large-scale production, hydrogen costs around $0.70/kg if it is consumed on-site (David, 2002). When hydrogen is sold on the market, the cost of liquefying the hydrogen and transporting it to the user adds considerably to production cost. The energy required to produce hydrogen via electrolysis (assuming 1.23 V) is about 33 kWh/kg. For 1 mole (2 g) of hydrogen, the energy is about 0.066 kWh/mole (David, 2002). The achieved efficiencies are more than 80%, and on this basis electrolytic hydrogen can be regarded as a storable form of electricity.

Hydrogen can be stored in a variety of forms:

- Cryogenic; this has the highest gravimetric energy density.
- High-pressure cylinders; pressures of 10,000 psi are quite normal.
- Metal hydride absorbs hydrogen, providing a very low pressure and extremely safe mechanism, but it is heavy and more expensive than cylinders.
- Chemical carriers offer an alternative, with anhydrous ammonia offering similar gravimetric and volumetric energy densities to ethanol and methanol.

Hydrogen can be used in internal combustion engines, fuel cells, turbines, cookers, gas boilers, roadside emergency lighting, traffic lights, or signaling where noise and pollution can be a considerable nuisance, but where traffic and pedestrian safety cannot be compromised.

1.5.5 Hydropower generation

Hydropower has a valuable role as a clean and renewable source of energy in meeting a variety of vital human needs. Water resources management and benefit sharing and among other points (safe drinking water and sanitation, water for food and rural development, water pollution and ecosystem conservation, disaster mitigation, and risk management) the recognition of the role of hydropower as one of the renewable and clean energy sources and that its potential should be realized in an environmentally sustainable and socially acceptable manner.

Water is a basic requirement for survival: for drinking, for food, for energy production, and for good health. As water is a commodity, which is finite and cannot be created, and in view of the increasing requirements as the world population grows, there is no alternative other than storing water for use when it is needed. The major challenges are to feed the increasing world population, to improve the standards of living in rural areas, and to develop and manage land and water in a sustainable way. Hydropower plants are classified by their rated capacity into one of four regimes: micro (<50 kW), mini (50–500 kW), small (500 kW–5 MW), and large (>5 MW) (John and James, 1989).

The total world installed hydro capacity today is around 730 GW, and 1500 GW more will be built during this century, principally in developing countries in Asia, Africa, and South America. The present production of hydroelectricity is only about 18% of the technically feasible potential (and 32% of the economically feasible potential); there is no doubt that a large amount of hydropower development lies ahead (IHA, 2003). Table 1.9, which is reproduced from (IHA, 2003), classified hydro plants in the world.

1.5.6 Wind energy

The utilization of energy from renewable sources, such as wind, is becoming increasingly attractive and is being widely used for the substitution of oil-produced energy, and eventually to minimize atmospheric degradation. Most of the world's energy consumption is greatly dependent on fossil fuel, which is exhaustible, and is being used extensively due to the continuous escalation in the world's population and development.

This valuable resource needs to be converted, and its alternatives need to be explored. In this perspective, utilization of renewables, such as wind energy, has gained considerable momentum since the oil crises of the 1970s. Wind energy is nondepleting, site-dependent, nonpolluting, and a potential source of the alternative energy option. Wind power could supply 12% of global electricity demand by 2020, according to a report by European Wind Energy Association and Greenpeace (EWEA, 2003). Wind energy can and will constitute a significant energy resource; it must be converted to a usable form (Fig. 1.1).

As Fig. 1.1 illustrates, information sharing is a four-stage process, and effective collaboration must also provide ways in which the other three stages of the "renewable" cycle: gather, convert, and utilize can be integrated. Efficiency in the renewable energy sector translates into lower gathering, conversion, and utilization (electricity) costs. A great level of installed capacity has already been achieved.

Fig. 1.2 clearly shows that the offshore wind sector is developed, and this indicates that wind is becoming a major factor in electricity supply with a range of significant technical, commercial, and financial hurdles to be overcome. The offshore

Table 1.9 World hydro potential and development.

Continent	Africa	Asia	Australia and Oceania	Europe	North and Central America	South America
Gross theoretical hydropower potential (GWh/year)	4×10^6	19.4×10^6	59.4×10^6	3.2×10^6	6×10^6	6.2×10^6
Technically feasible hydropower potential (GWh/year)	1.75×10^6	6.8×10^6	2×10^6	10^6	1.66×10^6	2.7×10^6
Economically feasible hydropower potential (GWh/year)	1.1×10^5	3.6×10^6	90×10^4	79×10^4	10^6	1.6×10^6
Installed hydro capacity (MW)	21×10^3	24.5×10^4	13.3×10^4	17.7×10^4	15.8×10^4	11.4×10^4
Production by hydro plants in 2002 or average (GWh/year)	83.4×10^3	80×10^4	43×10^3	568×10^3	694×10^3	55×10^4
Hydro capacity under construction (MW)	>3024	$>72.7 \times 10^3$	>177	$>23 \times 10^2$	58×10^2	$>17 \times 10^3$
Planned hydro capacity (MW)	77.5×10^3	$>17.5 \times 10^4$	>647	$>10^3$	$>15 \times 10^3$	$>59 \times 10^3$

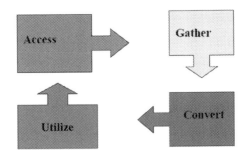

Figure 1.1 The renewable cycle.

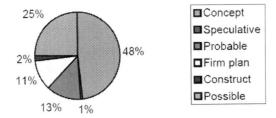

Figure 1.2 Global prospects by 2003–10.

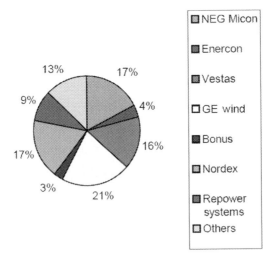

Figure 1.3 Turbines share for 2003–10.

wind industry has the potential for a very bright future and to emerge as a new industrial sector (Fig. 1.3).

The speed of turbine development means that more powerful models have superseded the original specification turbines in the time from concept to turbine order. Levels of activities are growing (Fig. 1.4) at a phenomenal rate, new prospects are

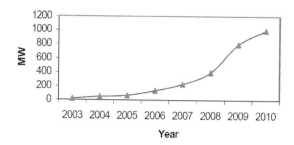

Figure 1.4 Average wind farm capacity 2003–10.

developing, new players are entering and existing players are growing in experience, technology is evolving, and political situation will appear to support the sector. Water is the most natural commodity for the existence of life in the remote desert areas. As a condition of settling and growing, the supply of energy comes into a second priority. The high cost and the difficulties of a main power lines extension, especially to low-populated regions, can divert the attention to the utilization of more reliable and independent sources of energy such as the renewable wind energy.

1.6 Energy and sustainable development

Sustainability has been defined as the extent to which progress and development should meet the need of the present without compromising the ability of the future generations to meet their own needs (Steele, 1997). This encompasses a variety of levels and scales ranging from economic development and agriculture to the management of human settlements and building practices. This general definition was further developed to include sustainable building practices and management of human settlements. The following issues were addressed during the Rio Earth Summit in 1992 (Sitarz, 1992):

- the use of local materials and indigenous building sources;
- incentive to promote the continuation of traditional techniques, with regional resources and self-help strategies;
- regulation of energy-efficient design principles;
- international information exchange on all aspects of construction related to the environment, among architects and contractors, particularly nonconventional resources;
- exploration of methods to encourage and facilitate the recycling and reuse of building materials, especially those requiring intensive energy use during manufacturing, and the use of clean technologies.

Action areas for producers:

- management and measurement tools—adopting environmental management systems appropriate for the business;
- performance assessment tools—making use of benchmarking to identify scope for impact reduction and greater eco-efficiency in all aspects of the business;

Table 1.10 Energy and sustainable environment.

Technological criteria	Energy and environment criteria	Social and economic criteria
Primary energy saving in regional scale	Sustainability according to greenhouse gas pollutant emissions	Labor impact
Technical maturity, reliability	Sustainable according to other pollutant emissions	Market maturity
Consistence of installation and maintenance requirements with local technical know-how	Land requirement	Compatibility with political, legislative, and administrative situation
Continuity and predictability of performance	Sustainability according to other environmental impacts	Cost of saved primary energy

- best practice tools—making use of free help and advice from government best practice programs (energy efficiency, environmental technology, and resource savings);
- innovation and ecodesign—rethinking the delivery of "value added" by the business, so that impact reduction and resource efficiency are firmly built in at the design stage;
- cleaner, leaner production processes—pursuing improvements and savings in waste minimization, energy and water consumption, transport and distribution, as well as reduced emissions. Tables 1.10–1.12 indicate energy conservation, sustainable development, and environment;
- supply chain management—specifying more demanding standards of sustainability from "upstream" suppliers, while supporting smaller firms to meet those higher standards.
- product stewardship—taking the broadest view of "producer responsibility" and working to reduce all the "downstream" effects of products after they have been sold on to customers;
- openness and transparency—publicly reporting on environmental performance against meaningful targets, actively using clear labels and declarations so that customers are fully informed, building stakeholder confidence by communicating sustainability aims to the workforce, the shareholders, and the local community (Fig. 1.5).

This is the step in a long journey to encourage a progressive economy, which continues to provide us with high living standards, but at the same time helps to reduce pollution, waste mountains, other environmental degradation, and environmental rationale for future policy-making and intervention to improve market mechanisms.

This vision will be accomplished by:

- "Decoupling" economic growth and environmental degradation. The basket of indicators illustrated shows the progress being made (Table 1.13). Decoupling air and water

Energy services: concepts, applications and historical background

Table 1.11 Classification of key variables defining facility sustainability.

Criteria	Intra-system impacts	Extra-system impacts
Stakeholder satisfaction	• Standard expectations met • Relative importance of standard expectations	• Covered by attending to extra-system resource base and ecosystem impacts
Resource base impacts	• Change in intra-system resource bases • Significance of change	• Resource flow into/out of facility system • Unit impact exerted by flow on source/sink system • Significance of unit impact
Ecosystem impacts	• Change in intra-system ecosystems • Significance of change	• Resource flows into/out of facility system • Unit impact exerted by how on source/sink system • Significance of unit impact

Table 1.12 Positive impact of durability, adaptability and energy conservation on economic, social, and environment systems.

Economic system	Social system	Environmental system
Durability	Preservation of cultural values	Preservation of resources
Meeting changing needs of economic development	Meeting changing needs of individuals and society	Reuse, recycling, and preservation of resources
Energy conservation and saving	Savings directed to meet other social needs	Preservation of resources, reduction of pollution, and global warming

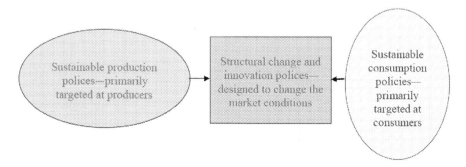

Figure 1.5 Link between resources and productivity.

Table 1.13 The basket of indicators for sustainable consumption and production.

Economy-wide decoupling indicators	Decoupling indicators for specific sectors
1. Greenhouse gas emissions 2. Air pollution 3. Water pollution (river water quality) 4. Commercial and industrial waste arisings and household waste not cycled *Resource use indicators* 5. Material use 6. Water abstraction 7. Homes built on land not previously developed and number of households	8. Emissions from electricity generation 9. Motor vehicle kilometers and related emissions 10. Agricultural output, fertilizer use, methane emissions, and farmland bird populations 11. Manufacturing output, energy consumption, and related emissions 12. Household consumption, expenditure energy, water consumption, and waste generated

pollution from growth, making good headway with CO_2 emissions from energy and transport. The environmental impact of our own individual behavior is more closely linked to consumption expenditure than the economy as a whole.
- Focusing policy on the most important environmental impacts associated with the use of particular resources, rather than on the total level of all resource use.
- Increasing the productivity of material and energy use that are economically efficient by encouraging patterns of supply and demand, which are more efficient in the use of natural resources. The aim is to promote innovation and competitiveness. Investment in areas such as energy efficiency, water efficiency, and waste minimization.
- Encouraging and enabling active and informed individual and corporate consumers.

1.7 Global warming

With the debate on climate change, the preference for real measured data has been changed. The analyses of climate scenarios need an hourly weather data series that allows for realistic changes in various weather parameters. By adapting parameters in a proper way, data series can be generated for the site. Weather generators should be useful for:

- calculating energy consumption (no extreme conditions are required),
- designing purposes (extremes are essential),
- predicting the effect of climate change such as increasing annually average of temperature.
    This results in the following requirements:
- Relevant climate variables should be generated (solar radiation: global, diffuse, direct solar direction, temperature, humidity, wind speed, and direction) according to the statistics of the real climate.

Table 1.14 EU member states greenhouse gas (GHG) emissions.

Country	1990	1999	Change 1990–99 (%)	Reduction target (%)
Austria	76.9	79.2	2.6	−13
Belgium	136.7	140.4	2.8	−7.5
Denmark	70.0	73.0	4.0	−21.0
Finland	77.1	76.2	−1.1	0.0
France	545.7	544.5	−0.2	0.0
Germany	1206.5	982.4	−18.7	−21.0
Greece	105.3	123.2	16.9	25.0
Ireland	53.5	65.3	22.1	13.0
Italy	518.3	541.1	4.4	−6.5
Luxembourg	10.8	6.1	−43.3	−28.0
Netherlands	215.8	230.1	6.1	−6.0
Portugal	64.6	79.3	22.4	27.0
Spain	305.8	380.2	23.2	15.0
Sweden	69.5	70.7	1.5	4.0
United Kingdom	741.9	637.9	−14.4%	−12.5
Total EU-15	4199	4030	−4.0	−8.0

- The average behavior should be in accordance with the real climate.
- Extremes should occur in the generated series in the way it will happen in a real warm period. This means that the generated series should be long enough to assure these extremes, and the series should be based on average values from nearby stations.

On some climate change issues (such as global warming), there is no disagreement among the scientists. The greenhouse effect is unquestionably real—it is essential for life on earth. Water vapor is the most important GHG; next is CO_2. Without a natural greenhouse effect, scientists estimate that the earth's average temperature would be $-18°C$ instead of its present $14°C$. There is also no scientific debate over the fact that human activity has increased the concentration of GHGs in the atmosphere (especially CO_2 from combustion of coal, oil, and gas). The greenhouse effect is also being amplified by increased concentrations of other gases, such as methane, nitrous oxide, and CFCs as a result of human emissions. Most scientists predict that rising global temperatures will raise the sea level and will increase the frequency of intense rain or snowstorms. Climate change scenarios sources of uncertainty and factors influencing the future climate are:

- the future emission rates of the GHGs (Table 1.14),
- the effect of these emissions on the GHGs concentrations in the atmosphere,
- the effect of this increase in concentration on the energy balance of the atmosphere,
- the effect of this change in energy balance on global and regional climate.

It has been known for a long time that urban centers have mean temperatures higher than their less-developed surroundings. The urban heat increases the average and peak air temperatures, which in turn affect the demand for heating and cooling.

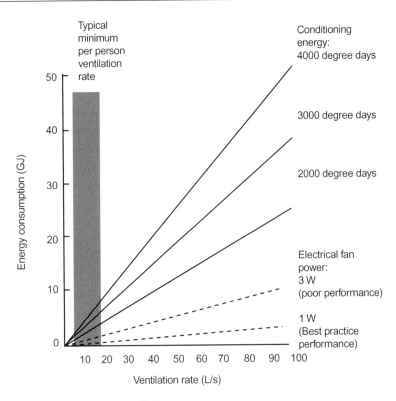

Figure 1.6 Energy impact of ventilation.

Higher temperatures can be beneficial in the heating season, lowering fuel use, but they exacerbate the energy demand for cooling in the summer times. In temperate climates, neither heating nor cooling may dominate the fuel use in a building, and the balance of the effect of the heat is less. As the provision of cooling is expensive with higher environmental cost, ways of using innovative alternative systems such as mop fan will be appreciated. The solar gains would affect energy consumption. Therefore lower or higher percentage of glazing or incorporating of shading devices might affect the balance between annual heating and cooling load. In addition to conditioning energy, the fan energy needed to provide mechanical ventilation can make a significant further contribution to energy demand. Much depends on the efficiency of design, both in relation to the performance of fans themselves and to the resistance to flow arising from the associated ductwork. Fig. 1.6 illustrates the typical fan and thermal conditioning needs for a variety of ventilation rates and climate conditions.

1.8 Recommendations

- Launching of public awareness campaigns among local investors particularly small-scale entrepreneurs and end users of RET to highlight the importance and benefits of renewable, particularly solar, wind, and biomass energies.

- Amendment of the encouragement of investment act to include furthers concessions, facilities, tax holidays, and preferential treatment to attract national and foreign capital investment.
- Allocation of a specific percentage of soft loans and grants obtained by governments to augment budgets of R&D related to manufacturing and commercialization of RET.
- Governments should give incentives to encourage the household sector to use renewable energy instead of conventional energy.
- Execute joint investments between the private sector and the financing entities to disseminate the renewable with technical support from the research and development entities.
- Availing of training opportunities to personnel at different levels in donor countries and other developing countries to make use of their wide experience in application and commercialization of RET particularly renewable energy.
- The governments should play a leading role in adopting renewable energy devices in public institutions, for example, schools, hospitals, government departments, and police stations for lighting, water pumping, water heating, communication, and refrigeration.
- To encourage the private sector to assemble, install, repair, and manufacture renewable energy devices via investment encouragement, more flexible licensing procedures.

1.9 Conclusion

There is strong scientific evidence that the average temperature of the earth's surface is rising. This was a result of the increased concentration of CO_2 and other GHGs in the atmosphere as released by burning fossil fuels. This global warming will eventually lead to substantial changes in the world's climate, which will, in turn, have a major impact on human life and the built environment. Therefore effort has to be made to reduce fossil energy use and to promote green energies, particularly in the building sector. Energy use reductions can be achieved by minimizing the energy demand, by rational energy use, by recovering heat, and by the use of more green energies. The study was a step toward achieving this goal. The adoption of green or sustainable approaches to the way in which society is run is seen as an important strategy in finding a solution to the energy problem. The key factors to reducing and controlling CO_2, which is the major contributor to global warming, are the use of alternative approaches to energy generation and the exploration of how these alternatives are used today and may be used in the future as green energy sources. Even with modest assumptions about the availability of land, comprehensive fuelwood farming programs offer significant energy, economic, and environmental benefits. These benefits would be dispersed in rural areas where they are greatly needed and can serve as linkages for further rural economic development. The nations as a whole would benefit from savings in foreign exchange, improved energy security, and socioeconomic improvements. With a ninefold increase in forest—plantation cover, the nation's resource base would be greatly improved.

References

ASHRAE, 1993. Energy Efficient Design of New Building Except New Low-Rise Residential Buildings. BSRIASHRAE Proposed Standards 90-2P-1993, Alternative GA.

American Society of Heating, Refrigerating, and Air Conditioning Engineers Inc., Atlanta, GA.

Barabaro, S., Coppolino, S., Leone, C., Sinagra, E., 1978. Global solar radiation in Italy. Sol. Energy 20, 431−438.

Baruah, D., 1995. Utilisation pattern of human and fuel energy in the plantation. J. Agric. Soil Sci. 8 (2), 189−192.

Bos, E., My, T., Vu, E., Bulatao, R., 1994. World Population Projection: 1994-95. World Bank by the John Hopkins University Press, Baltimore and London.

CAEEDAC, February 2000. A Descriptive Analysis of Energy Consumption in Agriculture and Food Sector in Canada. Final Report. Edmonton.

D'Apote, S.L., 1998. IEA biomass energy analysis and projections. In: Proceedings of Biomass Energy Conference: Data, analysis and Trends, 23−24 March 1998. OECD, Paris.

David, J.M., 2002. Developing hydrogen and fuel cell products. Energy World 303, 16−17.

Duchin, F., 1995. *Global Scenarios About Lifestyle and Technology, the Sustainable Future of the Global System.* United Nations University, Tokyo.

Duffie, J.A., Beckman, W.A., 1980. Solar Engineering of Thermal Processes. John Wiley & Sons, New York.

Dutt, B., 1982. Comparative efficiency of energy use in rice production. Energy 6, 25.

Energy Use in Offices, 2000. Energy Consumption Guide 19 (ECG019). Energy Efficiency Best Practice Programme. UK Government, London.

EWEA, 2003. Wind Force 12. Brussels.

Givoni, B., 1998. Climate Consideration in Building and Urban Design. Van Nostrand Reinhold, New York.

Hall, O., Scrase, J., 1998. Will biomass be the environmentally friendly fuel of the future? Biomass Bioenergy 15, 357−367.

IHA, 2003. World atlas & industry guide. Int. J. Hydropower Dams .

John, A., James, S., 1989. The Power of Place: Bringing Together Geographical and Sociological Imaginations. Unwin Hyman, Boston, MA.

Kammerud R., Ceballos E., Curtis B., Place W., Anderson B., 1984. Ventilation Cooling of Residential Buildings. ASHRAE Trans: 90 Part 1B. ASHRAE, Atlanta, GA.

OECD/IEA, 2004. Renewables for Power Generation: Status and Prospect. UK.

Pernille, M., 2004. Feature: Danish Lessons on District Heating. Energy Resource Sustainable Management and Environmental March/April 2004: 16−17.

Rees, W.E., 1999. The built environment and the ecosphere: a global perspective. Build. Res. Inf. 27 (4), 206−220.

Shaviv, E., 1989. The influence of the thermal mass on the thermal performance of buildings in summer and winter. In: Steemers, T.C., Palz, W. (Eds.), Science and Technology at the Service of Architecture. Kluwer Academic Publishers, Dordrecht, pp. 470−472.

Singh, J., 2000. Master of Science On Farm Energy Use Pattern in Different Cropping Systems in Haryana, India. International Institute of Management-University of Flensburg, Sustainable Energy Systems and Management, Germany.

Sitarz, D. (Ed.), 1992. Agenda 21: The Earth Summit Strategy to Save Our Planet. Earth Press, Boulder, CO.

Sivkov, S.I., 1964a. To the methods of computing possible radiation in Italy. Trans. Main Geophys. Obs. 160.

Sivkov, S.I., 1964b. On the computation of the possible and relative duration of sunshine. Trans. Main Geophys. Obs. 160.

Steele, J., 1997. Sustainable Architecture: Principles, Paradigms, and Case Studies. McGraw-Hill Inc, New York.

Thakur, C., Mistra, B., 1993. Energy requirements and energy gaps for production of major crops in India. Agric. Situat. India 48, 665−689.

United Nations, 2001. World Urbanisation Project: The 1999 Revision. The United Nations Population Division, New York.

Wu, J., Boggess, W., 1999. The optimal allocation of conservation funds. J. Environ. Econ. Manag. 38, 302−321.

Yaldiz, O., Ozturk, H., Zeren, Y., 1993. Energy usage in production of field crops in Turkey. In: 5th International Congress on Mechanisation and Energy Use in Agriculture, Kusadasi, Turkey.

Part 2

Energy financing schemas

The promotion of renewable energy communities in the European Union

2

Theodoros G. Iliopoulos
Hasselt University, Hasselt, Belgium

Chapter Outline

2.1 Overview 37
2.2 The link between the provision of energy services and the increase of energy efficiency 38
2.3 The efficiency gains stemming from distributed generation of energy production 39
2.4 The concept of renewable energy community 41
2.5 The promotion of renewable energy communities in EU law 44
2.6 The promotion of renewable energy communities in the draft National Energy and Climate Plans 47
2.7 Conclusion 50
References 51

2.1 Overview

This chapter examines the promotion of renewable energy communities in the European Union (EU), as a trend and a policy objective that aims to contribute to the adequate provision of energy services, to the increase of energy efficiency, as well as to the attainment of sustainable development and the low-carbon energy transition. Accordingly, the chapter presents the supranational legal framework for the renewable energy communities and examines the measures that have been and are expected to be employed by Member States to actually promote their development.

In terms of structure, Section 2.2 examines how the supranational legal order links the provision of energy services with the target of increasing energy efficiency. Next, Sections 2.3 and 2.4 focus on decentralization and renewable energy communities respectively as means for increasing energy efficiency. Section 2.5 focuses on the EU legal framework for the promotion of renewable energy communities, and Section 2.6 reviews how Member States plan to actually promote the development of renewable energy communities in the future. The analysis will be based on the national policy objectives expressed in the draft National Energy and Climate Plans (NECPs) that were submitted to the European Commission in 2019. Last, Conclusion concludes the chapter.

2.2 The link between the provision of energy services and the increase of energy efficiency

With the typical distinction between goods and services being conflated in the energy sector, it is becoming particularly challenging to define the concept of "energy services" (Nartova, 2017). Different definitions can be put forward, each one highlighting different energy policy objectives to be served. For instance, in the field of energy trade one could see energy services as comprising actions such as the production, the transportation, and the transmission and distribution of energy products (Nartova, 2017). In another example, in the area of tackling energy poverty, energy services have been defined as

> the benefits that energy carriers produce for human wellbeing. [...] Energy carriers, in turn, can be derived from a variety of primary energy sources; electricity for example can be generated from hydropower, petroleum, solar, or wind energy. From the point of view of the user, what matters is the energy service not the source. Whether in business, home, or community life, what matters are the reliability, affordability, and accessibility of the energy service (Hesselman, 2014; Modi et al., 2005).

A definition of "energy services" has also been adopted by the supranational legal order. Since this chapter focuses on the supranational framework, this is the definition that will be followed; given the autonomy of EU law, the content thus given to the notion of energy services is not dependent upon the understanding of others, such as international organizations or national laws (Pirker and Reitmeyer, 2015).

Accordingly, Article 2 (7) of the Energy Efficiency Directive 2012/27/EU, as amended by Directive (EU) 2018/2002, states that energy service means

> the physical benefit, utility or good derived from a combination of energy with energy-efficient technology or with action, which may include the operations, maintenance and control necessary to deliver the service, which is delivered on the basis of a contract and in normal circumstances has proven to result in verifiable and measurable or estimable energy efficiency improvement or primary energy savings.

Next, Article 2 (24) of the same Directive defines the concept of an energy service provider as "a natural or legal person who delivers energy services or other energy efficiency improvement measures in a final customer's facility or premises."

The same definitions are also used by the newly enacted Energy Union Governance Regulation 2018/1999/EU. Indeed, Articles 2 (47) and (56) define "energy service" and "energy service provider," respectively by referring to the definitions included in the Energy Efficiency Directive.

As per the earlier discussion, one can conclude that the EU legal order explicitly links the provision of energy services with the attainment of energy efficiency policy targets. Such a link is also affirmed by the fact that, according to Article 21, as well as to Annex I (3.2) and Annex III (3) of the Energy Union Governance Regulation EU 2018/1999, Member States are required to biannually report on energy efficiency to the Commission, and to describe therein inter alia their policy and the measures they take to promote energy services in the public sector.

Of course, this does not mean that energy services are only associated with energy efficiency, and that there is no other aspect that interests the EU legislator. Accordingly, it should be noted that the new Electricity Directive (EU) 2019/944 acknowledges in recital 59 that

> [e]nergy services are fundamental to safeguarding the well-being of the Union citizens. Adequate warmth, cooling and lighting, and energy to power appliances are essential services to guarantee a decent standard of living and citizens' health. Furthermore, access to those energy services enables Union citizens to fulfill their potential and enhances social inclusion. Energy poor households are unable to afford those energy services due to a combination of low income, high expenditure on energy and poor energy efficiency of their homes.

In a similar vein, Article 3 (3) (c) of the foregoing Energy Union Governance Regulation requires Member States to assess the number of households in energy poverty taking into account the necessary domestic energy services needed to guarantee basic standards of living in the relevant national context, existing social policy and other relevant policies, as well as indicative Commission guidance on relevant indicators for energy poverty.

In addition, the aforementioned Energy Efficiency Directive states in recital 20 that the provision of innovative energy services will increase competition because energy utilities can differentiate their product by providing complementary energy services, and recital 5 of the new Electricity Directive (EU) 2019/944 argues that this "will enable all consumers to fully participate in the energy transition, managing their consumption to deliver energy-efficient solutions which save them money and contribute to the overall reduction of energy consumption."

In conclusion, the provision of energy services in accordance with the supranational legal framework demands is inextricably linked with the increase of energy efficiency, but also with the empowerment of energy consumers. In this regard, the next part of this chapter will call attention to decentralization of energy production as a phenomenon that has the potential to serve both the efficiency and the consumer empowerment targets.

2.3 The efficiency gains stemming from distributed generation of energy production

It is not disputed that the supranational energy efficiency policy "aims to [stimulate consumers] to be more efficient in their energy use" (Lavrijssen, 2017). In pursuing this, regulators resort to demand-side management programs, to wit programs that aim to save energy through changing the patterns that consumers use energy

(Masters, 2004). This can be achieved through an overall reduction of energy consumption, which results in direct energy efficiency gains. But there are more demand-side management strategies. One could, for instance, reference a load management strategy that aspires to reduce consumption during peak hours and to increase consumption during off-peak hours. A diversion of consumption toward off-peak hours can be significantly facilitated by price signals that energy consumers can get thanks to technological developments. With smart meters, consumers have the possibility to get real-time price information about the energy they consume. And, of course, the price is more expensive at times of peak demand and, reversely, is low at times of low demand. Thus by responding to such price signals, consumers contribute to energy efficiency, while at the same time they have financial gains since their electricity bills are reduced (Hamilton et al., 2011). Efficiency gains can also be reaped with the help of other techniques, such as labeling energy-efficient appliances or properly informing consumers and raising their awareness about the merits of energy efficiency.

Distributed generation is another demand-side management technique (Kakran and Chanana, 2018); the term "distributed generation" refers to a "technology that allows power to be used and managed in a decentralized and small-scale manner, thereby siting generation close to load" (Cleveland and Morris, 2006). This is very much associated with self-consumption. Accordingly, a distributed generation model involves consumers generating on their own, fully or partly, the energy they need. A neologism has accurately characterized these actors *prosumers*, since they both produce and consume energy (Lavrijssen and Carrillo Parra, 2017; Parag and Sovacool, 2016). The rise of prosumers can make an electricity system more efficient, because the demand for energy from the central grid is reduced, and thus grid congestion is also reduced, especially during peak hours. At the same time, prosumers or self-consumers can be incentivized to actually self-consume the energy they produce; this will entail that energy consumption will be shifted to times when energy is abundantly generated by the decentralized installation. In addition, distributed generation is also linked with the promotion of renewable energy sources; indeed, the technologies that are used in a decentralized energy production system mostly exploit solar, wind or geothermal power, or biomass (European Commission, 2012). It is beneficial and in accordance with the low-carbon energy transition that such low-carbon energy sources replace fossil fuels. But solar and wind energy sources are intermittent sources, which means that generation does not have a steady flow, but depends on weather conditions, and more specifically on sun radiation and wind respectively. Yet once weather conditions are appropriate, energy is produced at a very low marginal cost and without depleting natural resources. Consequently, using the energy produced from solar and wind energy sources and not letting it unutilized have significant efficiency gains, as well as economic gains—not to mention its importance for the efforts to mitigate climate change.

Of course, distributed generation does not only relate to individual consumers or households. It also involves initiatives where groups of natural or legal persons own and develop one or more renewable energy installations. This is the case of renewable energy communities. Sections 2.4 and 2.5 examine the concept and the promotion of renewable energy communities under EU law respectively.

2.4 The concept of renewable energy community

Although one might intuitively understand what the term "renewable energy community" refers to, literature shows that it is actually an obscure term (Moroni et al., 2019; Wirth, 2014). Indeed, it has been proven significantly challenging to formulate a widely accepted definition that would contain all the key features of renewable energy communities in a satisfactory manner. To make things more complicated, many different terms have been brought into play. Accordingly, references have been made not only to renewable energy communities (Sokołowski, 2019), but also to energy communities (Moroni et al., 2019; Sokołowski, 2018), clean energy communities (Gui and MacGill, 2018), sustainable energy communities (Heaslip et al., 2016; Romero-Rubio and de Andrés Díaz, 2015), citizen energy communities (Hyysalo and Juntunen, 2011), civic energy communities (Verkade and Höffken, 2019; de Vries et al., 2016), renewable energy cooperatives (Heras-Saizarbitoria et al., 2018; Herbes et al., 2017; Huybrechts and Mertens, 2014). Each of these resembling terms highlights different aspects of similar phenomena. In this regard, the term "sustainable energy communities" goes beyond the objective to promote renewable energy sources and also encompasses the aim for a "rational use of energy" (Romero-Rubio and de Andrés Díaz, 2015); the term "civic energy communities" focuses on the local and at the same time innovative, unconventional character of the initiative (Verkade and Höffken, 2019; de Vries et al., 2016), while the word "cooperative" emphasizes a democratic decision-making model and the aspiration for a collective achievement of economic and noneconomic goals (Heras-Saizarbitoria et al., 2018; Herbes et al., 2017; Huybrechts and Mertens, 2014).

From all the different approaches suggested, one can remark the attempt of Moroni et al. (2019) to provide an objective, "general (nonideological) definition of this kind of community" and, next, to put forward a taxonomy that systematizes the different nuances of the broad concept of "energy community." Accordingly, the authors put forward the general term "energy-related communities", which denotes "groups of individuals who voluntarily accept certain rules for the purposes of shared common objectives (*only* or *also*) relating to energy; that is: (1) purchasing energy as collective groups, (2) and/or managing energy demand and supply, (3) and/or generating energy." Thus energy communities are "intentional" or "contractual" communities, and they can be classified into place-based ones, if they are connected with a specific area, or nonplace-based ones, that is, communities that are not linked with a certain area. Although the authors refer only to "individuals," their definition should be broadened so that energy communities be defined as entities addressing both natural and legal persons.

Turning to the supranational legal framework, the new Renewable Energy Directive (EU) 2018/2001, also known as "the RED II," and the new Electricity Directive (EU) 2019/944 adopted for the first time legal provisions that explicitly deal with the regulation of the energy communities (Sokołowski, 2018). But, similarly to the polyphonic situation in scientific literature, EU law also adopted a heterogeneous terminology.

Starting with the RED II, it refers to the concept of "renewable energy community," which is defined as a legal entity

1. which, in accordance with the applicable national law, is based on open and voluntary participation, is autonomous, and is effectively controlled by shareholders or members that are located in the proximity of the renewable energy projects that are owned and developed by that legal entity;
2. the shareholders or members of which are natural persons, small and medium enterprises (SMEs), or local authorities, including municipalities;
3. the primary purpose of which is to provide environmental, economic, or social community benefits for its shareholders or members or for the local areas where it operates, rather than financial profits.

As per the earlier discussion, the supranational renewable energy communities indeed encompass features that literature has attributed to such entities. They are intentional communities, open to membership, and they are characterized by an idiosyncratic governance model. They are normally multipurpose entities, in the sense that they aim to link their operation with the delivery of a multitude of benefits. It also seems that EU law ascribes to renewable energy communities a strong connection with the area in which they operate. The definition states that they are controlled by shareholders or members "that are located in the proximity of the renewable energy projects." In addition, the expected benefits are explicitly to be reaped primarily by the local areas where the renewable energy communities operate. Establishing such a link between a renewable energy community and a location is a clear policy objective, the *ratio* of which is partly provided by recital 70 of the RED II. According to it,

> [t]he participation of local citizens and local authorities in renewable energy projects through renewable energy communities has resulted in substantial added value in terms of local acceptance of renewable energy and access to additional private capital which results in local investment, more choice for consumers and greater participation by citizens in the energy transition. Such local involvement is all the more crucial in a context of increasing renewable energy capacity (see also Bauwens, 2016).

Last, the RED II does not require renewable energy communities to only engage in renewable energy projects. Of course, the foregoing definition states that they own and develop such projects. But such a clause does not necessarily prevent a renewable energy community from owning and developing a combined heat and power project. The question is, will entities that mostly exploit such combined heat and power systems, and are only slightly involved in renewable energy projects, still fall within the scope of the concept of a renewable energy community?

On the other hand, the EU legislator opted to refer to "citizen energy communities" in the Electricity Directive (EU) 2019/944. Nevertheless, the Directive does not contain a definition for this concept. However, taking a closer look at the relevant recitals reveals the key characteristics of a citizen energy community, and these are quite similar to the features of a renewable energy community. In this regard, according to recitals 43, 44, and 46 of the Electricity Directive 2019/944, citizen energy communities are also contractual communities that are open to membership to all categories of entities, with special reference being made to the right of households to participate in them. In addition, recital 46 starts with stating that "citizen energy communities constitute a new type of entity due to their membership structure, governance requirements, and purpose." This acknowledges that citizen energy communities have a distinctive governance model and that they aim to serve special purposes. But here—and this is an important difference from renewable energy communities—these purposes mostly relate to the active participation of smaller actors, such as households, to the energy market. Thus recital 43 highlights the need for ensuring "an enabling framework, fair treatment, a level playing field, and a well-defined catalogue of rights and obligations." Of course, similar rights shall also be granted to members of renewable energy communities, in accordance with Article 22 of the RED II. But it is the *ratio*, the purpose that makes the two legal instruments differ.

Moreover, recital 46 of the new Electricity Directive states that citizen energy communities

> should be allowed to operate on the market on a level playing field without distorting competition, and the rights and obligations applicable to the other electricity undertakings on the market should be applied to citizen energy communities in a nondiscriminatory and proportionate manner. Those rights and obligations should apply in accordance with the roles that they undertake, such as the roles of final customers, producers, suppliers or distribution system operators.

Next remark is that citizen energy communities do not have an explicit strong link with a location. Nevertheless, recital 44 hints that the law requires such a link to exist, since it provides that citizen energy communities "are considered to be a category of cooperation of citizens or local actors that should be subject to recognition and protection under Union law." In addition, citizen energy communities are not required to own or develop renewable energy projects. This is another important difference between the two types of energy communities. Nevertheless, in practice it is likely that such a citizen energy community does own and develop renewable energy projects, which means that it will qualify as a renewable energy community too—always provided that it primarily aims not to financial profits, but to delivering environmental, economic, or social community benefits. Consequently, the two types of energy communities are close to each other. In addition, both the two relevant Directives aim to create a legal framework that creates clear rights and obligations for persons involved in such energy communities, so that participation in them be encouraged.

After the terms have been clarified, the next part will examine how EU law promotes the development of renewable energy communities.

2.5 The promotion of renewable energy communities in EU law

As analyzed in the previous part, renewable energy communities own and develop renewable energy projects. Consequently, it is normal that the tools employed for the promotion of the use of renewable energy sources in general are also relevant to them. And these tools are the so-called "support schemes." The term is defined by Article 2 (5) of the RED II as meaning

> any instrument, scheme or mechanism applied by a Member State, or a group of Member States, that promotes the use of energy from renewable sources by reducing the cost of that energy, increasing the price at which it can be sold, or increasing, by means of a renewable energy obligation or otherwise, the volume of such energy purchased.

The same definition continues with an indicative catalogue of support schemes, which includes the most widespread ones, to wit "investment aid, tax exemptions or reductions, tax refunds, renewable energy obligation support schemes including those using green certificates, and direct price support schemes including feed-in tariffs and sliding or fixed premium payments."

Accordingly, renewable energy communities have been benefitting from various support schemes. For instance, Member States such as Denmark, the Netherlands, Spain, or the United Kingdom have enacted special tax regimes in order to incentivize participation in renewable energy communities (Verkade and Höffken, 2019; Heras-Saizarbitoria et al., 2018; Roberts et al., 2014). But it is direct price support schemes that have been the spearhead of the promotion of renewable energy communities, and mostly feed-in tariffs. Feed-in tariffs are long-term contracts that oblige the grid operator to purchase the renewable energy produced at a guaranteed fixed price (Pyrgou et al., 2016; Couture and Gagnon, 2010). This is a particularly attractive support scheme because it materially reduces the risk of investors. Indeed, the beneficiaries are relieved of worrying about future demand for energy or about price fluctuations, because regardless of the market conditions they will sell their whole production, and at a fixed price. This also allows investors to estimate their profits, as well as the time at which their investment will start yielding net profits, as accurately as possible. Feed-in tariffs have provided strong incentives to citizens to be engaged in renewable energy communities. Indeed, successful and well-established paradigms of renewable energy communities, such as the German one, have relied on feed-in tariffs. Other Member States that more recently intended to accelerate their progress in this field, such as Spain, have also resorted to such instruments (Herbes et al., 2017; Romero-Rubio and de Andrés Díaz, 2015).

Nevertheless, applying support schemes that disregarded market conditions proved nonviable in the long-term (Iliopoulos, 2016). Feed-in tariffs were so attractive that mobilized too many investments in renewable energy projects. Such an increase in supply would normally lead to a drop of energy prices. But because of the feed-in tariffs, prices were contractually fixed at the levels before a massive investment took place and stayed there. Consequently, authorities had to buy too expensive energy and from too many producers. Such a model could not be afforded and in the end, in early 2010s, Members States that based their national energy policy on feed-in tariffs imposed unilateral and abrupt changes in the contractual terms, such as reductions of the agreed tariffs.

At the same time, Members States started resorting to a more market-oriented variation of feed-in tariffs, namely feed-in premiums. In a feed-in premium scheme, the renewable energy generators are only guaranteed to receive the market price plus a bonus payment, an add-on fee. EU law has recently affirmed this turn to feed-in premiums. Article 4 (3) of the RED II states that "with regard to direct price support schemes, support shall be granted in the form of a market premium," with exemptions, such as feed-in tariffs, allowed only for small-scale installations and demonstration projects.

But, of course, this model is less attractive for investors compared to feed-in tariffs. Consequently, as might be expected, the combination of the reduction of fixed tariffs and the introduction of premiums had brought a certain deceleration in the promotion of renewable energy sources, as well as in the interest in participation in renewable energy communities. (Herbes et al., 2017; Iliopoulos, 2016; Parkhill et al., 2015; Romero-Rubio and de Andrés Díaz, 2015). This deceleration might prove an interim one if the new supranational legal framework, in conjunction with the reformed national regimes, manage to give a new impetus to the deployment of renewable energy sources.

Another price instrument that is particularly relevant for the promotion of renewable energy sources is net metering and, especially, virtual net metering. Net metering regimes allow prosumers to feed the electricity they produce into the grid. In return, they receive a remuneration that normally equals the retail electricity price. This remuneration normally is a credit on the next electricity bill that is reduced accordingly (Jacobs, 2017; Butenko, 2016). As for virtual net metering, it allows the electricity produced in one site and fed into the grid to be used for remunerating more persons, regardless of whether the beneficiaries are connected with the site that produces energy or not. Of course, virtual net metering is particularly important for the promotion of renewable energy communities, which involve more persons collectively owning, developing, and administering renewable energy projects.

However, net metering regimes are still under development and have not been widespread yet. As for virtual net metering, it only finds a limited application in the EU, with Greece, Italy, Lithuania, and Spain having enacted such regimes. This can be partly attributed to the fact that the European Commission, and possibly Member States too, is rather skeptical about the mid- and long-term viability of net metering regimes (European Commission, 2015). This is because beneficiaries are remunerated at the retail price, although the good they offer, that is, electricity, would be purchased by the suppliers at a wholesale price. Such a system raises economic efficiency issues and might spur a chain of events that will very much resemble what happened in the past with feed-in tariffs (Iliopoulos, 2019).

As accurate as such reservations are, they should not prevent Member States from enacting net metering, but should stimulate legislators to carefully design such support schemes so that renewable energy sources be promoted without overproduction and overcompensation posing grid stability risks and financial stability risks, respectively. An interesting approach is given by the Cypriot net metering model. By way of a safety net, Cypriot authorities announce annual net metering plans, and they set a ceiling on the aggregate capacity that can benefit from the support granted. Thus they are in a position to assess the grid stability and the financial situation on a rolling basis. The conclusions drawn are then taken into account before the next annual net metering plan is issued. This allows a constant monitoring of the system and diminishes the possibility that authorities and prosumers are caught by surprise.

EU law does not contain clear rules about net metering. But it seems that net metering, as a direct price support scheme, should only apply to small-scale installations and to demonstration projects, in accordance with the abovementioned Article 4 (3) of the RED II.

In terms of renewable energy communities, it is not always easy for them to enter the available support schemes. For instance, labyrinthine administrative requirements might be a significant problem for communities that, unlike bigger companies, do not have special departments that know how to deal with procedures and how to communicate with authorities. Similarly, if the support is granted in a tendering procedure, it might be hard for communities to compete with larger market actors in order to be awarded with the contract. The EU legislator is aware of such difficulties and, hence, the RED II aims to ensure that they can be addressed. Accordingly, recital 26 of the Directive states that

> *Member States should ensure that renewable energy communities can participate in available support schemes on an equal footing with large participants. To that end, Member States should be allowed to take measures, such as providing information, providing technical and financial support, reducing administrative requirements, including community-focused bidding criteria, creating tailored bidding windows for renewable energy communities, or allowing renewable energy communities to be remunerated through direct support where they comply with requirements of small installations.*

This recital gives Member States a road map about what should be done so that support to renewable energy communities is facilitated. This recital is also reflected in the law, in Article 22 of the RED II, which specifically deals with renewable energy communities. In this regard, Article 22 (7) uses a coercive language and provides that "Member States shall take into account specificities of renewable energy communities when designing support schemes in order to allow them to compete for support on an equal footing with other market participants." Therefore, Member States are required to tackle the factors that prevent renewable energy communities from benefitting from support schemes, but they have the discretion to identify what these factors are.

Nevertheless, Article 22 (7) should be read in conjunction with Article 22 (4) of the RED II, which is broader, as it refers to numerous factors that are crucial for the promotion and development of renewable energy communities. And, in essence, this paragraph establishes the minimum content that an enabling legal framework for renewable energy communities shall have. From the different requirements set down by the law, it is indicatively highlighted here that Member States shall ensure that renewable energy communities can benefit from tools that facilitate access to finance and information, and that they and their members are subject to fair, proportionate and transparent treatment, procedures, and charges, without facing unjustified regulatory and administrative barriers. This is very important, because access to finance has been reported to be one of the major issues that persons interested in energy communities need to overcome, while lengthy and complicated procedures or high charges further exacerbate the problem (Herbes et al., 2017; Romero-Rubio and de Andrés Díaz, 2015).

In addition, the RED II also bestows renewable energy communities and their members with certain rights. These rights are not directly linked with the support schemes and the financing of renewable energy projects, but they complement such instruments because they allow citizens to actively participate in the energy markets and to take important initiatives. The core relevant provision can be found in Article 22 (1); it provides that Member States shall ensure that consumers, and especially households "are entitled to participate in a renewable energy community while maintaining their rights or obligations as final customers, and without being subject to unjustified or discriminatory conditions or procedures that would prevent their participation in [it]." In paragraph 2, it is further specified that Member States shall ensure that renewable energy communities are entitled to produce, consume, store, and sell renewable energy, but also to share the energy they produce between their members, as well as to access energy markets. Of course, Member States retain an important degree of discretion about the exact terms, under which the foregoing requirements will be put into practice.

For the sake of completion, it should be noted that the new Electricity Directive also contains certain rights that energy communities and their members shall enjoy. Of course, this Directive only refers to citizen energy communities, but, as already explained, it is highly likely that a citizen energy community will also fall within the scope of the term "renewable energy community." To a large extent the two Directives confer matching rights on the communities, such as the right to a nondiscriminatory and proportionate treatment or the right to access to all electricity markets.

From the earlier discussions, it is becoming clear that EU law encourages the rise of renewable energy communities and requires Member States to form their energy policies having such initiatives into account and to take measures that will facilitate their actual development. In this regard, the next part looks at the draft NECPs, as submitted to the Commission in 2019 and examines the measures for renewable energy communities therein contained.

2.6 The promotion of renewable energy communities in the draft National Energy and Climate Plans

The NECPs set out the national energy policy objectives as well as the measures that will be taken so that these objectives are met. Accordingly, examining the draft

NECPs that were submitted to the Commission in 2019 reveals interesting information about the zeal of Member States to actually promote renewable energy communities and about the instruments on which such promotion will be founded.

One can see that about one-third of the Member States highlight the rise of renewable energy communities as an important feature of their energy policy, and they refer to specific measures that will serve this development. Perhaps the most elaborate plans vis-à-vis renewable energy communities are, in alphabetical order, the Greek, the Irish, and the Spanish one.

Greece makes extensive references to the merits the promotion of renewable energy communities has and sets down a quantitative target for renewable energy communities: the intention of policy makers is that their capacity will exceed 500 MW by 2030. Furthermore, the Greek draft NECP also provides a detailed presentation of the measures that will be supporting the operation of renewable energy communities. These include not only special financing mechanisms and support schemes, such as virtual net metering that has been already enacted since 2017, but also other incentives and facilitating measures, such as preferential participation in tendering procedures or prioritized examination of the authorization application that such communities file to the authorities.

Turning to the Irish draft NECP, it states, in verbatim, that "Ireland strongly supports the ambitions behind the establishment of the rights and entitlements associated with both renewable self-consumers and renewable energy communities within the [RED II]." It also contains specific measures, such as the enactment of investment opportunities for communities and citizens, the enactment of pilot tariffs schemes plus a grant for solar micro-generation and storage, or the implementation of two programs for local energy communities, namely the "better energy communities" for funding installation of heat pumps, solar thermal and solar PV, and the "sustainable energy communities network" for developing technical and project management skills.

Next, Spain presents a catalogue of numbered energy policy targets and measures, with some of them addressing renewable energy communities. Accordingly, measure 1.3, entitled "development of self-consumption using renewables and distributed generation," explicitly aims to promote the development of local energy communities and sets down certain mechanisms, such as the introduction of indicative targets for self-consumption, the provisions of soft financing schemes, the involvement of experts in the management of self-consumption projects, and the simplification of administrative procedures so that locals also partake in such projects.

The Italian draft NECP is also rather concrete with regard to renewable energy communities. More specifically, it emphasizes Italy's intention to promote "individual and collective self-consumption" and states specific instruments, such as the application of net metering and the simplification of the authorization procedures; in addition, it declares that an even more elaborate plan will be announced soon. Similarly, the Netherlands refer to specific measures for the promotion of renewable energy communities, to wit the application of a preferential tax regime, while the government will assess "whether a development facility can be set up that allows energy cooperatives to fund development costs."

Interesting are also the Austrian, the Bulgarian, and the French draft NECPs. Austria states that a Renewable Energy Expansion Act will be in force in 2020, and it will be incentivizing decentralization and the operation of renewable energy communities. But the draft NECP provides little information about funding mechanisms, while the relevant programs therein included refer to the self-marketing of such projects, to cooperation structures, to exchanges of best practices, and so on. This might signal an interesting approach not to found the promotion of renewable energy communities on financing instruments. In a similar vein, Bulgaria, which also emphasizes her aspiration to materially promote renewable energy communities, refers to the "Energy from Renewable Sources Act" that already contains certain favorable provisions for self-consumers. Next, it is planned that regional measures will be adopted for facilitating the exchange of best practices, for training to strengthen the regulatory, technical, and financial expertise, and for informing about the available financing opportunities. As for the French NECP, it mentions the support for the development of renewable energy communities as crucial for the increase in supply of electricity from renewable energy sources. And while it contains a number of measures that could assist in this direction, it does not explicitly target renewable energy communities; the exception is the program "Villes solaire" that relates to the promotion of photovoltaic panels and explicitly addresses renewable energy communities; but there is no further elaboration on it.

However, the majority of the Member States does not give a clear picture about the measures that will be taken to support energy communities. Most of them acknowledge the importance of renewable energy communities, and they state their willingness to promote their development, but they do not provide many details about the instruments that will be serving this objective. A few Member States, such as Hungary, Luxembourg, or Romania briefly indicate general measures that might contribute to this target, such as special tenders for SMEs, feed-in tariffs for small-scale solar production at roof level or tax reliefs for self-consumers—but it is not planned that such measures will be designed especially for renewable energy communities. Another interesting draft NECP is the one submitted by Lithuania, which has set down specific targets for self-consumption, but has been relatively silent on how these targets will be met. Interesting are also the draft NECPs of the Nordic countries. Finland and Sweden place the emphasis on informative instruments, whereas Denmark underlines the long tradition in the country for energy communities, especially in the district heating sector and in renewable energy projects. Other Member States only express their intention to put forward more elaborate plans about policies for renewable energy communities in the future. Last, several Member States do not make any reference to the promotion of renewable energy communities. In this regard, one could highlight the Maltese draft NECP, which does the exact opposite. It states that "[i]n view of the structure of the Maltese electricity system in which there is no electricity supply market (Enemalta is designated as an exclusive supplier in Malta), it is not foreseen that renewable energy communities will develop". Of course, all Member States have to transpose the RED II and the new Electricity Directive into national legislation and to comply with EU law.

2.7 Conclusion

The rise of renewable energy communities is linked with a number of positive developments in the field of energy policy. Indeed, renewable energy communities can be regarded as a vehicle for increasing energy efficiency, contributing to the low-carbon energy transition target as well as empowering energy consumers, since they can significantly promote decentralization and self-consumption. Consequently, Member States have already started employing instruments that will be supporting the development of such initiatives. These instruments mostly match the support schemes for renewable energy sources, to wit the typical instruments that have been used for the promotion of renewable energy sources. Nevertheless, the typical form under which the support schemes have been operated needs certain adaptations. This is not only because a renewable energy community is a special kind of a renewable energy project that needs special treatment, but also because the lessons learned from national support frameworks show that the traditional support schemes need reforms so that they can be viable in the long-term.

In front of this picture, the EU legislator has recently enacted several legal acts that deal with energy policy and with energy communities in specific. The relevant provisions aim not only to ensure that Member States will indeed enact support schemes that will be specially designed to accommodate to the needs of energy communities, but also to ensure that Member States will facilitate participation in energy communities through an enabling legal framework. Thus the new Renewable Energy Directive and the new Electricity Directive contain certain requirements regarding the support schemes' design, and they also contain a number of rights that self-consumers and energy communities shall be bestowed with.

Nevertheless, this is a supranational legal framework, the effectiveness of which depends to a large extent on the specific measures that will be taken at the national level. Accordingly, the Energy Union Governance Regulation requires Member States to submit to the Commission detailed plans that present the national energy policy and the instruments that are to be enacted so that EU energy objectives are attained. Member States have already submitted the draft NECPs. These drafts show a relatively irresolute stance vis-à-vis the exact measures to be taken to deliver the development of renewable energy communities. Few Member States have already put forward a detailed plan, while a number of them have only made general and vague references to broader policy objectives.

In short, the EU legal framework already ensures that self-consumers will enjoy certain rights and that the new legal landscape will be more favorable for energy communities. But will such a landscape be a truly enabling one? This also depends on the implementation of the Directives by Member States. At the moment, most of them seem zealous toward this direction, but it is the concrete measures to be taken that will give a determinate answer.

References

Bauwens, T., 2016. Explaining the diversity of motivations behind community renewable energy. Energy Policy 93, 278–290.
Butenko, A., 2016. Sharing Energy. Eur. J. Risk Regul. 7 (4), 701–716.
Cleveland, C., Morris, C., 2006. Dictionary of Energy. Elsevier, Amsterdam, p. 120.
Couture, T., Gagnon, Y., 2010. An analysis of feed-in tariff remuneration models: Implications for renewable energy investment. Energy Policy 38 (2), 955–965.
de Vries, G., Boon, W., Peine, A., 2016. User-led innovation in civic energy communities. Environ. Innov. Societal Transit. 19, 51–65.
Directive (EU) 2019/944 of the European Parliament and of the Council of 5 June 2019 on Common Rules for the Internal Market for Electricity and Amending Directive 2012/27/EU OJ L 158, 14.6.2019, pp. 125–199.
Directive (EU) 2018/2001 of the European Parliament and of the Council of 11 December 2018 on the Promotion of the Use of Energy from Renewable Sources OJ L 328, 21.12.2018, pp. 82–209.
Directive (EU) 2018/2002 of the European Parliament and of the Council of 11 December 2018 Amending Directive 2012/27/EU on Energy Efficiency, OJ L 328, 21.12.2018, pp. 210–230.
Directive 2012/27/EU of the European Parliament and of the Council of 25 October 2012 on Energy Efficiency, Amending Directives 2009/125/EC and 2010/30/EU and Repealing Directives 2004/8/EC and 2006/32/EC OJ L 315, 14.11.2012, pp. 1–56.
European Commission, 2012. Communication from the Commission to the European Parliament, the Council, the European Economic and Social Committee and the Committee of the Regions. Renewable Energy: A Major Player in the European Energy Mmarket. COM 271 Final, p. 9.
European Commission, 2015. Commission Staff Working Document. Best practices on Renewable Energy Self-consumption. SWD 141 Final, p. 10.
Gui, E., MacGill, I., 2018. Typology of future clean energy communities: An exploratory structure, opportunities, and challenges. Energy Research & Social Science 35, 94–107.
Hamilton, B., Thomas, C., Park, S.J., Choi, J.-G., 2011. The customer side of the meter. In: Sioshansi, F. (Ed.), Smart Grid: Integrating Renewable, Distributed and Efficient Energy, first ed. Elsevier Inc, Oxford.
Heaslip, E., Costello, G.J., Lohan, J., 2016. Assessing good-practice frameworks for the development of sustainable energy communities in Europe: lessons from Denmark and Ireland. J. Sustain. Dev. Energy Water Environ. Syst. 4 (3), 307–319.
Heras-Saizarbitoria, I., Sáez, L., Allur, E., Morandeira, J., 2018. The emergence of renewable energy cooperatives in Spain: a review. Renew. Sustain. Energy Rev. 94, 1036–1043.
Herbes, C., Brummer, V., Rognli, J., Blazejewski, S., Gericke, N., 2017. Responding to policy change: new business models for renewable energy cooperatives – Barriers perceived by cooperatives' members. Energy Policy 109, 82–95.
Hesselman, M., 2014. Realizing 'universal access to modern energy services. In: McCann, A., Rooij, M., Hallo de Wolf, A., Neerhof, A. (Eds.), When Private Actors Contribute to Public Interests, first ed. Eleven International Publishing, The Hague, p. 111.
Huybrechts, B., Mertens, S., 2014. The relevance of the cooperative model in the field of renewable energy. Ann. Public. Cooperative Econ. 85 (2), 193–212.

Hyysalo, S., Juntunen, J., 2011. User innovation and peer assistance in small-scale renewable energy technologies. In: Davidson, D., Gross, M. (Eds.), The Oxford Handbook of Energy and Society, first ed. Oxford University Press, New York.

Iliopoulos, T., 2016. Renewable energy regulation: feed-in tariff schemes under recession conditions? Eur. Netw. Law Regul. 4 (2), 110−117.

Iliopoulos, T., 2019. Regulating smart distributed generation electricity systems in the European Union. In: Reins, L. (Ed.), Regulating New Technologies in Uncertain Times, first ed. T.M.C. Asser Press, The Hague.

Jacobs, S., 2017. The energy prosumer. Ecol. Law Q. 43 (3), 519−579.

Kakran, S., Chanana, S., 2018. Smart operations of smart grids integrated with distributed generation: a review. Renew. Sustain. Energy Rev. 81, 524−535.

Lavrijssen, S., 2017. Power to the energy consumers. Eur. Energy Environ. Law Rev. 26 (6), 172−185. 175.

Lavrijssen, S., Carrillo Parra, A., 2017. Radical prosumer innovations in the electricity sector and the impact on prosumer regulation. Sustainability 9 (7), 1207.

Masters, G., 2004. Renewable and Efficient Electric Power Systems, first ed. John Wiley & Sons, Inc., Hoboken, New Jersey, pp. 292−293.

Modi, V., McDade, S., Lallement, D., Saghir, J., 2005 Energy Services for the Millennium Development Goals. The International Bank for Reconstruction and Development/The World Bank and the United Nations Development Programme, Washington, DC, p. 9.

Moroni, S., Alberti, V., Antoniucci, V., Bisello, A., 2019. Energy communities in the transition to a low-carbon future: a taxonomical approach and some policy dilemmas. J. Environ. Manag. 236, 45−53.

Nartova, O., 2017. Energy Services. In: Cottier, T., Nadakavukaren Schefer, K., Baumann, J., Imeli, B., Powell, J., Gilgen, R. (Eds.), Elgar Encyclopedia of International Economic Law. Edward Elgar Publishing, Cheltenham, p. 448.

Parag, Y., Sovacool, B., 2016. Electricity market design for the prosumer era. Nat. Energy 1 (4).

Parkhill, K., Shirani, F., Butler, C., Henwood, K., Groves, C., Pidgeon, N., 2015. 'We are a community [but] that takes a certain a mount of energy': exploring shared visions, social action, and resilience in place-based community-led energy initiatives. Environ. Sci. Policy 53, 60−69.

Pirker, B., Reitmeyer, S., 2015. Between discursive and exclusive autonomy − opinion 2/13, the protection of fundamental rights and the autonomy of EU Law. Camb. Yearb. Eur. Leg. Stud. 17, 168−188.

Pyrgou, A., Kylili, A., Fokaides, P., 2016. The future of the Feed-in Tariff (FiT) scheme in Europe: the case of photovoltaics. Energy Policy 95, 94−102.

Regulation (EU) 2018/1999 of the European Parliament and of the Council of 11 December 2018 on the Governance of the Energy Union and Climate Action, Amending Regulations (EC) No 663/2009 and (EC) No 715/2009 of the European Parliament and of the Council, Directives 94/22/EC, 98/70/EC, 2009/31/EC, 2009/73/EC, 2010/31/EU, 2012/27/EU and 2013/30/EU of the European Parliament and of the Council, Council Directives 2009/119/EC and (EU) 2015/652 and Repealing Regulation (EU) No 525/2013 of the European Parliament and of the Council OJ L 328, 21.12.2018, pp. 1−77.

Roberts, J., Bodman, F., Rybski, R., 2014. Community Power: Model Legal Frameworks fo Citizen-Owned Renewable Energy. ClientEarth, London.

Romero-Rubio, C., de Andrés Díaz, J., 2015. Sustainable energy communities: a study contrasting Spain and Germany. Energy Policy 85, 397−409.

Sokołowski, M., 2019. Renewable Energy Communities in the Law of the EU, Australia, and New Zealand. European Energy and Environmental Law Review 28 (2), 34−46.

Sokołowski, M., 2019. European Law on the Energy Communities: a Long Way to a Direct Legal Framework. European Energy and Environmental Law Review 27 (2), 60–70.

Verkade, N., Höffken, J., 2019. Collective energy practices: a practice-based approach to civic energy communities and the energy system. Sustainability 11 (11), 3230.

Wirth, S., 2014. Communities matter: institutional preconditions for community renewable energy. Energy Policy 70, 236–246.

Financial schemes for energy efficiency projects: lessons learnt from in-country demonstrations

Charikleia Karakosta, Aikaterini Papapostolou, George Vasileiou and John Psarras
Energy Policy Unit (EPU-NTUA), Decision Support Systems Laboratory, School of Electrical and Computer Engineering, National Technical University of Athens, Athens, Greece

Chapter Outline

3.1 Introduction 55
3.2 The proposed methodology 58
3.3 Innovative financing schemes 59
    3.3.1 Crowdfunding 60
    3.3.2 Energy performance contracting 60
    3.3.3 Green bonds 60
    3.3.4 Guarantee funds 61
    3.3.5 Revolving funds 61
    3.3.6 Soft loans 61
    3.3.7 Third-party financing 62
3.4 Case study countries 62
    3.4.1 Bulgaria 62
    3.4.2 Greece 65
    3.4.3 Lithuania 66
    3.4.4 Spain 67
3.5 Key actors identification 67
3.6 Knowledge transfer 68
    3.6.1 Peer-to-Peer learning 69
    3.6.2 Capacity building activities 72
3.7 Conclusions 72
References 73

3.1 Introduction

Indeed, one of the main actions to tackle climate change and its impact is energy efficiency (Doukas et al., 2011, 2014; EC, 2014; Karakosta and Askounis, 2010; Marinakis et al., 2013; Miller and Carriveau, 2018). European Commission (EC) in 2008 has reached a political agreement, which includes a binding energy efficiency target of 32.5%, for 2030, for the European Union (EU), with a clause for an upward revision by 2023. Moreover, the EU has set itself a long-term goal of reducing greenhouse gas emissions by 80%–95% compared to 1990 levels by 2050.

Moving forward, the EC adopted the Paris Agreement (United Nations, 2015) on climate change and the United Nations 2030 Agenda for Sustainable Development (UN, 2015) in 2015, and thus many of its priorities (EC, 2014, 2015, 2018a) for 2014–20 feed into the Union's energy and climate goals and work toward delivering the 2030 Agenda's for Sustainable Development Goals. At the end of 2016, the Commission appointed a high-level expert group on sustainable finance (EC, 2018c). In 2018, the expert group published its final report (EC, 2018d, 2019a), offering a comprehensive vision on how to build a sustainable finance strategy for the EU.

One of the open questions today in mainstreaming energy efficiency finance is: what are the effective ways to drive new finance for energy efficiency investments (Becqué et al., 2016; G20 Energy Efficiency Investment Toolkit, 2017)? In this respect, literature review indicates, among others, a lack of evidence on the performance of energy efficiency investment benefits, as well as of commonly agreed procedures and standards for energy efficiency investment (Cooremans and Schönenberger, 2019; EEFIG Underwriting Toolkit, 2017; Karakosta et al., 2011; Painuly et al., 2003; Sarkar and Singh, 2010; Zhan et al., 2018). There are also many problems that a potential investor needs to put up with the initial investment (Schlein et al., 2017), and the project developers (small in most cases) are struggling to finance their needs. This is particularly true during the first stages of investments generation and preselection/preevaluation (Doukas, 2018; Triple-A, 2019).

The "gap" can be identified in the concept development phase of energy efficiency investments. Indeed, project developers spent a huge amount of hours auditing one plant's potential energy savings, but in most cases, never actually carrying out the project, because they cannot convince investors to give the capital needed to do the work (BPIE, 2010). Private investors often lack the knowledge to understand how project developers do business, especially at an early state of project identification (Boza-Kiss et al., 2017). At the same time, project developers do not have the expertise or resources to make a convincing case for the investors (BPIE, 2014).

Indeed, partly due to the heterogeneity of energy efficiency investments and due to the immaturity of the market for such investments, the relative costs for project development, finance documentation, processing, and aggregation (together "transaction costs") are high, making entry into this business unattractive for many financial institutions (EEFIG, 2015). Usually, the investment institutions lack the technical knowledge and experience in understanding which project constitutes an "energy efficiency investment"; this often creates a lack of trust in such investments, acting as a barrier to including energy efficiency projects in the investment portfolio, even though they are often robust, with a guaranteed return. These institutions seek strong evidence of profitability in such projects before they are willing to support them, and this is particularly important for the small-scale projects.

The main challenge lies on the mobilization of capital through a targeted use of funds; thus, innovative decision support schemes and standardization tools for the respective key actors are required. Crucial is the identification of which energy efficiency investments can be considered as attractive and eligible, fostering sustainable growth, while also having an extremely strong capacity to meet their commitments (Bankers Almanac, 2018; Canes, 2017; Doukas, 2018; Morgan Stanley, 2005; Triple-A, 2019), already from the first stages of investment generation and preselection/preevaluation (Triple-A, 2019).

To achieve the necessary investment levels, the energy sector must generate a robust pathway of energy efficiency investments (Brown et al., 2019). However, the preparation of energy efficiency investments requires knowledge of specialized consultants with technical, as well as financial knowledge. In this respect, capacity building, awareness, communication, and marketing are the priorities to support the demand for energy efficiency investments (Hinge et al., 2013).

In this context, this chapter targets to present a very practical result-oriented approach, seeking to address the aforementioned challenges by providing a methodology using in-country demonstrations and replicability of energy efficiency investments that are realistic and feasible in the national and sectoral context, as well as on how they could be financed in practice in the short or medium term.

The aim of this chapter is to present a methodological approach on energy efficiency investments and robust financing programs/models in targeted countries that will act as first-of-a-kind demonstration for the banks, institutional investors, and asset managers. The experience gathered in each in-country demonstration will also facilitate and accelerate the replicability. The findings could then be used to lead to concrete recommendations toward a successfully energy efficiency financing.

The initial step toward this direction is an extended literature review on conventional financing schemes, as well as innovative schemes that have emerged recently, such as crowdfunding, energy performance contracting (EPC), green bonds, and so on, in order to explore a variety of options, advantages, disadvantages, regarding the energy efficiency financing. In the meanwhile, the specific characteristics of several countries are explored so as to select the targeted case studies and apply the proposed approach. These steps are followed by an extensive stakeholder engagement in order to identify the most relevant key actors that will benefit from the whole process according to their needs and priorities. Finally, the overall aim of the approach is to enable peer-to-peer learning, as well as capacity building among targeted stakeholders, in order to successfully communicate the knowledge and best practices toward financing and implementation of energy efficiency projects.

The research conducted in this study is based on the aim of the Horizon 2020 European project "Triple-A." Triple-A initiated in September 2019 and its overall aim is to assist financial institutions and project developers increase their deployment of capital in energy efficiency, making investments more transparent predictable and attractive.

Apart from the introductory section, this chapter is structured into five sections. More particularly, in Section 3.2 the proposed methodology is described through the introduction of the main methodological steps. These steps are described in detail one by one in the remaining sections. Section 3.3 includes a literature review on the key innovative financing schemes, whereas in Section 3.4 the main characteristics of the case study countries are presented with regards to their performance in energy efficiency financing. The process of key actors' identification is further elaborated in Section 3.5, whereas in Section 3.6 the two methods for knowledge and best practices transfer and exchange are introduced. Finally, Section 3.7 includes the conclusions, lessons learnt, and recommendation for the further research based on the proposed approach.

3.2 The proposed methodology

The objective of the proposed methodology is mainly to ensure capacity building and knowledge exchange among all interested parties in national, regional, and cross-country level, especially banks and investors and other relevant stakeholders, with regards to energy efficiency financing and, in particular, on underwriting sustainable energy investments. The ultimate goal is to exchange lessons learnt and explore energy efficiency investments, aiming to reduce the respective time and effort required at the crucial phase of the investments conceptualization, as well as to increase transparency and efficiency of respective decision-making. Recommendations on what energy efficiency investments are realistic and feasible in the national and sectoral context, as well as on how they could be financed in practice in the short or medium term, are some of the outputs of the proposed framework.

Relevant national and regional stakeholders engaged in the energy efficiency financing procedure have varying levels of knowledge, skills, and capacity for energy efficiency financing planning and implementation (Karakosta, 2016). As such, it is important to meet diverse needs of different case studies, to learn from each other, and to strengthen collaboration and engagement among all the involved parties (Karakosta and Flamos, 2016; Karakosta et al., 2018; Papadopoulou et al., 2011, 2013). Beyond benchmarking the performances and sharing of best practices, this engagement will enable national stakeholders to take possible advantages of synergies and economies on scale, especially through bundling or aggregating projects to increase bankability of projects and develop business cases.

To address the aforementioned objectives, a methodological approach has been designed in order to share the knowledge and to build capacity toward a successfully energy efficiency financing in selected country case studies.

The methodology starts with two parallel steps. On the one hand, all the innovative financing schemes should be carefully reviewed in order to explore the pros and cons of their selection for specific energy efficiency projects in targeted sectors (private/public buildings, public lighting, and so on). On the other hand, several countries should be examined in order to select the most suitable as case study countries. The case study countries should be strategically selected so as to promote diversity across a number of factors and thereby to enhance the results of the approach. To this end, an in country-analysis should take place, as far as the performance of several countries in energy efficiency financing is concerned.

After reviewing the available financing schemes and the case study countries characteristics, the matching process takes place in order to focus more particularly in financing schemes that are successfully applied in the case study countries that are under examination and in specific energy sectors.

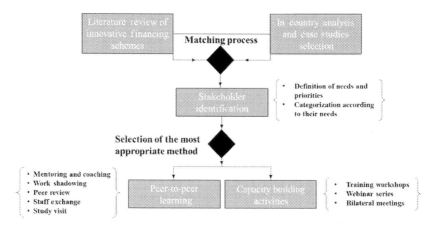

Figure 3.1 The proposed methodology.

The next important step of the approach is the identification of relevant stakeholders and key actors that will be engaged during the whole process. In this context, the different type of stakeholders' needs should be listed taking into account in which way they would like to be engaged in the approach and benefit from their participation. Are they interested in showcasing achievements of their projects at national or regional level? Do they want to get additional feedback on their projects by confronting their experience with peers working on the same issues? Or, in the opposite hand, they aim to learn from other projects implemented and apply what they learned in their own context, while getting tailor-made assistance adapted to their learning objectives and needs. If the right questions have been raised, then it becomes relatively easy to identify the type of stakeholder categories and profiles to be solicited, because of their influential role, background experience, specific expertise, and the degree of involvement in the particular subject/area that is meant to be discussed.

The final step refers to the selection of the most appropriate methods to be followed for the knowledge and best practices exchange, namely the peer-to-peer learning and the capacity building. The two approaches differ in the way the message will be conveyed, how energy efficiency investments will be promoted, and in which way the relevant stakeholders will be engaged. Fig. 3.1 shows the proposed methodology, presenting each step to be followed, so as to facilitate the mainstreaming of energy efficiency projects, whereas the remaining sections analyze each step separately in order to assist in their implementation.

3.3 Innovative financing schemes

In the following sections, a literature review on the key innovative financing schemes for energy efficiency projects is presented, highlighting the main advantages and barriers for each financing tool. The review includes financing schemes that have been already used for several years, as well as innovative schemes that have emerged recently. Although every scheme is characterized by its own special features, they are often combined, resulting in reduced risks.

3.3.1 Crowdfunding

The term "crowdfunding" refers to the mobilization of funding for projects from a large number of investors ("the crowd") via the utilization of Internet-based platforms and online processes.

Crowdfunding is generally divided into four different modalities (GIZ, 2013, 2016):

1. Donations—the oldest form of crowdfunding, using the Internet to fundraise for projects, causes, and organizations.
2. Rewards—in exchange for a contribution, the crowd investor receives a nonfinancial return, such as new music CD, the production of which was crowdfunded, or vouchers to make purchases in a specific shop.
3. Debt—the crowd investor provides a loan to a project or to another person (e.g., peer-to-peer lending) and expects in exchange interest payments and the return of the principal.
4. Equity—the crowd investor acquires a share in a company and expects dividends and/or a value increase in return. Here the crowd participates in upside and downside risks of the business.

3.3.2 Energy performance contracting

EPC is an innovative financing scheme offered by a contractor energy service companies (usually ESCOs) to clients (e.g., a municipality), who are in need of energy efficiency improvements but have limited financial means or technical capacities to implement such projects on their own (EnPC Intrans, 2015). The innovation of EPC lies in the fact that an ESCO finances the project and implements energy efficiency investments (PROSPECT, 2018b). In principle, the ESCO plans and conducts the project and receives only service fees—and gets the return of investment—from the client using the savings from energy costs (Eurostat, 2017; Töppel and Tränkler, 2019). Clients will eventually benefit from energy and cost savings after the end of the contract (Novikova et al., 2017).

3.3.3 Green bonds

Green bonds are bonds where proceeds are exclusively applied to finance or refinance green projects. Green bonds can be issued by:

- city governments
- utilities: water, transport, energy, and so on.
- corporations that are developing, building, or managing green assets for issuers
- states or development banks (PROSPECT, 2018c).

Green bonds are a recent financial instrument providing support to green investments. International organizations and banks are currently developing standards in order to regulate and foster the growth of this emerging market (OECD, 2015).

3.3.4 Guarantee funds

Guarantee funds facilitate the engagement of financial institutions and allow sharing of credit risk or distribution of loss in energy efficiency investments. In this respect, they constitute a means of transferring credit risk from a creditor to another entity (guarantor) that is capable and ready to deal with part of the risk and/or cover the loss in case the loans default or if the debtor fails to meet the conditions of the loan (ESMAP, 2014; PROSPECT, 2018a).

3.3.5 Revolving funds

Revolving funds are pools of capital comprising of cost savings from energy efficiency and renewable energy projects or the interest paid by the sustainability measures financed by the fund (PROSPECT, 2018a). In this respect, cost savings or interest revenues are used for the continuous financing of new investments in similar projects, leading to the development of a sustainable funding cycle. Revolving funds can be helpful when dealing with the long payback time of the projects. Their long-term effectiveness can be enhanced, when coupled with other financing schemes, especially soft (BPIE, 2010) or ESCOs (de T'Serclaes, 2007).

There are two types of revolving funds:

1. External revolving funds often developed and managed by a selected fund manager (with its compensation tied to the fund's performance) or by a utility or specially created organization. In such cases, the revolving fund can lend to multiple entities, which are obliged to repay the loan in an agreed date (ESMAP, 2014).
2. Internal revolving funds, developed by a single entity, "which provides the initial capital and may also manage the fund itself" (ESMAP, 2014). Their structure varies on local factors and needs (C40, 2016). This type of revolving funds requires an initial capital contribution.

3.3.6 Soft loans

Soft loans can enable the reduction of the loan interest rate below the market interest rate and may even provide zero interest rates at the beginning of the loan agreement (ACE, 2013; EEFIG, 2014; FEDARENE, 2015), as well as grant concessions, such as longer repayment periods, with certain conditions to meet (EEFIG, 2014; FEDARENE, 2015). They can be applied in all types of buildings and are commonly offered by governments, both in the form of tenders and direct negotiations. Due to the longer maturity provided by soft loans, homeowners who take the loan for energy-efficient renovation work in their homes can be more flexible regarding their monthly installments (Energy Cities, 2017). Many public international financing institutions and national governments are utilizing such programs to trigger the market and to close the lending gap left by the passive local and traditional banking sector actors (Makinson, 2006; PROSPECT, 2018a).

3.3.7 Third-party financing

Third-party financing is the most common financing model for EPC in developed EPC markets and is more common in the residential sector. More specifically, it can be described as debt financing; building owners assure their funding from a third party, usually investors or banks, instead of getting financial resources from internal funds or ESCOs (Energy Charter Secretariat, 2003). In addition, the ESCOs offer guaranteed savings, leading to a positive project cash flow, thus minimizing the risk of repayment. The interest costs during the period of construction and installation are included in the project financing agreement, as well as the risk is assumed by the third-party lender (The Edison Foundation, 2010).

The advantages and barriers of each financing scheme are presented in Table 3.1, as they ensued from the review, so as to provide a clear picture of their feasibility in relation to different types of projects and facilitate the understanding of stakeholders.

3.4 Case study countries

The case study countries were strategically selected so as to promote diversity across a number of factors and thereby enhance the results of the approach, including:

- a country of upper middle income economy, that has registered steady growth in the recent past, although it still remains one of the least developed countries in Europe, but appears to be already well ahead in reaching its targets for reducing greenhouse gas emissions and for increasing the share of renewable energy (Republic of Bulgaria);
- a weak economy going through one of the longest and most severe recessions any EU Member State has ever faced, therefore cultivating an ideal background for assessing the investment potential amid these conditions and difficulties hindering financing and implementing EU legislation (Greece);
- a country that experienced one of the fastest economic recovery in Europe where the efficient use of energy resources and energy is one of its key long-term strategic objectives in the energy sector (Lithuania);
- a diversified economy with a well-established legislative framework with regards to energy efficiency (Spain).

The current situation of the incentive framework in energy efficiency financing in the case study countries is described in the following sections.

3.4.1 Bulgaria

The national energy efficiency target for 2020 is 716 ktoe/year energy savings in final energy consumption and 1590 ktoe/year in primary energy consumption of which 169 ktoe/year is in the transformation, transmission, and distribution processes in the energy sector (Republic of Bulgaria, 2014). The results from meeting the national target under Directive 2006/32/EC (EC, 2006) show that by 2016 the

Table 3.1 Advantages and barriers of financing schemes.

Financing scheme	Advantages	Barriers
Crowdfunding	• New funding sources • Empowering responsible investors • Greater diversification and smaller amounts per investor • Faster decision and transaction processing, through standardized online processes GIZ (2016)	• Legal uncertainty • Limited institutional capacity of crowdfunding platforms and lack of track record, evidence, and benchmarks as for the priorities and preferences of the "crowd" • Lack of tools to manage foreign exchange risk in cross-border crowdfunding for energy efficiency • Lack of widely recognized project quality assurance mechanisms • Competition from highly subsidized programs GIZ (2016)
EPC	• Investment risks are transferred to the ESCO • No investment or up-front capital required • ESCO provides the guaranteed energy savings and required energy and maintenance services Novikova et al. (2017), PROSPECT (2018b)	• Complex procedures and restrictive regulations • Little interest • Lack of finance, understanding, and personnel • Complex processes and distributed responsibility in municipalities • Calculation risk • Energy prices • Feasibility • High cost, lack of loans Novikova et al. (2017), PROSPECT (2018b)
Green bonds	• Demonstrating and implementing issuer's approach • Diversification of a bond issuer's investor base • Potential to increase issuance size • Greater proportion of "buy and hold" investors • Reputational benefits • Articulation and enhanced credibility of sustainability strategy	• Costs of meeting green bond requirements • Difficulties for international investors to access local markets • Lack of domestic green investors • Lack of awareness of the benefits of green bonds • Lack of univocal international guidelines and standards • Lack of green bond ratings, indices, and listings

(Continued)

Table 3.1 (Continued)

Financing scheme	Advantages	Barriers
	• Access to "economies of scale" • Improved international governance structures • Municipalities can reach constituencies physically located close to the green project they intend to support and can provide them with opportunities to invest in programs that have direct proximal impact Banga (2019), OECD (2015), PROSPECT (2018c)	• Lack of supply of labeled green bonds Banga (2019), PROSPECT (2018c)
Guarantee funds	• Leveraging of public funds • Perception of risk among commercial lenders • Risk reduction • Availability EEFIG (2014), ESMAP (2014), Makinson (2006), PROSPECT (2018a)	• Excessive bureaucracy • Time needed to structure and negotiate • Moral hazard • Lack of know-how and expertise EEFIG (2014), ESMAP (2014), FEDARENE (2015), PROSPECT (2018a)
Revolving funds	• Recycles capital for future use • Efficient allocation of public funds • Long-term sustainability of public investment • Direct and clear impact on the lack of liquidity • Enables future commercial financing BPIE (2010), Energy Cities (2013), FEDARENE (2015), de T'Serclaes (2007), PROSPECT (2018a), World Bank (2014)	• Parallel negotiations • Depends strongly on the economic situation of each country • Complex architecture, time-consuming preparation • Final beneficiaries remain reluctant C40 (2016), Energy Cities (2013), FEDARENE (2015), PROSPECT (2018a)
Soft loans	• Scalability • Funds can revolve • Longer duration • 1:1 Refinance to commercial banks • Positive impact on public budgets • Reduced interest rates • Higher leverage effect	• Unclear energy refurbishment process • Transaction costs • Increased regulations/provisions for (promotional) banks • Time-consuming and strictly supervised procedures

(*Continued*)

Table 3.1 (Continued)

Financing scheme	Advantages	Barriers
Third-party financing	• Easy to roll out • Large flexibility • Use of cohesion funds BPIE (2010), EEFIG (2014), ESMAP (2014), KFW (2011), PROSPECT (2018a) • Functions typically performed by a lending institution do not have to be adopted by the utility or program administrator • The third-party lender assumes the risk The Edison Foundation (2010)	• Lack of interest, information, and coordination EEFIG (2014), Energy Cities (2017), FEDARENE (2015), PROSPECT (2018a) • High risk of job losses • Energy service companies suffer from a lack of credibility and trust by industrial energy users • Lack of knowledge • Falling electricity prices and increased uncertainty in liberalized energy markets • Sufficient know-how and capital of large industrial customers • Industrial sector companies want to keep control over the production process prefer to own all their equipment • Low contract volume of smaller companies • Defensive stance of energy managers • Lack of trust toward energy service companies Energy Charter Secretariat (2003)

country exceeded the provisional target for the period 2008−16 by 3.4% (Republic of Bulgaria, 2017). According to Directive 2012/27/EU (EC, 2012), the national energy savings target for the period 2014−16 is 8325.6 GWh/year and the implementation is 3532.2 or 42.4% of the target fulfilment rate (Republic of Bulgaria, 2017).

3.4.2 Greece

Since 2000, Greece has done significant improvement related to the energy consumption legislation framework. According to a recent report of the International Energy Agency (IEA, 2017), in 2016, approximately, 50% of the final energy consumption in Greece was regulated by existing mandatory codes and standards,

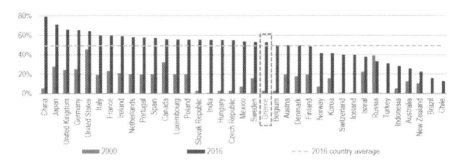

Figure 3.2 Coverage potential of existing mandatory codes and standards.
Source: Data from International Energy Agency (IEA), 2017. Energy Efficiency 2017. OECD/IEA, Paris, p. 143.

including energy efficiency matters, whereas it was much less than 10% in 2000 (Fig. 3.2).

The existence of energy efficiency legislation allowed banks and other investing bodies to activate in the sector. So, the four major Greek Banks (Alpha Bank, Eurobank, National Bank of Greece, Piraeus Bank) provide energy efficiency mortgages and consumer loans, whereas two of them provide energy efficiency renovation loans additionally for small businesses. In the Greek market, there are not available specialized energy efficiency loans for medium or large companies, even though all major banks support such projects (EIB, 2019a). Additionally, Public Investment Bank (TPD), a public bank institution, provides loans for public lighting energy efficiency improvement (TPD, 2019).

Finally, energy efficiency projects are supported by various EU and national grants and subsided schemes, such as the upcoming Infrastructure Fund of Funds ("InfraFoF") (InfraFoF, 2019).

3.4.3 Lithuania

The efficient use of energy resources and energy is one of Lithuania's key long-term strategic objectives in the energy sector. In the light of Energy Efficiency Directive 2012/27/EU (EC, 2012), Lithuania has to ensure that the cumulative amount of 11.67 TWh (Republic of Lithuania, 2019) of energy will be saved by 2020 and is drafting required documentation to set in place an Energy Efficiency Obligation Scheme or adopt alternative measures, enable and promote faster development of Energy Service Companies, and hasten the rate of public and multiapartment building renovations and modernization of street lightning.

By the end of 2014, the Government of Lithuania has established and adopted the program for improvement of energy efficiency in public buildings. The program aims to renovate public buildings (total area of 700,000 m^2) and to achieve 60 savings of final energy (GWh) (Republic of Lithuania, 2017). In addition, Energy Efficiency Fund was established in February 2015 from the European structural and investment funds in order to reach goals set by the program and to stimulate modernization of street lightning. Allocated funds to Energy Efficiency Fund amount to 79 million € (EC, 2019b).

In 2004, the Government of Lithuania has adopted multiapartment buildings' renovation (modernization) program (VIPA, 2016), which aimed to encourage the apartment owners to comprehensively upgrade multiapartment buildings and residential districts in order to achieve higher living standards, rational use of energy resources, compensation for reduced budgetary expenditure on heating costs, to ensure effective use of housing, to improve the living environment and the quality of life of residents.

3.4.4 Spain

The Spanish energy efficiency services market has been growing slowly but steadily during the last years. Among other reasons, this growth is also due to the regulatory insecurity originated with regards to the renewable energy national support programs, which were introduced intensely in 2008 and 4 years later in 2012, were abruptly stopped causing damages to the country reputation and both, national and international investments.

Yet, there is a well-established legislative framework in Spain with regards to energy efficiency. In addition, the government is trying to regain the confidence in the sector through promotional programs, institutional frameworks, and financial credit lines to achieve the national savings targets. Jessica Fund (120 million €) (EIB, 2019b) and PAREER Plan (200 million €) (IEA, 2019; Solar Thermal World, 2019) are two of the most important programs that financed sustainable urban development projects through the implementation of energy efficiency or renewable energy measures (EC, 2018b).

Recently, on April 2, 2018, a pool of experts published a report about the current situation of decarbonization in Spain, including an analysis and proposed measures (CETE, 2018) to be used by the Ministry for the Ecological Transition in the elaboration of the Strategic Plan on Energy and Climate. This report highlights the financing process as the main barrier for the energy efficiency investments since these imply long payback times and significant guarantees. Hence, new financing mechanisms are required to scale up energy efficiency investments in order to achieve the new targets imposed by 2030.

The case studies' context is briefly depicted in Fig. 3.3. For the GDP per capita, we used the World Bank's respective data (World Bank, 2018). For the final energy consumption, we used Eurostat (Eurostat, 2018) and for the deviation from the Article 7 energy savings target, the respective qualitative information was based on Marina Economidou (Economidou et al., 2017).

3.5 Key actors identification

A list of the relevant groups, organizations, experts, and everyone that has an interest in the specific objectives of the proposed approach is compiled. This list is developed in view of a balanced professional, institutional, and geographic

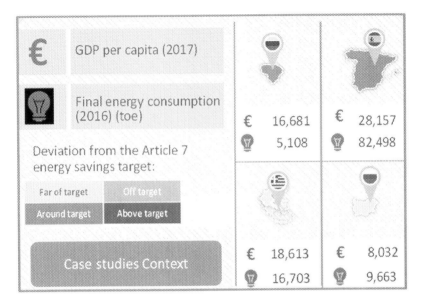

Figure 3.3 Case study context.

representation of stakeholders (Karakosta et al., 2007, 2015). Ranking stakeholders' relevance, according to their expertise, interest, power, influence, commitment, and interest levels, will help to target each stakeholder and each stakeholder segment properly and efficiently (Karakosta and Papapostolou, 2019a, b).

The key stakeholders that participate in the entire energy efficiency investments value chain, with their interactions, are presented in Fig. 3.4. They are the ones that will provide the required knowledge to achieve the objective of implementing energy efficiency investments, aiming to reduce the respective time and effort required at the crucial phase of the investments conceptualization, as well as to increase transparency and efficiency of respective decision-making.

First, the financing bodies and the companies/project developers are the major actors of the proposed methodology. There are also additional target groups that are involved in the development, testing, and implementation phase of innovative schemes for energy efficiency financing, such as policy makers and policy support institutes, researchers, and academics in business and techno-economic fields, and other groups, as described in the following table. The target group descriptions and benefits that emerge through their participation in the proposed methodological steps are presented in Table 3.2.

3.6 Knowledge transfer

In this section, the two methods proposed for the knowledge and best practices sharing and exchanging are elaborated.

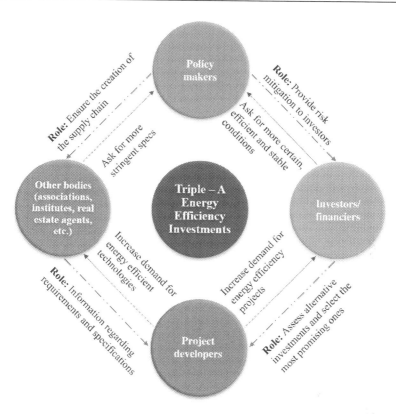

Figure 3.4 Key stakeholders' interactions.

3.6.1 Peer-to-Peer learning

Peer-to-peer learning could be conducted by national or regional key stakeholders (financing bodies, companies, project developers) to other relevant experts delivering the knowledge regarding successfully financed and implemented energy efficiency projects. Oftentimes, peer-to-peer learning involves pairing a more experienced part (i.e., stakeholders that have invested or gained finance for projects in the past) with a less experienced one that needs support. These can be effective for the high- and middle-level participants. However, for each profile a specific appropriate training approach will be designed and followed, considering not only the level but also the field of expertise of the participants. To maximize knowledge sharing and learning experiences, peer-to-peer groups do not have to be site specific. As there are groups that may have similar needs and issues, such as countries or regions with similar characteristics (e.g., geographical or economical contexts), they can be grouped together to build stronger teams and better understand each other's needs. The learning program will be enriched with a fair amount of case study examples. Lastly, the parameters of the learning program based on the peer-

Table 3.2 Key actors descriptions and benefits.

Key actors	Description	Benefits
Financing bodies	Commercial/green investment banks, EEFIG—energy efficiency financial institutions group members (EEFIG is an open dialogue and work platform for public and private financial institutions, industry representatives, and sector experts established in 2013 by DG Energy and UNEP Finance Initiative), institutional investors (e.g., pension funds) and their financial advisors, insurance companies, brokerages, investment funds (national and international) and their managers, unit investment trusts, and developers/managers of financial products	They will provide key parameters on the available funding, and their perceptions on how to assess alternative investments and select the most promising one.
Companies/project developers	Energy companies, ESCOs looking for additional finance, accredits professionals, management investment companies, and construction companies	These target groups will be supported on where to go, how to present their project ideas, how to involve private sector, and how to apply for funds.
Policy makers and policy support institutes	EC Directorates and Units, Governments and local authorities, Ministries who provide incentives and set the scene in a national and European level, as well as policy support institutes	Knowledge transfer to policy makers in order to update policy frameworks with new financing mechanisms that promote energy efficiency investments.
Researchers and academia in business and Techno-economic fields	Individuals engaged in research initiatives and/or working in research/academic institutes will bring their expertise in innovative energy efficiency financing	Design innovative financing schemes; inspiration for future research initiatives based on the methodology outcomes.
Others	Technology suppliers, property valuers, real estate agents, technical chambers, notaries, associations' individuals (e.g., architects, engineers)	Identification of project ideas' requirements and specification, based on their expertise and experience; diffusion and exploitation of the project's results.

Table 3.3 Overview of the most common and effective peer-to-peer learning tools and methods.

Mentoring and coaching	Mentoring involves a partnership between a "mentor" and a "mentee" wherein the mentee can collaborate with the mentor for improved capacity and enhanced knowledge.
Work shadowing	Work shadowing involves a "learning" participant and an "experienced" participant wherein the trainee can observe and engage with the expert in certain work tasks. Involvement of decision makers can create networks that enable synergies and economies of scale.
Peer review	Peer review is not only a process of assessing the performance of peers, but also sharing of knowledge and experiences.
Staff exchange	Staff exchange aims to enhance cooperation between peers through a better understanding of the culture in which they operate.
Site visits	Site visits are intended for peers to acquire firsthand knowledge on the ground work. Work shadowing can accompany site visits as participants who can build good interpersonal relationships while on the job, describing their working methods and reflecting on their own actions.

to-peer learning objectives and activities will be set. A detailed curriculum of the peer-to-peer learning program could be prepared and adapted to the specific peers, based on the outcomes of an analysis of received relevant requests or ideas resulting (mainly) from the identified needs of trainees.

The main aim of this method is to develop and execute a complete and easily replicable peer-to-peer learning program addressing national and regional stakeholders. The program will be focused on the development of financing schemes for implementing the energy efficiency and sustainable energy and climate projects. Given that the topic on financing is quite broad, the program could be structured in a variety of specific sectors, namely public buildings, private buildings, transport, public lighting, and so on.

The main idea is to create effective and productive peer-to-peer groups among national or regional stakeholders in order to ensure the exchange of experience and expertise. The program should entail tools specifically tailored for peer-to-peer learning. These tools, such as mentoring, work shadowing, and site visits, could enable the proper matching of peer groups, which will feed information on the development of financing for sustainable energy plans (Table 3.3).

Through the peer-to-peer learning, partnerships could be built that stimulate mutual understanding of each other's issues, situations, and challenges with the aim of exploring new ideas, options, and solutions. The methodology aims to build partnerships within countries and regions and their relevant stakeholders targeting at exchanging of experiences for developing finance for energy efficiency.

3.6.2 Capacity building activities

Capacity building activities will be set up to explore how energy efficiency investments and robust financing programs/models in case study country are developed through the participation of interested stakeholders. To ensure workable stakeholder participation, groups of key stakeholders will be brought together for regional training workshops tailored to country's needs. Training and know-how transfer in general could be implemented in a reliable cooperation environment during the training workshops for the presentation of case studies with regards to project investment in targeted countries.

In particular, this method includes the organization and implementation of one day regional training workshops in each involved country to exchange knowledge/experience and stimulate the interest and participation of key stakeholders. They will also facilitate a dynamic dialogue mechanism to share common tools and instruments on a national/regional level when introducing energy efficiency investments.

The outcomes of these workshops will provide appropriate risk mitigation options, including, among others, the possibility to aggregate "similar" projects and to create financial products that can be traded in secondary markets (e.g., securitizations and bonds and insurance products that focus on removing specific risks).

Following key activities to be included:

- *Development of the training content and material*: A detailed plan for the training courses will be drafted including agendas, venues, dates, trainers to deliver training, participants to be targeted, and responsibilities for the participants. The development of the training material will, in particular, bring in training concepts and topics, which have previously been proven successful. The materials will be further adjusted to local specifics and needs and translated if needed. The course will include key recommendations to assessment process on what energy efficiency investments are realistic and feasible in the country context; means of financing the projects in practice in the short or medium term; financing methods and approaches; evaluation and verification of the results.
- *Conducting the training courses*: This activity will include: arranging infrastructure and services for the training—training courses will be organized regionally with support of national stakeholders; sending invitations; recruiting participants—confirming participation partners will assist in reaching to the target group and recruiting participants to the training courses.
- *Evaluating the training*: Evaluation of the training courses will be carried out to make final fine-tuning of the course including its key elements (regional network, trainers, evidence of trainees, and so on).

Finally, apart from the regional workshops, webinar series will be organized across participating countries addressing: asset owners (public institutions, corporate entities, multilateral banks), project developers (ESCOs, construction companies, manufacturers). Bilateral meetings/calls will be also organized with stakeholders that wish to develop projects and passed an initial screening.

3.7 Conclusions

The identification and implementation of suitable energy efficiency projects is critical in order to foster the confidence of investment institutions regarding the inclusion of such projects in their investment portfolios, where reliable decision support frameworks are required for stakeholders. To this end, the communication of the knowledge and best practices toward financing and implementation of such projects is of paramount importance for relevant stakeholder, banks, institutional investors, and asset managers.

The proposed methodological framework supports the aforementioned actors during the early stages of project development, enabling them to identify project and financing scheme ideas, which can provide an admissible return on investment at low risk through capacity and peer- to-peer learning activities in selected case study countries. A key advantage of the proposed framework is that stakeholders are involved early on, in order to create a sense of ownership regarding the whole procedure, while they gain knowledge through already implemented cases.

The development of a methodology for in-country demonstrations, as well as the selection of case study countries, promotes the diversity of economies involved and fosters the thorough examination of the status quo and possibilities of different countries, including leading, innovative, recovering, and weak economies.

The proposed methodology proved to offer many advantages, since it enables to:

- build capacity of countries and regions in financing sustainable energy plans through capacity building and peer-to-peer learning activities,
- support key stakeholders in utilizing the rich experience available, which is yet difficult to gain solely on their own;
- link together stakeholders at the national and regional levels and their respective associations along in an intra-European network through an innovative peer-to-peer leaning experience on financing sustainable energy projects and measures.

In addition, the proposed methodological framework could be extended, so as to provide key recommendations on what energy efficiency investments are realistic and feasible in the country context, as well as on underwriting sustainable energy investments.

References

Association for the Conservation of Energy (ACE), 2013. Financing energy efficiency in buildings: an international review of best practice and innovation. Report to the World Energy Council. https://wec-policies.enerdata.net/Documents/casesstudies/Financing_energy_efficiency_buildings.pdf (accessed 12.06.19.).

Banga, J., 2019. The green bond market: a potential source of climate finance for developing countries. J. Sustain. Finance Invest. 9 (1), 17–32.

Bankers Almanac, 2018. Standard & poor's definitions. https://www.bankersalmanac.com/addcon/infobank/credit_ratings/standardandpoors.aspx (accessed 05.07.19.).

Becqué, R., Mackres, E., Layke, J., Aden, N., Liu, S., Managan, K., et al., 2016. Accelerating Building Efficiency: Eight Actions for Urban Leaders. World Resources Institute, Washington, DC.

Boza-Kiss, B., Bertoldi, P., Economidou, M., 2017. Energy Service Companies in the EU. Publications Office of the European Union, Luxembourg.

Brown, D., Sorrell, S., Kivimaa, P., 2019. Worth the risk? An evaluation of alternative finance mechanisms for residential retrofit. Energy policy 128, 418–430.

Building Performance Institute Europe (BPIE), 2010. Financing Energy Efficiency (EE) in Buildings. Brussels, Belgium, p. 35.

Building Performance Institute Europe (BPIE), 2014. Investing in the European Buildings Infrastructure – An Opportunity for the EU's New Investment Package. Discussion Paper, Brussels, Belgium, p. 4.

C40, 2016. C40 Cities Good Practice Guide – City Climate Funds, p. 33.

Canes, M.E., 2017. The inefficient financing of federal agency energy projects. Energy Policy 111, 28–31.

Comisión de Expertos de Transición Energética (CETE), 2018, Análisis y propuestas para la descarbonización. Spain, p. 548.

Cooremans, C., Schönenberger, A., 2019. Energy management: a key driver of energy-efficiency investment? J. Clean. Prod. 230, 264–275.

de T'Serclaes, P., 2007. Financing Energy Efficient Homes: Existing Policy Responses to Financial Barriers. OECD/IEA, Paris, France, p. 52, IES Information Paper.

Doukas, H., 2018. On the appraisal of "Triple-A" energy efficiency investments. Energy Sources B Econ.Plann Policy 13 (7), 320–327.

Doukas, H., Flamos, A., Psarras, J., 2011. Risks on the security of oil and gas supply. Energy Sources B Econ. Plann. Policy 6 (4), 417–425.

Doukas, H., Karakosta, C., Flamos, A., Psarras, J., 2014. Foresight for energy policy: techniques and methods employed in Greece. Energy Sources B Econ. Plann. Policy 9 (2), 109–119.

Economidou, M., Kona, A., Bertoldi, P., 2017. Implementation of the energy efficiency directive: progress, challenges and lessons learned. In: Proceedings of the ECEEE 2017 Summer Study, 29 May–3 June 2017, Belambra Presqu'île de Giens, France.

EEFIG Underwriting Toolkit, 2017. https://valueandrisk.eefig.eu (accessed 11.06.19.).

Energy Charter Secretariat, 2003. Third Party Financing. Brussels, Belgium, p. 150.

Energy Cities, 2013. Intracting—Internal Performance Contracting. Brussels, Belgium, p. 34.

Energy Cities, 2017. Financing the Energy Renovation of Residential Buildings through Soft Loans and Third-Party Investment Schemes. Infinite Solutions Guidebook. Brussels, Belgium, p. 84.

Energy Efficiency Financial Institution Group (EEFIG), 2014. Energy Efficiency Investments: Energy Efficiency – The First Fuel for the EU Economy Part 1: Buildings, p. 50.

Energy Efficiency Financial Institutions Groups (EFFIG), 2015. Energy Efficiency – The First Fuel for the EU Economy. How to Drive New Finance for Energy Efficiency Investments, p. 50.

Energy Sector Management Assistance Program (ESMAP), 2014. Improving Energy Efficiency in Buildings. Mayoral Guidance Note; No. 3. Energy Efficient Cities; Energy Sector Management Assistance Program (ESMAP); Knowledge Series 019/14. World Bank, Washington, DC, p. 28.

EnPC Intrans, 2015. Adapted Business Models for Energy Performance Contracting in the Public Sector. Karlsruhe, Germany, p. 49.

European Commission (EC), 2006. Directive 2006/32/EC of the European Parliament and of the Council of 5 April 2006 on Energy End-Use Efficiency and Energy Services and Repealing Council Directive 93/76/EEC. Official Journal of the European Union, Luxemburg 22.

European Commission (EC), 2012. Directive (EU) 2012/27/EU of the European Parliament and of the Council of 25 October 2012 on Energy Efficiency, Amending Directives 2009/125/EC and 2010/30/EU and Repealing Directives 2004/8/EC and 2006/32/EC. Official Journal of the European Union, Luxemburg 56.
European Commission (EC), 2014. A Policy Framework for Climate and Energy in the Period from 2020 to 2030; COM(2014) 15 Final. Brussels, Belgium, p. 18.
European Commission (EC), 2015. Communication from the Commission to the European Parliament, the Council, the European Economic and Social Committee, the Committee of the Regions and the European Investment Bank: A Framework Strategy for a Resilient Energy Union with a Forward-Looking Climate Change Policy. COM (2015) 80 Final. Brussels, Belgium, p. 21.
European Commission (EC), 2018a. A Clean Planet for All. A European Strategic Long-Term Vision for a Prosperous, Modern, Competitive and Climate Neutral Economy. COM(2018) 773 Final. Brussels, Belgium, 2014, p. 25.
European Commission (EC), 2018b. Annual report for Spain. https://ec.europa.eu/energy/sites/ener/files/documents/es_annual_report_2018_en.pdf (accessed 20.06.19.).
European Commission (EC), 2018c. Action Plan: Financing Sustainable Growth. COM (2018) 97 Final. Brussels, Belgium, p. 20.
European Commission (EC), 2018d. Financing a Sustainable European Economy. Final Report 2018 by the High-Level Expert Group on Sustainable Finance. Brussels, Belgium, p. 100.
European Commission (EC), 2019a. Financing energy efficiency. https://ec.europa.eu/energy/en/topics/energy-efficiency/financing-energy-efficiency (accessed 11.07.19.).
European Commission (EC), 2019b. City street lighting modernization programme. https://ec.europa.eu/eipp/desktop/en/projects/project-18.html (accessed 10.06.19.).
European Investment Bank (EIB), 2019a. Greece: New EUR 100 million EIB and Piraeus Bank initiative to cut energy bills. https://www.eib.org/en/press/all/2018-075-new-eur-100-million-eib-and-piraeus-bank-initiative-to-cut-energy-bills-in-greece.htm (accessed 08.07.19.).
European Investment Bank (EIB), 2019b. Spain: creation of first JESSICA urban development fund in Andalusia. https://www.eib.org/en/press/all/2011-027-spain-creation-of-first-jessica-urban-development-fund-in-andalusia (accessed 08.07.19.).
Eurostat, 2017. The recording of energy performance contracts in government accounts. https://ec.europa.eu/eurostat/documents/1015035/7959867/Eurostat-Guidance-Note-Recording-Energy-Perform-Contracts-Gov-Accounts.pdf/ (accessed 08.07.19.).
Eurostat, 2018. Eurostat, 2018. Final energy consumption. https://ec.europa.eu/eurostat/en/web/products-datasets/-/T2020_34 (accessed 17.09.19)
FEDARENE (European Federation of Agencies and Regions for Energy and the Environment, 2015. Innovative Financing Schemes in Local and Regional Energy Efficiency Policies. Brussels, Belgium.
G20 Energy Efficiency Investment Toolkit, 2017. Published by the G20 Energy Efficiency Finance Task Group under the content direction of the International Energy Agency (IEA), the UN Environment Finance Initiative (UNEP FI) and the International Partnership for Energy Efficiency Collaboration (IPEEC).
GIZ, 2013. Assessing Framework Conditions for Energy Service Companies in Developing and Emerging Countries. Berlin, Germany, p. 74.
GIZ, 2016. CF4EE—Crowdfunding for Energy Efficiency. Berlin, Germany, p. 65.
Hinge, A., Beber, H., Laski, J., Nishida, Y., 2013. Building efficiency policies in world leading cities: what are the impacts?. Proceedings of the European Council for an Energy-

Efficient Economy. 2013 Summer Study, 3-8 June 2013, Belambra Les Criques, Presqu'île de Giens, Toulon/Hyères, France.

Infrastructure Fund of Funds, Greece (InfraFoF), 2019. https://www.eib.org/en/products/blending/esif/eoi/mha-1484.htm?f = search&media = search (accessed 15.06.19.).

International Energy Agency (IEA), 2017. Energy Efficiency 2017. OECD/IEA, Paris, p. 143.

International Energy Agency (IEA), 2019. PAREER Programme (Aid Programme for Energy Rehabilitation in Buildings in the Household and Hotel Sectors). https://www.iea.org/policiesandmeasures/pams/spain/ (accessed 05.06.19.).

Karakosta, C., 2016. A holistic approach for addressing the issue of effective technology transfer in the frame of climate change. Energies 9 (7), 503.

Karakosta, C., Askounis, D., 2010. Challenges for energy efficiency under programmatic CDM: case study of a CFL project in Chile. Int. J. Energy Environ. 1 (1), 149−160.

Karakosta, C., Flamos, A., 2016. Managing climate policy information facilitating knowledge transfer to policy makers. Energies 9 (6), 454.

Karakosta, C., Papapostolou, A., 2019a. Impact analysis of multiple future paths towards a clean energy sector: a stakeholder participatory approach. In:Proceedings of the third International Scientific Conference on Economics and Management—EMAN 2019, 28 March 2019, Ljubljana, Slovenia.

Karakosta, C., Papapostolou, A., 2019b. Impact analysis of multiple future paths towards a clean energy sector: a stakeholder participatory approach. In: Akkucuk, U. (Ed.), Recent Developments on Creating Sustainable Value in the Global Economy. IGI Global, p. 350. , ISBN: 9781799811961, In press.

Karakosta, C., Doukas, H., Flamos, A., Psarras J., 2007. Sustainable technology transfer through the clean development mechanism: a collective approach grounded in participatory in-country processes. In: Proceedings of ENERTECH 2007, Second International Conference on Renewable Energy Sources and Energy Efficiency, 18−21 October 2007, Athens, Greece.

Karakosta, C., Doukas, H., Psarras, J., 2011. CDM sustainable technology transfer grounded in participatory in-country processes in Israel. Int. J. Sustain. Soc. 3 (3), 225−242.

Karakosta, C., Dede P., Flamos, A., 2015. Identification of knowledge needs on climate policy implications through a participatory process. In: Proceedings of the Eighth International Scientific Conference on Energy and Climate Change, Contributing to deep decarbonization, 7−9 October 2015, Athens, Greece, ISBN: 978-618-82339-2-8.

Karakosta, C., Flamos, A., Forouli, A., 2018. Identification of climate policy knowledge needs: a stakeholders consultation approach. Int. J. Clim. Change Strateg. Manag. 10 (5), 772−795.

KFW, 2011. Impact on Public Budgets of KFW Promotional Programmes in the Field of "EnergyEfficient Building and Rehabilitation. Frnakfurt am Main, Germany, p. 14.

Makinson, S., 2006. Public Finance Mechanisms to Increase Investment in Energy Efficiency: A Report for Policymakers and Public Finance Agencies. BASE—Basel Agency for Sustainable Energy for the UNEP Sustainable Energy Finance Initiative (SEFI), Basel, p. 56.

Marinakis, V., Doukas, H., Karakosta, C., Psarras, J., 2013. An integrated system for buildings' energy-efficient automation: application in the tertiary sector. Appl. Energy 101, 6−14.

Miller, L., Carriveau, R., 2018. A review of energy storage financing—Learning from and partnering with the renewable energy industry. J. Energy Storage 19, 311−319.

Morgan Stanley, 2005. Individual investors - an educational look at bond credit ratings. Morganstanleyindividual.com (accessed 10.06.19.).
Novikova, A., Stelmakh, K., Hessling, M., 2017. Financing models for energy-efficient street lighting. In: ECEE Summer Study Proceedings.
OECD, 2015. Green Bonds: Mobilising the Debt Capital Markets for a Low-Carbon Transition. Bloomberg Philanthropies, New York, p. 28.
Painuly, J.P., Park, H., Lee, M.K., Noh, J., 2003. Promoting energy efficiency financing and ESCOs in developing countries: mechanisms and barriers. J. Clean. Prod. 11 (6), 659−665.
Papadopoulou, A., Doukas, H., Karakosta, C., Makarouni, I., Ferroukhi, R., Luciani, G., et al., 2011. Tools and mechanisms fostering EU GCC cooperation on energy efficiency. In: World Renewable Energy Congress-Sweden, 8−13 May 2011, Linköping; Sweden (No. 057, pp. 2308-2315). Linköping University Electronic Press, Linköping, Sweden.
Papadopoulou, A.G., Al Hosany, N., Karakosta, C., Psarras, J., 2013. Building synergies between EU and GCC on energy efficiency. Int. J. Energy Sect. Manag. 7 (1), 6−28.
PROSPECT, 2018a. Private Buildings Module. European Commission H2020 Project Number 752126. European Commission, Brussels, Belgium.
PROSPECT, 2018b. Public Buildings Module. European Commission H2020 Project Number 752126. European Commission, Brussels, Belgium.
PROSPECT, 2018c. Transport Module. European Commission H2020 Project Number 752126. European Commission, Brussels, Belgium.
Public Investment Bank (TPD), 2019. https://www.tpd.gr/ (accessed 10.06.19.).
Republic of Bulgaria, 2014. National Energy Efficiency Action Plan 2014-2020. Sofia, Bulgaria, p. 93.
Republic of Bulgaria, 2017. National Energy Efficiency Action Plan 2014-2020, updated 2017. Sofia, Bulgaria, p. 85.
Republic of Lithuania, Ministry of Energy, 2017. Order Approving the Energy Efficiency Action Plan for 2017−2019. Vilnius, Lithuania, p. 18.
Republic of Lithuania, Ministry of Energy, 2019. https://enmin.lrv.lt/en/sectoral-policy/energy-efficiency-sector (accessed 20.06.19.).
Sarkar, A., Singh, J., 2010. Financing energy efficiency in developing countries—lessons learned and remaining challenges. Energy Policy 38 (10), 5560−5571.
Schlein, B., Szum, C., Zhou, N., Ge, J., He, H., 2017. Lessons from Europe, North America, and Asia: financing models that are facilitating building energy efficiency at scale. In: 2017 ECEEE Conference, Hyeres, France.
Solar Thermal World, 2019. Spain: 20% direct energy efficiency subsidy − up to EUR 200 million. https://www.solarthermalworld.org/content/spain-20-direct-energy-efficiency-subsidy-eur-200-million (accessed 07.07.19.).
The Edison Foundation, Institute for Electric Efficiency, 2010. Alternative Financing Mechanisms for Energy Efficiency. IEE Brief, Washington, DC, p. 14.
Töppel, J., Tränkler, T., 2019. Modeling energy efficiency insurances and energy performance contracts for a quantitative comparison of risk mitigation potential. Energy Econ. 80, 842−859.
Triple-A, 2019. Enhancing at an Early Stage the Investment Value Chain of Energy Efficiency Projects. European Commission, Horizon 2020 Research and Innovation Programme, Brussels, Belgium.
United Nations (UN), 2015. Transforming Our World: The 2030 Agenda for Sustainable Development. New York.

United Nations Framework Convention on Climate Change (UNFCCC), 2018. The Paris Agreement. C40 Cities Finance Facility, 2016. C40 Cities Good Practice Guide – City Climate Funds, p. 33, www.c40.org.

VIPA—Public Investment Development Agency, 2016. State of Art and SWOT Analysis. Vilnius, Lithuania, p. 36.

World Bank, 2014. Establishing and Operationalizing an Energy Efficiency Revolving Fund – Guidance Note. World Bank Group, Washington, DC, p. 36.

World Bank (2018). GDP per capita. https://data.worldbank.org/indicator/ny.gdp.pcap.cd (accessed 07.07.19.).

Zhan, J., Li, S., Chen, X., 2018. The impact of financing mechanism on supply chain sustainability and efficiency. J. Clean. Prod. 205, 407–418.

Part 3

Energy systems in buildings

Energy in buildings and districts

Jacopo Vivian
University of Padua, Padua, Italy

Chapter Outline

4.1 Introduction 81
4.2 Thermal comfort 84
4.3 User behavior 88
4.4 Weather conditions under climate change and growing urbanization 91
4.5 Envelope and materials 94
4.6 From passive to nearly zero-energy building design 96
4.7 Smart buildings and home automation 99
4.8 From smart buildings to smart districts and cities 100
4.9 Concluding discussion 103
References 104

4.1 Introduction

Buildings play a key role in modern society as most of human activities occur therein, with direct consequences on both health of the occupants and consumption of energy and related greenhouse gas emissions. According to the IEA, the buildings and building construction sectors combined are responsible for 36% of final energy consumption and nearly 40% of total CO_2 emissions worldwide (Abergel et al., 2018). Since the problem of climate change has a global scale, the analysis and the solutions need to adopt a global perspective. Fig. 4.1 shows some worldwide trends using projections to 2050: (A) the share of urban versus rural population, (B) share of urban population between developed and developing countries, and (C) average population density in developed and developing countries as reported by Ameen et al (2015). The first picture shows the massive movement of people from rural areas to cities. Second, we see that the major increase in urban population in the upcoming years comes from developing countries. Exactly in these countries, the urban population increase will not be accompanied by a proportioned increase of living space. As a result, the metropolitan cities of the next future will have a very high population density, especially in developing countries. Most of these countries are located in regions with tropical or subtropical climates in Southeast Asia and Africa. The combined effect between the growth of income and the global warming will boost the energy demand for space cooling by 2050 in a

Energy Services Fundamentals and Financing. DOI: https://doi.org/10.1016/B978-0-12-820592-1.00004-X
© 2021 Elsevier Inc. All rights reserved.

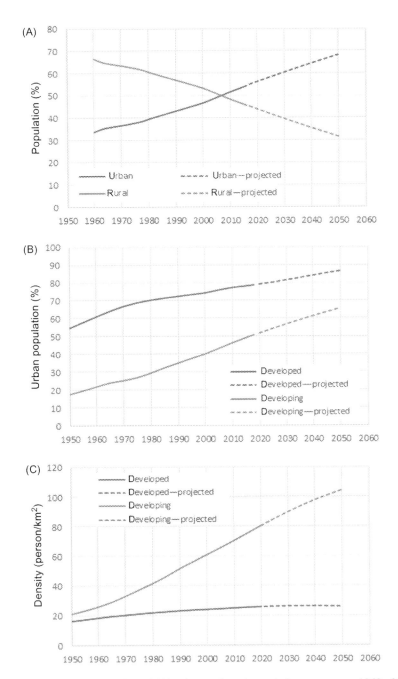

Figure 4.1 Historical and projected (A) urban and rural population percentage 1960–2050, (B) rates of urbanization, and (C) population densities in developed and developing countries 1950–2050[1] (Ameen et al., 2015).

---

[1] The data used for generating graphs (A), (B), and (C) have been obtained from online public databases World Bank (n.d.), United Nations, Department of Economic and Social Affairs (2018), and United Nations, Department of Economic and Social Affairs (2019), respectively.

range between 25% and 58% before adaptation according to a recent study (van Ruijven et al., 2019).

Since most of the energy demand will be concentrated in cities, this chapter is focused on infrastructure-based services rather than on isolated microgrids that are most suitable for rural areas. In Europe, the Energy Performance of Buildings Directive (EPBD) 2010/31/EU established that all of the new constructions, from December 31, 2020, will have to reach the standard nearly zero-energy buildings (nZEBs), which implies a large-scale deployment of this kind of buildings (European Commission, 2010). The 2015 Paris Agreement on climate change following the COP21 Conference boosted the European Union's efforts to decarbonize its building stock (Savaresi, 2016). As a result, the European Commission presented the clean energy package in November 2016, a set of measures aimed at regulating the transition toward a sustainable energy system. The first document to be approved was the new EPBD 2018/844 (European Commission, 2018), amending previous EPBD 2010/31/EU (European Commission, 2010) and the 2012/27/EU Directive on energy efficiency (European Commission, 2012). According to the new Directive, to achieve a highly energy-efficient and decarbonized building stock and to ensure the transformation of existing buildings into nZEBs, in particular by an increase in deep renovations, member states should provide clear guidelines and outline measurable, targeted actions, as well as promote equal access to financing. Among other novelties, the Directive 2018/844 introduces a framework to assess the smart readiness of buildings, which rates their ability to adapt their operation to the needs of the occupant and the grid and to improve their energy efficiency and overall performance. For the first time, the EPBD does not only look at the performance of single buildings, but also on their ability to interact and provide additional services to the energy system without affecting the indoor environment. This regulatory step marks the advent of the digital age in the world of buildings. The close updates of the legislation in this sector demonstrate the deep evolution of the building design approach in terms of targets, technology functions, overall performances and domain experienced in the last years (Vigna et al., 2018). The evolutionary path of building transformation started with passive buildings intended to minimize the energy demand through passive solutions (building envelope domain), then evolved into the nZEBs aimed at obtaining an energy balance (consumption-production) through on-site generation from renewable energy sources (building as energy system domain), and will now find its latest evolution in the energy matching required by smart buildings in order to improve resilient building behavior coupled with grid interaction. Fig. 4.2 illustrates the conceptual path from passive to smart buildings.

This chapter provides an overview on the main aspects of the energy performance of buildings from the perspective of building energy analysts and building designers. A set of important concepts have been selected and analyzed in the next sections. Section 4.2 provides an overview on the thermal comfort theory and then focuses on the mechanisms of adaptation. Section 4.3 is focused on the user behavior, due to its significant impact on the energy consumption of buildings. Such impact is going to grow in the next years due to the increasing share of low-energy

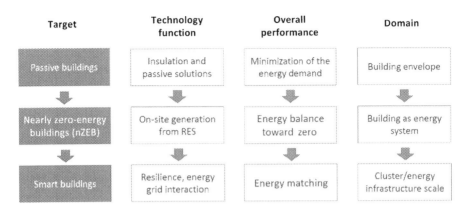

Figure 4.2 Evolution of building design approach (Vigna et al., 2018).

buildings and due to the penetration of user-interactive technologies. Section 4.4 provides an analysis on driving forces for the thermal behavior of buildings—the outdoor weather conditions—and how they will be affected by the aforementioned trends of global warming and growing urbanization. Section 4.5 is focused on the thermohygrometric properties of structural components for new building envelopes and energy refurbishments. After these theoretical premises, the chapter provides an overview on the main concepts and the last trends of building design and operation, following the conceptual path depicted in Fig. 4.2. Section 4.6 talks about the evolution from passive to nZEBs, then Section 4.7 is focused on the technological upgrade needed to shift from nZEB to "smart buildings." Finally, in Section 4.8 buildings will be analyzed as parts of wider systems, namely the district or the city. Section 4.9 links these concepts and draws some general conclusions.

4.2 Thermal comfort

The thermohygrometric comfort in the indoor environment plays an important role in our society due to its impact on our consumption of energy, on our wealth, as well as on our productivity during any of our activities. The condition of thermohygrometric comfort is defined as that condition of mind that expresses satisfaction with the thermal environment (International Organization for Standardization, 2005) or as that sensation of thermal neutrality for which the individual does not feel the need to correct the environment by the behavior (Hensen, 1991). In other words, the thermal discomfort arises as soon as the individual moves away from the condition of neutrality, or feels the need to change environmental parameters.

What gives rise to discomfort? The first answer, driven by our everyday experience, would be the air temperature and humidity, that is, the physical state of the environment that surrounds us. This answer does not consider the individual, who is central in both the aforementioned definitions of thermal comfort. The subjective sensation of heat or cold arises first of all from human physiology. The human body has an internal temperature that is approximately constant and a surface temperature that varies according to the conditions of the environment in which it is located. Two mechanisms of thermoregulation (vasomotor thermoregulation and sweating) combine to vary the surface temperature of the skin in such a way as to keep the internal temperature constant, so as not to damage the organic functions of the human body. As the skin exchanges heat by both convection with the air and radiation with the surfaces close to us, physiologists introduced the concept of operative temperature, that is, an equivalent temperature taking into account the mean radiant temperature of the surrounding environment (Winslow et al., 1937). The concept was later extended to determine a standard effective temperature (SET), that is, an operative temperature for given conditions of relative humidity (50%) and air movement. The SET was found to accurately predict thermal discomfort of people wearing standard clothing in a standard environment on a scale between 1 and 7 (Gonzalez et al., 1974). In the same period, Fanger utilized the predicted mean vote (PMV) to measure the mean thermal sensation of a large group of persons using a scale between -3 (cold) and $+3$ (hot), where 0 corresponds to both the preferred thermal condition (minimum predicted percentage of dissatisfied) and to a condition of thermal neutrality. This index was found to accurately predict the thermal sensation for any given combination of activity level, clothing, and the four thermal environmental parameters: dry-bulb air temperature, air velocity, mean radiant temperature, and relative humidity (Fanger, 1982). As a result, Fanger proposed a correlation between the PMV and the percentage of people dissatisfied as shown in Eq. (4.1).

$$PPD = 100 - 95\ e^{(-0.03353\ PMV^4 - 0.2179\ PMV^2)} \tag{4.1}$$

The method was based on the assumption that the thermal comfort depends only on the energy balance of the human body interacting with the surrounding environment.

Fanger's method has the advantage to better describe the behavioral aspect compared to the SET method, and to clearly define the boundaries of thermal comfort —see Fig. 4.3. Nowadays, both the ASHRAE 55 Standard (American Society of Heating, Refrigerating and Air-Conditioning Engineers, 2017) on thermal comfort (previously based on the SET method) and the International Standard ISO 7730 adopt the Fanger's method (International Organization for Standardization, 2005). As Fig. 4.3 shows, the humidity ratio does not affect the boundaries of thermal comfort significantly, a part from the recommended upper boundary at 12 g/kg. Many other studies have focused on asymmetric radiant fields, draughts, and contact with cold or warm surfaces, as particular cases of local discomfort (Fanger et al., 1985). According to de Dear and Brager, Fanger's model of thermal comfort views occupants as passive recipients of thermal stimuli driven by the physics of the body's thermal balance with its immediate environment, and mediated by autonomic physiological responses (de Dear and Brager, 1998). They proposed an alternative approach called adaptive thermal comfort model. This model relies on the

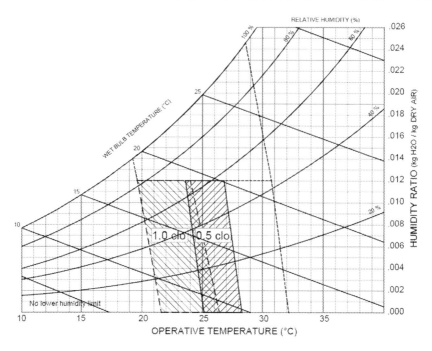

Figure 4.3 Thermal comfort limits in the ASHRAE psychrometric chart (American Society of Heating, Refrigerating and Air-Conditioning Engineers, 2017).

assumption that building occupants play an active role in creating their own thermal preferences. In short, satisfaction occurs through appropriate adaptation to the indoor climatic environment. One of the first studies to shift the attention from the indoor parameters to the people reaction was carried out by Nicol and Humphreys in schools and office buildings (Nicol and Humphreys, 1973). Fig. 4.4 shows the difference between the static model proposed by Fanger and the adaptive model proposed by de Dear and Brager. The subjective nature of thermal comfort does not only arise from physiological differences between individuals, but also from their personal experience and expectations. The latter play an important role in determining how the building occupants adapt to the indoor environment by means of three main mechanisms: physiological acclimatization, psychological adaptation, and behavioral regulation.

In practical terms, one hypothesis of the adaptive model is that people in warm-climate zones prefer warmer indoor temperatures than people living in cold-climate zones (de Dear and Brager, 1998). The research project RP-884 showed that the hypothesis was correct in case of buildings without air-conditioning systems, and that the adaptive model was more suitable to predict the preferred temperature in such buildings compared to the standard thermal comfort model (de Dear and Brager, 1998). Moreover, it was found that the neutral temperature does not necessarily coincide with the preferred temperature. In fact, people living in conditioned

Energy in buildings and districts

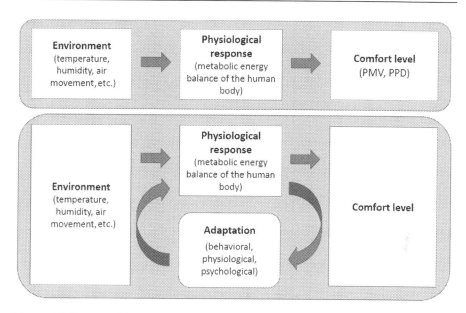

Figure 4.4 Conceptual framework of classical and adaptive thermal comfort models.

spaces preferred a thermal sensation corresponding to "slightly cool" in summer and "slightly warm" in winter, as a result of their thermal expectation. All these aspect may find a justification with the adaptive thermal comfort model. Nicol and Humphreys claimed that a variable set point depending on the running mean outdoor temperature does not increase discomfort and allows significant reductions in energy use in buildings (Nicol and Humphreys, 2002). The research works about thermal comfort strongly relies on the availability of indoor environmental monitoring datasets. A recent research project coordinated by the University of California at Berkeley has systematically collected and harmonized raw data of thermal comfort field studies around the world in the last two decades (Földváry Ličina et al., 2018). After the quality assurance process, there was a total of 81,846 rows of data of paired subjective comfort votes and objective instrumental measurements of thermal comfort parameters. An additional 25,617 rows of data from the original ASHRAE RP-884 database are included, bringing the total number of entries to 107,463. Furthermore, a web-based interactive thermal comfort visualization tool has been developed that allows end-users to quickly and interactively explore the data. The database, called ASHRAE Global Thermal Comfort Database II, is intended to support diverse inquiries about thermal comfort in field settings (Földváry Ličina et al., 2018).

What are the implications of the adaptive approach to thermal comfort? Under the umbrella of the IEA EBC Annex 69, different research groups around the world are now trying to build a common framework for the adaptive thermal comfort theory, which is now implemented in several Standards such as the aforementioned ASHRAE 55 (American Society of Heating, Refrigerating and Air-Conditioning

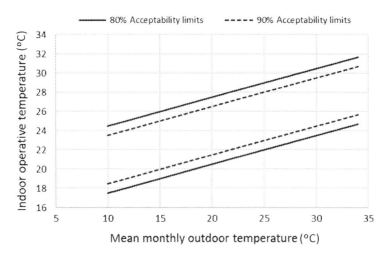

Figure 4.5 Acceptable operative temperature for naturally conditioned spaces according to ASHRAE Standard 55 (American Society of Heating, Refrigerating and Air-Conditioning Engineers, 2017).

Engineers, 2017), EN 16798 (European Committee for Standardization, 2019), and other national Standards (Carlucci et al., 2018) for buildings without mechanical cooling systems—see Fig. 4.5. Moreover, they are trying to extend the validity of adaptive thermal comfort to buildings with mixed-mode HVAC systems, that is, cooling/heating together with natural ventilation (Kim et al., 2019). This could lead to significant energy savings as a result of different building design and operational strategies, and indoor environment standards compared to engineering practice nowadays.

4.3 User behavior

The previous section has shown that the thermal sensation, which is driven by physical, physiological and psychological impulses, influences the thermal comfort in the indoor environment. Comfort-driven interactions of the building occupants with HVAC systems such as changing the temperature set point on a thermostat or switching off a heat emitter directly influence the thermal behavior of buildings and, in turn, their energy demand. While these actions are clearly related to the thermal comfort perceived by the occupants, many other actions such as closing the blinds, opening windows and doors, cooking, or having a shower could be driven by other reasons. Therefore a broader perspective would consider not only the thermal sensation, but also the other human senses such as sight and hearing. For instance, people are usually positively affected by natural light and disturbed by glare and noise. Indoor environmental quality (IEQ) indicators help to measure thermal, acoustic, visual comfort, and indoor air quality. In general, human interactions with buildings are not necessarily driven by IEQ aspects. All the human activities occurring within buildings—having a shower, working, sleeping, cooking, watching TV, etc.—have an impact on the heating and cooling load of buildings and on the IEQ indicators (relative humidity, air freshness, etc.). Even the mere presence of people in buildings can significantly affect their energy demand. The increasing share of refurbished and newly constructed low-energy buildings is going to increase the impact of user behavior on the building energy demand. Depending on the building type and degree of automation, occupants remain as one of the greatest influences on building energy use and among the main reasons for discrepancies between predicted and measured energy performance (IEA EBC Annex 79, 2019). This discrepancy is the so-called energy performance gap, which was highlighted in several studies,—for example, Housez et al. (2014) and Calì et al. (2016). Moreover, emerging evidence suggests that occupants are often dissatisfied with home automation systems and may intervene. Providing greater control to occupants increases their acceptance and preference for a wider range of indoor environmental conditions (IEA EBC Annex 79, 2019). The IEA EBC Annex 79 set out to improve knowledge about occupants interaction with building technologies, with a specific focus on comfort-driven actions (IEA EBC Annex 79, 2019). Data mining and machine learning techniques deployed over large datasets, as well as ad hoc experiments, will help to define the relationship between occupants and building systems. This understanding could facilitate adaptive opportunities and promote energy-saving behavior. Accordingly, a new modeling framework should be deployed assuming that occupants are active decision-making agents who respond to indoor environmental conditions, as shown in Fig. 4.6.

Nowadays, Standards adopt deterministic schedules to simulate set points of the technical systems, ventilation rates, domestic hot water (DHW) demand, and internal heat gains with a hourly resolution. In the ISO 18523, three daily schedules are proposed for both residential and nonresidential buildings, and the latter have been divided into 12 categories (International Organization for Standardization, n.d.). Similar schedules have been proposed by ASHRAE 90 (Goel et al., 2017). All these variables are extrapolated from reference occupancy schedules. Despite the uncertainty inherent to the user behavior, these schedules as well as those used by most of the research papers concerning with the energy performance of buildings are deterministic profiles. Recently, stochastic occupancy profiles based on time-use surveys (TUS) have been generated using Markov chains (Richardson et al., 2008). Already in the 1980s Van Raaij and Verhallen discovered some behavioural patterns (conservers, spenders, cool and warm dwellers) and claimed they could play an important role in the effectiveness of energy conservation measures (Van Raaij and Verhallen, 1983). For instance, they discovered that spenders could consume up to 38% more energy than conservers, as the former maintain a high temperature and a high level of ventilation in their homes. They had a lower educational level and are more often at home compared to other energy users. For warm dwellers, comfort was the main reason to heat more (Van Raaij and Verhallen, 1983). Market economy links the behavioral aspects of energy use commented above to the so-called rebound effect, that is, the trend whereby improved efficiencies are offset by

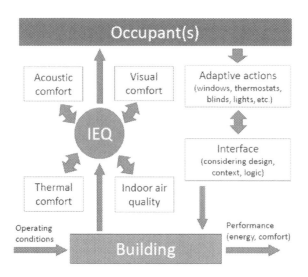

Figure 4.6 Advanced modeling of occupant behavior in buildings (IEA EBC Annex 79, 2019).

a higher spending attitude (Hens et al., 2010). As thermal comfort is clearly driving energy demand for heating, an occupancy-based control could help reduce the energy demand by conditioning only those spaces that are actually occupied. For instance, a monitoring campaign of occupancy in a building with 16 offices revealed that only six of them were simultaneously occupied for 99.8% of the time throughout the year (O'Brien et al., 2017). A literature review from the University of Southampton revealed that occupancy-based smart controls may offer savings between 10% and 15% in building energy consumption, as long as an accurate prediction of occupancy is provided. The study highlighted a lack of evidence on the occupancy and heating patterns (Gauthier et al., 2016). It is well-known that the switching behavior of the electric lights is individual but not arbitrary, that is, while switching thresholds vary within a group of subjects, individuals use their controls consciously and consistently (Reinhart and Voss, 2003). Burak Gunay et al. claimed that the order and availability of manual control behaviors may represent a limitation for the generalization of a particular adaptive behavior (Gunay et al., 2013). They questioned the existence/usefulness of generalized behavior model and suggested that automation systems, instead of undertaking a traditional set point-based control strategy for the whole building, can learn from occupants' interventions over their sensors or controlled devices by using, for example, logistic regression models to predict the likelihood of adaptive occupant behavior. With regard to data collection methods, time-use surveys (TUS) are so far the main method for inferring domestic occupancy patterns at a regional or national scale (Gauthier et al., 2016). TUS have validity issues make wireless sensor networks the best method for monitoring occupancy, given the low-cost software and hardware of Internet of things (IoT) devices already available on the market. Occupancy may also be inferred from metered electricity data, providing that the sampling rate is short enough. This is a promising method for both domestic and commercial buildings (Gauthier et al., 2016). In a recent work, clustering techniques applied to Wi-Fi device count time series data were demonstrated to accurately forecast occupancy over the 24-hour prediction horizon (Hobson et al., 2019).

4.4 Weather conditions under climate change and growing urbanization

The previous sections have been focused on the indoor thermal environment and on the impact of user behavior on the energy performance of buildings. From the perspective of the energy analysts, the weather represents the boundary conditions "on the other side of the wall." For investors, policy makers, and other stakeholders, weather conditions are a key indicator for estimating the energy demand for space heating and cooling on an aggregated scale. Decisions about the design of buildings and the energy supply at community level must take into account a time horizon of at least 30 years. So far, little attention has been paid to the uncertainty that characterizes the weather conditions in the coming years in the light of the ongoing global warming and of the growing urbanization.

According to a recent report of the IEA, the energy needs for space cooling will triple by 2050, with nearly 70% of the increase from the residential sector, driven by economic and population growth in the hottest parts of the world. The lion's share of the projected growth in energy use for space cooling by 2050 comes from the emerging economies, with just three countries—India, China, and Indonesia—contributing half of global cooling energy demand growth (International Energy Agency, 2018). Fig. 4.7 shows a world map of the expected increase in cooling degree days. As shown by the figure, the aforementioned countries will be among those parts of the world that will suffer the most from global warming. Although the future energy demand is likely to increase due to climate change, the magnitude of such increase depends on many interacting sources of uncertainty. A recent study that combines econometrically estimated responses of energy use to different socioeconomic and climate scenarios shows that vigorous (moderate) warming increases global climate-exposed energy demand before adaptation around 2050 by 25%–58% (11%–27%), on top of a factor 1.7–2.8 increase above present-day due to socioeconomic developments (van Ruijven et al., 2019). According to the study, the energy demand will rise by more than 25% in the tropics and southern regions of the United States, Europe, and China (van Ruijven et al., 2019). Since all models for calculating the building energy demand for space heating and cooling rely on weather data, it is of utmost importance to define how to treat the uncertainty of the considered weather data. Those weather files that take into consideration both typical and extreme conditions, such as heat waves, are the most reliable files for providing representative boundary conditions to test the energy robustness of buildings under future climate uncertainties (Moazami et al., 2019).

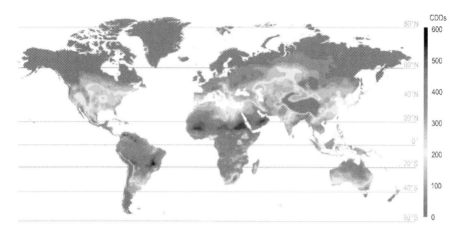

Figure 4.7 Increase in cooling degree days (CDDs) relative to historical CDDs, 2016–50[2] (International Energy Agency, 2018).

The movement of people from rural to urban areas has become a consolidated global trend. As people increasingly live and will live in large urban centers, it is interesting to study the local climatic conditions within these places. In the last decades, several scientists observed a significant temperature difference between city centers and their surrounding areas, especially at night—see for example, Bornstein (1968). This is known as the urban heat island (UHI) effect, and can be explained with the energy balance at the city surface. Taha performed meteorological simulations and his results suggest that cities can feasibly reverse heat islands and offset their impacts on energy use simply by increasing the albedo of roofing and paving materials and reforesting urban areas (Taha, 1997). Although the UHI was highly investigated, there are fewer reports on the urban cool island (UCI) phenomenon, that is, where the air temperature of the surrounding rural area is warmer than that of the urban area. In contrast to the UHI, the UCI phenomenon always occurs during the day, and with relatively weak intensity. The low solar radiation received especially by high rise and compact cities in moments with low anthropogenic heat can lead to the UCI, as reported in Hong Kong (Yang et al., 2017).

The rapid urbanization in developing countries cannot be analyzed only from a physical standpoint. In fact, the actual living conditions of people are clearly influenced by deep social and economical inequalities (OECD, 2018). While the rich

---

[2] This map is without prejudice to the status of or sovereignty over any territory, to the delimitation of international frontiers and boundaries, and to the name of any territory, city, or area.
Notes: The CDDs shown here are relative to a base temperature of 18°C; see Chapter 3 of reference International Energy Agency (2018) for an explanation of the baseline scenario.
Sources: Derived using NCAR (2004), Community Climate System Model, Version 3.0, www.cesm.ucar.edu/models/ccsm3.0/; NCAR (2012), GIS Program Climate Change Scenarios, Version 2.0, www.gisclimatechange.org; CIESIN (2017), Gridded Population of the World, Version 4 (GPWv4): Population Count Adjusted to Match 2015 Revision of UN WPP Country Totals, Revision 10, https://doi.org/10.7927/H4JQ0XZW.

Energy in buildings and districts

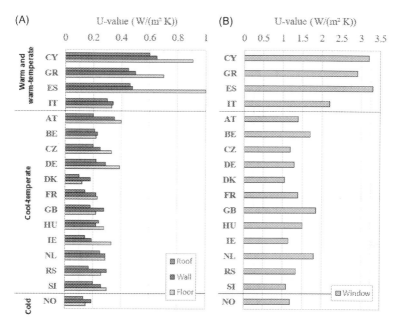

Figure 4.8 Current minimum requirements[3] on U-values of (A) opaque and (B) glazed building components in new buildings across EU countries (Stein et al., 2016).

people can afford living in comfortable and even energy-efficient places, the poor people are often pushed toward informal settlements characterized by poor quality of life having one or more of the following deprivations: lack of access to improved water sources and sanitation facilities, lack of sufficient living area, lack of housing durability, and lack of security of tenure (Nutkiewicz et al., 2018). What is the role of energy modelers in such complex cases where thermal comfort and energy efficiency do not seem among the top priorities? A possible answer comes from India, where the central government has adopted a policy for the redevelopment of informal settlements. A recent study used energy modeling techniques to define the best configurations for the "redevelopment" of the slum of Dharavi in Mumbai (India), a slum in the center of Mumbai (India) that is estimated to give a home to approximately 1 million people. The redevelopment means replacing parts of the slum with high rise social housing buildings. This chapter shows that the early design choices about the building morphology as well as measures to improve the ventilation after the construction may have a significant impact on thermal comfort parameters, thus indicating opportunities for significantly enhancing the living conditions of the slum inhabitants after the refurbishment (Nutkiewicz et al., 2018).

---

[3] Fig. 4.8 provides an overview of the average component U-values for all exemplary new buildings (latest construction year class) included in the TABULA database (Stein et al., 2016).

4.5 Envelope and materials

The European Directive EPBD 2010/31/EU obliges each member state to implement policies to improve the efficiency of buildings, until new buildings have almost zero energy consumption by end of 2020. The minimum requirements prescribed by the national laws of member countries were progressively updated in the last years in order to reach the target values at the mentioned deadline. Fig. 4.8 shows the minimum U-value requirements for different European countries belonging to different climate zones according to a report in 2016 (Stein et al., 2016). However, a comprehensive reduction of emissions in the building sector can be achieved only by acting on the existing stock. Therefore both new building constructions and the refurbishment of the existing building stock are important factors to achieve the efficiency targets. As far as refurbishment of existing buildings is concerned, the main type of intervention consists in increasing the thermal resistance of external walls by applying a layer of insulating material. The most common materials used for this purpose are extruded and expanded polystyrene, polyurethane (PUR), stone wool, and cellulose fiber. The thermohygrometric properties of these and other insulation materials are shown in Table 4.1.

The thermal insulation of the perimeter walls can be carried out from the external side, from the internal side, or in cavities. The insulation applied to the internal surface involves the creation of a plasterboard counter-wall and the use of insulating panels. This type of intervention involves a reduction in the living space and possible interstitial condensation problems. The insulation of the cavity in the external walls of existing buildings is carried out by drilling holes in the wall and blowing insulation materials in loose form (cellulose fibers, glass wool flakes, polystyrene in granular form, vermiculite, etc.) or PUR foams. The most critical aspect of this type of intervention is the uniformity of filling of the cavity. The insulation of the external side of the walls is the most effective solution, as it allows the correction of thermal bridges (due to the presence of concrete columns and floors, for example). This type of insulation includes external cladding systems and ventilated facades. The thermal insulation system consists of the application, on the entire external vertical surface of the building, of insulating panels that are then covered with a thin thickness of finish made with special plasters. This solution is particularly suitable for the restoration of vertical surfaces whose cladding is in an advanced stage of deterioration. Further advantages of the external insulation is that it avoids losing part of the thermal buffering potential of the building structure, which is especially useful during summer to mitigate the effect of solar loads, and it reduces the risk of interstitial condensation.

Nowadays there is a growing use of prefabricated and/or modular solutions for both new construction and refurbishments. As a result, the proportions of building components and systems preassembled or fabricated off-site are increasing. Prefabrication reduces variation in product quality and process timing and reduces cycle times for production and installation. The shift toward prefabricated and/or modular components can be regarded as a process of "industrialization" of the

Table 4.1 Thermohygrometric performance of some insulation materials.

Material	Thermal conductivity, λ (W/(m² K))	Density, ρ (kg/m³)	Specific heat, c_p (J/(kg K))	Vapor diffusion resistance factor, μ (–)	Embodied energy, EE (MJ/m²)
Polystyrene (expanded polystyrene)	0.037	17	1570	29–58	88.6
Hydrophobic mineral wool	0.037	100	790	1.3–2.6	–
Multipor (MP)	0.047	125	1070	2–11	–
Polystyrene (extruded polystyene)	0.036	10	800	80–250	–
Rigid polyurethane (PUR)	0.024	25	800	20	101
Rockwool	0.040	40	840	1	17
Cellulose fiber	0.040	40	–	–	3.3
Wood fiberboard (WFB)	0.047	54	2130	2.7	–
Flax fibers	0.054	23	1443	1.3	–
Hemp fibers	0.054	38	1906	1.8	–
Jute fibers	0.052	33	1819	1.8	–
Sheep wool	0.045	25	1940	1.9	2.5

Source: From Jerman, M., Palomar, I., Kočí, V., Černý, R., 2019. Thermal and hygric properties of biomaterials suitable for interior thermal insulation systems in historical and traditional buildings. Build. Environ. 154, 81–88. https://doi.org/10.1016/j.buildenv.2019.03.020; Hossain, A., Mourshed, M., 2018. Retrofitting buildings: embodied & operational energy use in English housing stock. Proceedings 2, 1135. https://doi.org/10.3390/proceedings2151135.

building envelope, thus boosting both the construction productivity and reducing the environmental impact of the construction process (Andaloro et al., 2019). Prefabricated envelopes may differ in terms of main support materials (e.g., steel, aluminum, concrete, and timber), load transfer mode (load bearing or nonload bearing), shell arrangement (single skin or multilayered), and the degree of prefabrication versus on-site construction (Gasparri and Aitchison, 2019).

The ventilated facade makes it possible to insulate the buildings thermally and acoustically and at the same time to protect them from humidity and atmospheric agents. The insulating panels are fixed to the external surface of the masonry structures with special plugs. The external finishing elements are spaced from the insulation by means of a system of metal profiles and brackets, directly anchored to the wall structure behind. The ventilated cavity between the insulation layer and the facade finish improves the summer performance of the building, as the heat from the solar radiation incident on the facade is disposed of through the chimney effect. The ventilated facade also protects the external walls from rainwater, while maintaining the wall's breathability. This technology is mainly used in tertiary and commercial buildings and represents an excellent solution for the restyling of the building.

The environmental impact of building materials and products can be quantified by means of indicators such as their embodied energy (EE) and embodied carbon (EC). These indicators are becoming a dominant construction design variable as buildings move toward the nZEB standard. Moran et al. (2017) found that the life cycle of EE and EC ranged from 30% to 33% and from 41% to 100%, respectively, for eight case study buildings in Ireland, thus stressing the importance of an appropriate selection of "eco-friendly" materials at the design stage (Moran et al., 2017). The heat losses of a simple pane of glass are on an average five times higher than those of an uninsulated wall. Double and triple glazing can significantly reduce the heat loss to the external environment. The air in the cavity/cavities can also be replaced by noble gases such as argon and krypton, which guarantee greater thermal resistance. To further increase the thermal performance of the glazed package, it is also common to use low-emission glass instead of simple float glass. A thin layer of metal oxides (e.g., silver) is deposited on the internal (external) surface of the glass facing the outdoor (indoor) environment, which reduces the heat transfer by radiation through the window.

4.6 From passive to nearly zero-energy building design

As it was mentioned in Section 4.1, in the last years building design has experiences a shift from the passive house approach to the nZEB approach. This section aims at describing the principles behind the two design approaches.

We could say that the passive house concept builds upon this logical basis: "the less we lose, the less we need." It was therefore originally conceived for cold climates. Passive design strategies aim at maximizing heat gains and minimizing heat loss through the envelope. Therefore passive buildings have a very low thermal transmittance ($U \leq 0.15$ W/(m^2 K) for wall, floors, and roofs and $U \leq 0.85$ W/(m^2 K) for windows), glazed elements with good solar transmittance ($g \geq 0.5$) possibly facing toward the south (in the northern hemisphere), and have a very compact form to reduce the surface area to volume ratio ($S/V \leq 0.7$ m^{-1}). Virtually, all construction methods can be successfully utilized to this end. Masonry (cavity wall and monolithic), timber frame, off-site prefabricated elements, insulated concrete formwork, steel, straw bale, and many hybrid constructions have been successfully used in Passivhaus buildings (Schnieders et al., 2015). The thermal insulation material should be placed on the exterior side of the walls, and any other measure should be taken to prevent thermal bridges to increase the heat transmission to the outside. In order to reduce the heating demand and prevent warm moisture laden air from entering the fabric, the building must have a very good airtightness levels. The latter can be achieved by using an airtight membrane or barrier within each of the building elements. Shading systems play a fundamental role to avoid overheating in summer. A number of further strategies are available to reduce overheating risk, including the use of conventional cross ventilation, free cooling at nights, and heat recovery systems in case of mechanical ventilation yielding a heat recovery efficiency $\geq$ 75% and a specific fan power of $\leq$ 0.45 Wh/m^3.

The EPBD 2018/844 (European Commission, 2018) defines a nZEB as "...a building that has a very high energy performance, as determined in accordance with Annex I. The nearly zero or a very low amount of energy required should be covered to a very significant extent by energy from renewable sources, including energy from renewable sources produced on-site or nearby." Attia et al. (2017) resumed the design process of nZEB as an integrative approach looking to:

1. reduce energy needs for heating and cooling by optimizing the envelope and integrating passive heating and cooling techniques;
2. improve energy efficiency of active systems;
3. incorporate renewable energy.

It is evident how the nZEB concept involves making two steps further compared to the passive house concept, that was based mainly on optimizing the envelope and adopting passive strategies to minimize the heat loss. Although the concept is clear, researchers and building designers have been concerned with the lack of a common methodology to assess the energy performance and of clear threshold values to accompany the definition of nZEB. The first document to address this issue was produced by the European Commission based on a study from Ecofys (European Commission, 2016). This document suggests some benchmark ranges for the energy performance of nZEB offices and single-family housesin different European climates, as reported in Table 4.2.

However, according to the new EPBD the decisions about the calculation methodology and the energy performance targets have been delegated to the individual member states. According to the new EPBD 2018/844, the energy performance of a building shall be determined on the basis of calculated or actual energy use and shall reflect typical energy use for (1) space heating, (2) space cooling, (3) DHW,

Table 4.2 Benchmark ranges for the energy performance of nZEB in different EU climatic zones as suggested in the EC recommendations (European Commission, 2016).

Climate—building	Net primary energy (kWh/m^2)	Primary energy use (kWh/m^2)	On-site production from RES (kWh/m^2)
Mediterranean—office	20–30	80–90	60
Mediterranean—SFH	0–15	50–65	50
Oceanic—office	40–55	85–100	45
Oceanic—SFH	15–30	50–65	35
Continental—office	40–55	85–100	45
Continental—SFH	20–40	50–70	30
Nordic—office	55–70	85–100	30
Nordic—SFH	40–65	65–90	25

SFH, Single-family house.

(4) ventilation, (5) built-in lighting, and (6) other technical building systems. The energy performance of a building shall be expressed by a numeric indicator of primary energy use in kWh/(m^2 year) for the purpose of both energy performance certification and compliance with minimum energy performance requirements.

Fig. 4.9 shows a simple primary energy balance of the building. The primary energy use should be calculated for each of the aforementioned end uses with appropriate primary energy conversion factors provided by the national Standards. The same factors should be used in case the electrical or thermal energy produced on-site is exported to the grid. A recent study revealed that the best cost-effective solutions to low-energy building designs vary considerably by climate (D'Agostino and Parker, 2018). In warmer, sunny locations, appliance efficiency measures and light colored surfaces are of utmost importance as heating loads are not significantly increased, whereas cooling loads may be reduced. In colder climates, insulation and building tightness appear much more important as thermal improvements are strongly dependent on heating and cooling loads in a city. As the nZEBs target can be missed if buildings are designed without considering climate change, the possibility that Europe may experience warmer periods in summer was considered in the calculations. According to the results, the energy excess in summer decreases over time due to the increased cooling loads. The same does not occur in winter, when the imbalance between generation and consumption is not reduced in the same proportion (D'Agostino and Parker, 2018). Stazi et al. (2014) compared three building envelopes for a multifamily building in Central Italy. They evaluated the use of masonry, wood−cement, and wood envelopes with a multidisciplinary approach considering energy saving, comfort, and environmental sustainability. They found that masonry is the best solution for summer comfort and wood−cement for winter performance. The wooden solution is preferred for low environmental impact and cost effectiveness (Stazi et al., 2014). The adoption of a lightweight envelope, even if not convenient with regard to comfort aspects for the overheating phenomena, is preferable for lower exploitation of environmental resources and economic convenience (Stazi et al., 2014).

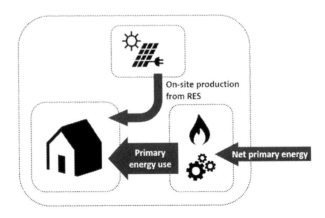

Figure 4.9 Primary energy balance of a building.

The results of another simulation study showed that comfort parameter settings have a higher impact on the air-conditioning energy demand for a nZEB than for a traditional dwelling (Guillén-Lambea et al., 2017). By adopting extended comfort ranges, significant energy savings could be achieved in countries with temperate climates. For this reason, the authors recommended the use of adaptive comfort standards to nZEB dwellings in Southern Europe. The aforementioned considerations suggest that the design of low-energy buildings must be climate dependent and must consider the adaptive capabilities of future occupants to achieve the desired energy performance targets.

4.7 Smart buildings and home automation

According to the author, everything that makes the home autonomous and interactive contributes to the transition from nZEB to smart buildings. The autonomy of the building necessarily passes through its control systems. We can distinguish between low-level control systems and the so-called building energy management systems (BEMS). The former is responsible for controlling individual components, such as the control unit that starts the engine of a pump following an increase in the pressure difference in a hydraulic system, or the thermostat that starts a boiler or any other heat generator following a drop in the water temperature in a tank or the air temperature in a room. These low-level controllers enable the automated operation of HVAC devices in buildings, thus avoiding the necessity of a continuous manual intervention from the building occupants. BEMS, instead, operate on a higher level: they coordinate the actions of low-level controllers so that the building improve its performance either in terms of (1) costs, (2) energy efficiency, or (3) thermal comfort. Usually, higher energy efficiency means lower energy bills. So the first two objectives involve aligned control actions. On the other hand, the achievement of these objectives (minimization of costs or energy expenditure) usually conflicts with the improvement of indoor thermal comfort. The second pillar for the design of a smart building is its capability to interact with both the occupants and the energy grids to which the latter is connected. The latter concept is strongly related to the so-called energy flexibility of the building, that is, "…its ability to manage its demand and generation according to local climatic conditions, user needs, and grid requirements," as defined by the research group of the IEA EBC Annex 67 (Jensen et al., 2017). A simulation study by Vivian et al. (2020) showed that the energy flexibility potential of a building is mainly related to the level of insulation of the envelope during the heating season, and to both insulation and the time of the event during the cooling season. The combination of these variables and the climatic conditions during the event determine the amount of energy that can be shifted over time. Moreover, part of the thermal energy shifted is lost through the envelope after the events, resulting in lower efficiencies for old buildings. In general, upward and downward modulation events are preferable just before and just after the peak load periods, respectively. This chapter shows that severe weather conditions and intermittent set point schedules lead to exceptions to this general rule (Vivian et al., 2020). In addition, the results show that the cost associated with flexibility is an increase in consumption for upward modulation events and a reduction in thermal comfort conditions for downward modulation events. The greater the flexibility potential of the buildings, the higher the costs, both in heating and cooling modes (Vivian et al., 2020). More insights on the applications of energy flexible buildings are discussed in the next section. As already mentioned in Section 4.3, home automation systems are often bypassed by users if they do not have the opportunity to interact with them according to their needs. Therefore the user interaction is another key feature of any successful BEMS. To this end, the user interface should be clear and easy-to-use. The principles of adaptive comfort described in Section 4.2 could help widen the boundaries of thermal comfort and therefore give the BEMS more room for optimizing the HVAC operation and at the same time allow the users to gain control over the indoor environment. The Standard EN 15232 describes two procedures to calculate the contribution of building automation and controls (BAC) to the energy performance of buildings (European Committee for Standardization, 2017). The output of the first method is a list of automation, control, and management function types that is used to run a detailed calculation of building energy performance based on other EPBD standards. The second method provides BAC efficiency factors needed to calculate the final energy demand of a building according to a given building automation and control classification on a yearly basis (European Committee for Standardization, 2017). The smart readiness of buildings, introduced by the new EPBC 2018/844 and still to be clearly defined by an appropriate indicator, assesses their ability to adapt their operation to the needs of the occupant and the grid and to improve their energy efficiency and overall performance. Therefore this indicator will probably measure the energy flexibility of the whole building, including its HVAC systems and the corresponding BAC.

Summarizing, according to the smart building concept, BEMS will optimize the HVAC operations, thus making the building both autonomous and interactive. This technological evolution of buildings does not come for free. A technological infrastructure is indeed needed to allow the bidirectional communication between the building and its interfaces to the occupants and to the energy grids. Therefore the connectivity is an important prerequisite for the transition to the smart buildings scenario. To this end, the market offers several low-cost hardware components such as wireless sensor networks, microcontrollers, and single-board computers that can enable the interaction with the objects of everyday life—including HVAC systems—through suitable communication protocols, thus becoming an integral part of the Internet (Atzori et al., 2010). This communication paradigm is called IoT, and home automation is in fact one of its most promising application areas (Jia et al., 2019).

4.8 From smart buildings to smart districts and cities

The Heat Roadmap Europe has clearly indicated that district heating and cooling is mostly indicated for dense urban areas, whereas individual heat pumps are considered as the most suitable technology for the decarbonization of peripheral urban areas and rural areas. A widespread use of heat pumps for heating and cooling European households confirms a general trend of electrification of end uses, which includes other sectors such as transport. On the other hand, it is clear from the previous section that the smart building paradigm fosters a scenario where buildings are no longer independent energy consumers. Indeed, their inherent flexibility allows them not only to produce energy locally but also to adapt their energy consumption to the needs of the energy system, thus providing auxiliary services to the electricity grid. From the perspective of the electric utility, any activity designed "to influence customer uses of electricity in ways that will produce desired changes in the utility's load shape" is called demand-side management (DSM) (Gellings and Chamberlin, 1993). Among the possible DSM activities, active demand response (ADR) programs are defined as the changes in electric energy use implemented directly or indirectly by end users from their normal consumption patterns as function of certain signals (Arteconi et al., 2013). One of the most promising applications of ADR consists of using the thermal capacitance of buildings and thermal energy storage (TES) systems to shift the load of electrically driven HVAC devices, such as heat pumps (Arteconi et al., 2013). Using the building structures alone or combined with existing heat storage systems is potentially disruptive, as no additional investment costs are needed but those for the "intelligence" and "connectivity" of existing HVAC control devices, as discussed in the previous section. A higher energy flexibility of buildings will favor their participation to ADR programs based on the requirements of the system operator (SO) or of external third-party actor, often named aggregator (Hurtado et al., 2017). The control in ADR programs can be either direct or indirect. In the direct control, the SO (or the aggregator) can directly change the consumption pattern, whereas indirect control implies that the user has full control over its appliances, and can freely choose whether to participate or not in the ADR program (Shen et al., 2014). The benefit of ADR actions should not be intended as energy savings for single participants, but rather as the possibility to shift the energy consumptions of the whole set of participants toward time windows with cheaper and/or cleaner energy production. These strategies are what Fischer and Madani (2017) called price focused and renewable energy focused services—see Fig. 4.10. According to their review, heat pumps could also contribute to grid relieving through voltage control —especially in case of distribution grids with high PV penetration—, congestion management, and frequency stabilization. On the other hand, distribution networks with a high share of heat pumps could experience problems due to high peak loads, possibly due to (1) simultaneous switch-on or shut-off a heat pump pool that responds to a unique control signal and (2) low PV feed-in (Wagener, 2017).

The provision of these auxiliary energy services to the grid is a possible remedy for the increase of electrical loads and distributed energy generation systems. Indeed, demand response would therefore reduce the needs for infrastructural investments for grid reinforcement. The aggregation of responsive energy consumers is only one among many initiatives based on the idea that organized groups, sometimes called "energy communities," can better tackle with the challenges posed by the energy transition compared to single individuals. This consideration is

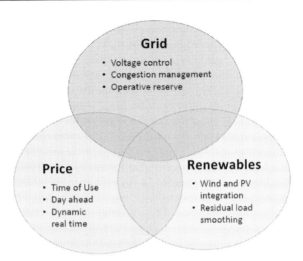

Figure 4.10 Schematic diagram of the application of heat pumps in a smart grid context (Fischer and Madani, 2017).

not limited to technical issues such as those mentioned earlier. In fact, individuals forming an "energy community" may have very different purposes, such as sharing the costs of an investment, or collectively purchasing clean energy from renewable energy production sites. In these cases, the projects are driven by the common needs of groups of final energy users. Moroni et al. (2018) have identified four groups of energy communities: place (nonplace) if the members (do not) belong to a specific territory, and energy-only (multiissue) if the goods and services managed by the community are (not) exclusively related to energy. The common framework between participants in nonplace energy communities such as virtual power plants is the market. An interesting application for both place and market-based energy communities is the PV energy sharing. According to this concept, different households can use the electricity produced by one or more "shared" PV system(s). This concept is implemented differently from country to country, depending on the national legislation. Fina et al. (2019) presented the differences between the legislation in force in Austria and Germany, and then showed that the profitability of shared PV depends on the settlement where the latter is applied (historical, rural, mixed, and residential areas). In general, settlements including buildings with different available roof surfaces and different power load profiles benefit the most from the PV-sharing energy community, as a result of a high exchange rate between local suppliers and consumers (Fina et al., 2019). An example of a "place-based" energy community is represented by zero and positive energy districts (Gollner et al., 2019). The latter is the portions of the city including several buildings with different uses where the exported primary energy is higher than the imported primary energy. As for nZEBs, a null or positive primary energy balance entails a significant on-site production from renewable energy sources. Since this kind of projects requires a very high investment costs, they are usually carried out by both private and public stakeholders, such as construction firms, energy service companies, municipalities, and so on. In these cases, participation of the final energy users in the decision-making processes helps increase their social acceptance. A well-known example of energy planning at urban level comes from the city of Graz, Austria. Here, the municipality has decided to reduce the air pollution by changing the energy supply mix of the city. A participated process involving citizens and local experts has been carried out, and around 40 action proposals came out as a result. The municipality decided to increase the share of buildings connected to the municipal district heating grid and at the same time to increase the number of heat suppliers. This process is required to modify regional laws on buildings and to interact with private stakeholders such as local industries for increasing the share of waste heat supply. Moreover, it was decided to produce part of the heat with solar thermal collectors assisted by a seasonal TES and by biomass boilers limited to the amount of biomass that could be supplied by local wood suppliers (Reiter et al., 2016). The project, called Solar Graz, shows the importance of the local authorities in the transition toward a sustainable energy supply at urban level.

4.9 Concluding discussion

This chapter offers an overview on different theoretical and practical aspects concerning the energy performance of buildings, focusing the attention on the last trends in research and technology and on the parallel evolution of the corresponding Standards.

Thermal comfort has been regarded as one of the main drivers of energy consumption in buildings. According to the recently proposed adaptive thermal comfort theory, thermal comfort cannot be simply explained by the metabolic energy balance of the human body in a certain environment, because people adapts to local conditions according to subjective impulses of physiological, psychological, and behavioral nature. The building occupants therefore hide a great potential for reducing the energy consumption in buildings. This is especially true for low-energy buildings, where the human interaction with the building has a dramatic impact on their energy needs compared to existing buildings where the energy demand is mainly caused by transmission losses through the envelope. Therefore building physics research has started to move out of its traditional domain, looking at solutions together with experts of the human behavior. In this context, the availability of data is a key asset, and the use of machine learning techniques seem to be powerful tools for the research community, that is, trying—among other things—to detect the patterns of occupants' presence and their interactions with buildings (HVAC systems, shading devices, etc.).

The global warming will boost the demand for space cooling worldwide, especially in big cities of the developing world. Here, the rapid urbanization is already asking for technical solutions on a urban scale rather than on a building scale in order to improve the thermal comfort of both indoor and outdoor spaces frequented by millions of people, with direct consequences on their health and well-being.

In industrialized countries, one of the main challenges will be the refurbishment of the building stock and the increase of on-site renewable energy production that can be mainly achieved through PV modules. The decarbonization of the heating sector necessarily passes through a widespread use of heat pumps and district heating systems. The use of bio-based materials and prefabrication of building components seem to be among the most suitable solutions to improve both the sustainability and the productivity of the construction sector. The choices concerning both the envelope characteristics and the technical systems to reach nZEB standards depend on the considered climate.

Due to the increasing number of distributed energy resources and to the foreseen electrification of end uses (mainly due to heat pumps, new cooling capacity and electric vehicles), buildings will be the sites of significant power production and consumption. The shift toward the smart grid paradigm requires new ways of managing energy within buildings. The availability of low-cost devices and open protocols for digital devices and particularly home automation systems will foster the technological advancement in this direction.

Besides being energy efficient, those buildings that will be able to participate in the electricity markets or those that will take part in the so-called energy communities in a broader sense will gain benefit from added value compared to passive consumers. To this scope, aggregation of prosumers using new business models seems to be the most promising option to make buildings play an active role in the energy system.

Finally, the infrastructure for energy supply at urban or district level must be implemented by the local administration with the necessary political support and by the ability to raise both public and private funds. Best practices include the definition of participatory pathways to broaden or strengthen the consensus among citizens.

References

Abergel, T., Dean, B., Dulac, J., Hamilton, I., 2018. Global Status Report—Towards a Zero-Emission, Efficient and Resilient Buildings and Construction Sector. International Energy Agency (IEA).

Ameen, R.F.M., Mourshed, M., Li, H., 2015. A critical review of environmental assessment tools for sustainable urban design. Environ. Impact Assess. Rev. 55, 110−125. Available from: https://doi.org/10.1016/j.eiar.2015.07.006.

American Society of Heating, Refrigerating and Air-Conditioning Engineers (ASHRAE), 2017. ANSI/ASHRAE Standard 55 − Thermal Environmental Conditions for Human Occupancy. Atlanta, GA.

Andaloro, A., Gasparri, E., Avesani, S., Aitchison, M., 2019. Market survey of timber prefabricated envelopes for new and existing buildings. In: Proceedings of Powerskin Conference 2019, Munich, Germany.

Arteconi, A., Hewitt, N.J., Polonara, F., 2013. Domestic demand-side management (DSM): role of heat pumps and thermal energy storage (TES) systems. Appl. Therm. Eng. 51, 155−165. Available from: https://doi.org/10.1016/j.applthermaleng.2012.09.023.

Attia, S., Eleftheriou, P., Xeni, F., Morlot, R., Ménézo, C., Kostopoulos, V., et al., 2017. Overview and future challenges of nearly zero energy buildings (nZEB) design in

Southern Europe. Energy Build. 155, 439−458. Available from: https://doi.org/10.1016/j.enbuild.2017.09.043.

Atzori, L., Iera, A., Morabito, G., 2010. The Internet of things: a survey. Computer Netw. 54, 2787−2805. Available from: https://doi.org/10.1016/j.comnet.2010.05.010.

Bornstein, R.D., 1968. Observations of the urban heat island effect in New York City. J. Appl. Meteorol. 7, 575−582. Available from: https://doi.org/10.1175/1520-0450(1968) 007 < 0575:OOTUHI > 2.0.CO;2.

Calì, D., Osterhage, T., Streblow, R., Müller, D., 2016. Energy performance gap in refurbished German dwellings: lesson learned from a field test. Energy Build. 127, 1146−1158. Available from: https://doi.org/10.1016/j.enbuild.2016.05.020.

Carlucci, S., Bai, L., de Dear, R., Yang, L., 2018. Review of adaptive thermal comfort models in built environmental regulatory documents. Build. Environ. 137, 73−89. Available from: https://doi.org/10.1016/j.buildenv.2018.03.053.

D'Agostino, D., Parker, D., 2018. Data on cost-optimal nearly zero energy buildings (NZEBs) across Europe. Data Brief 17, 1168−1174. Available from: https://doi.org/10.1016/j.dib.2018.02.038.

de Dear, R.J., Brager, G.S., 1998. Developing an adaptive model of thermal comfort and preference. ASHRAE Trans. 101.

European Commission, 2010. Directive 2010/31/EU of the European Parliament and of the Council on the Energy Performance of Buildings. Brussels.

European Commission, 2012. Directive 2012/27/EU of the European Parliament and of the Council on Energy Efficiency. Brussels.

European Commission, 2016. Commission Recommendation (EU) 2016/1318 of 29 July 2016 on Guidelines for the Promotion of Nearly Zero-Energy Buildings and Best Practices to Ensure that, by 2020, all New Buildings Are Nearly Zero-Energy Buildings. Brussels.

European Commission, 2018. Directive (EU) 2018/844 of the European Parliament and of the Council of 30 May 2018 Amending Directive 2010/31/EU on the Energy Performance of Buildings and Directive 2012/27/EU on Energy Efficiency. Brussels.

European Committee for Standardization, 2017. EN 15232-1:2017 Energy Performance of Buildings - Part 1: Impact of Building Automation, Controls and Building Management Modules M10-4,5,6,7,8,9,10. Brussels.

European Committee for Standardization, 2019. EN 16798-1: Energy Performance of Buildings - Ventilation for Buildings. Part 1: Indoor Environmental Input Parameters for Design and Assessment of Energy Performance of Buildings Addressing Indoor Air Quality, Thermal Environment, Lighting and Acoustics - Module M1-6. Brussels.

Fanger, P.O., 1982. Thermal Comfort: Analysis and Applications in Environmental Engineering. R.E. Krieger Pub. Co, Malabar, FL.

Fanger, P.O., Ipsen, B.M., Langkilde, G., Olessen, B.W., Christensen, N.K., Tanabe, S., 1985. Comfort limits for asymmetric thermal radiation. Energy Build. 8, 225−236. Available from: https://doi.org/10.1016/0378-7788(85)90006-4.

Fina, B., Auer, H., Friedl, W., 2019. Profitability of PV sharing in energy communities: use cases for different settlement patterns. Energy 189, 116148. Available from: https://doi.org/10.1016/j.energy.2019.116148.

Fischer, D., Madani, H., 2017. On heat pumps in smart grids: a review. Renew. Sustain. Energy Rev. 70, 342−357. Available from: https://doi.org/10.1016/j.rser.2016.11.182.

Földváry Ličina, V., Cheung, T., Zhang, H., de Dear, R., Parkinson, T., Arens, E., et al., 2018. Development of the ASHRAE global thermal comfort database II. Build. Environ. 142, 502−512. Available from: https://doi.org/10.1016/j.buildenv.2018.06.022.

Gasparri, E., Aitchison, M., 2019. Unitised timber envelopes. A novel approach to the design of prefabricated mass timber envelopes for multi-storey buildings. J. Build. Eng. 26, 100898. Available from: https://doi.org/10.1016/j.jobe.2019.100898.

Gauthier, S., Aragon, V., James, P., Anderson, B., 2016. Occupancy Patterns Scoping Review Project. University of Southampton, Southampton.

Gellings, C.W., Chamberlin, J.H., 1993. Demand-Side Management: Concepts and Methods, second ed. Fairmont Press, Lilburn, GA.

Goel, S., Rosenberg, M.I., Eley, C., 2017. ANSI/ASHRAE/IES Standard 90.1-2016 Performance Rating Method Reference Manual. https://doi.org/10.2172/1398228

Gollner, C., Hinterberger, R., Noll, M., Meyer, S., Schwarz, H.-G., 2019. Booklet of Positive Energy Districts in Europe (Preview): A Compilation of Projects Towards Sustainable Urbanization and the Energy Transition. JPI Urban Europe, Vienna.

Gonzalez, R.R., Nishi, Y., Gagge, A.P., 1974. Experimental evaluation of standard effective temperature a new biometeorological index of man's thermal discomfort. Int. J. Biometeorol. 18, 1–15. Available from: https://doi.org/10.1007/BF01450660.

Guillén-Lambea, S., Rodríguez-Soria, B., Marín, J.M., 2017. Comfort settings and energy demand for residential nZEB in warm climates. Appl. Energy 202, 471–486. Available from: https://doi.org/10.1016/j.apenergy.2017.05.163.

Gunay, H.B., O'Brien, W., Beausoleil-Morrison, I., 2013. A critical review of observation studies, modeling, and simulation of adaptive occupant behaviors in offices. Build. Environ. 70, 31–47. Available from: https://doi.org/10.1016/j.buildenv.2013.07.020.

Hens, H., Parijs, W., Deurinck, M., 2010. Energy consumption for heating and rebound effects. Energy Build. 42, 105–110. Available from: https://doi.org/10.1016/j.enbuild.2009.07.017.

Hensen, J.L.M., 1991. On the Thermal Interaction of Building Structure and Heating and Ventilating System (Thesis). Eindhoven University of Technology, Eindhoven.

Hobson, B., Lowcay, D., Gunay, B., Ashouri, A., Newsham, G.R., Opportunistic occupancy-count estimation using sensor fusion: a case study, Build. Environ. 159, 2019, 106154. Available from: https://doi.org/10.1016/j.buildenv.2019.05.032

Hossain, A., Mourshed, M., 2018. Retrofitting buildings: embodied & operational energy use in English housing stock. Proceedings 2, 1135. Available from: https://doi.org/10.3390/proceedings2151135.

Housez, P.P., Pont, U., Mahdavi, A., 2014. A comparison of projected and actual energy performance of buildings after thermal retrofit measures. J. Build. Phys. 38, 138–155. Available from: https://doi.org/10.1177/1744259114532611.

Hurtado, L.A., Rhodes, J.D., Nguyen, P.H., Kamphuis, I.G., Webber, M.E., 2017. Quantifying demand flexibility based on structural thermal storage and comfort management of non-residential buildings: a comparison between hot and cold climate zones. Appl. Energy 195, 1047–1054. Available from: https://doi.org/10.1016/j.apenergy.2017.03.004.

IEA EBC Annex 79, 2019. Occupant-Centric Building Design and Operation. Paris.

International Energy Agency (IEA), 2018. The Future of Cooling: Opportunities for Energy-Efficient Air Conditioning. OECD/IEA.

International Organization for Standardization, n.d. ISO 18523-1:2016 Energy Performance of Buildings - Schedule and Condition of Building, Zone and Space Usage for Energy Calculation. Part 1: Non-residential Buildings. Geneva.

International Organization for Standardization, 2005. ISO 7730: Ergonomics of the Thermal Environment − Analytical Determination and Interpretation of Thermal Comfort Using Calculation of the PMV and PPD Indices and Local Thermal Comfort Criteria. Geneva.

Jensen, S.Ø., Marszal-Pomianowska, A., Lollini, R., Pasut, W., Knotzer, A., Engelmann, P., et al., 2017. IEA EBC Annex 67 Energy Flexible Buildings. Energy Build. 155, 25−34. Available from: https://doi.org/10.1016/j.enbuild.2017.08.044.

Jerman, M., Palomar, I., Kočí, V., Černý, R., 2019. Thermal and hygric properties of biomaterials suitable for interior thermal insulation systems in historical and traditional buildings. Build. Environ. 154, 81−88. Available from: https://doi.org/10.1016/j.buildenv.2019.03.020.

Jia, M., Komeily, A., Wang, Y., Srinivasan, R.S., 2019. Adopting Internet of things for the development of smart buildings: a review of enabling technologies and applications. Autom. Constr. 101, 111−126. Available from: https://doi.org/10.1016/j.autcon.2019.01.023.

Kim, J., Tartarini, F., Parkinson, T., Cooper, P., de Dear, R., 2019. Thermal comfort in a mixed-mode building: are occupants more adaptive? Energy Build. 203, 109436. Available from: https://doi.org/10.1016/j.enbuild.2019.109436.

Moazami, A., Nik, V.M., Carlucci, S., Geving, S., 2019. Impacts of future weather data typology on building energy performance − investigating long-term patterns of climate change and extreme weather conditions. Appl. Energy 238, 696−720. Available from: https://doi.org/10.1016/j.apenergy.2019.01.085.

Moran, P., Goggins, J., Hajdukiewicz, M., 2017. Super-insulate or use renewable technology? Life cycle cost, energy and global warming potential analysis of nearly zero energy buildings (NZEB) in a temperate oceanic climate. Energy Build. 139, 590−607. Available from: https://doi.org/10.1016/j.enbuild.2017.01.029.

Moroni, S., Alberti, V., Antoniucci, V., Bisello, A., 2018. Energy communities in a distributed-energy scenario: four different kinds of community arrangements. In: Bisello, A., Vettorato, D., Laconte, P., Costa, S. (Eds.), Smart and Sustainable Planning for Cities and Regions. Springer International Publishing, Cham, pp. 429−437. Available from: https://doi.org/10.1007/978-3-319-75774-2_29.

Nicol, J.F., Humphreys, M.A., 2002. Adaptive thermal comfort and sustainable thermal standards for buildings. Energy Build. 34, 563−572. Available from: https://doi.org/10.1016/S0378-7788(02)00006-3.

Nicol, J.F., Humphreys, M.A., 1973. Thermal comfort as part of a self-regulating system. Build. Res. Pract. 1, 174−179. Available from: https://doi.org/10.1080/09613217308550237.

Nutkiewicz, A., Jain, R.K., Bardhan, R., 2018. Energy modeling of urban informal settlement redevelopment: exploring design parameters for optimal thermal comfort in Dharavi, Mumbai, India. Appl. Energy 231, 433−445. Available from: https://doi.org/10.1016/j.apenergy.2018.09.002.

O'Brien, W., Gunay, H.B., Tahmasebi, F., Mahdavi, A., 2017. A preliminary study of representing the inter-occupant diversity in occupant modelling. J. Build. Perform. Simul. 10, 509−526. Available from: https://doi.org/10.1080/19401493.2016.1261943.

OECD, 2018. Divided Cities: Understanding Intra-Urban Inequalities. Paris.

Reinhart, C., Voss, K., 2003. Monitoring manual control of electric lighting and blinds. Light. Res. Technol. 35, 243−258. Available from: https://doi.org/10.1191/1365782803li064oa.

Reiter, P., Poier, H., Holter, C., 2016. BIG Solar Graz: solar district heating in Graz − 500,000 m^2 for 20% solar fraction. Energy Procedia 91, 578−584. Available from: https://doi.org/10.1016/j.egypro.2016.06.204.

Richardson, I., Thomson, M., Infield, D., 2008. A high-resolution domestic building occupancy model for energy demand simulations. Energy Build. 40, 1560−1566. Available from: https://doi.org/10.1016/j.enbuild.2008.02.006.

Savaresi, A., 2016. The Paris Agreement: a new beginning? J. Energy Nat. Resour. Law 34, 16–26. Available from: https://doi.org/10.1080/02646811.2016.1133983.
Schnieders, J., Feist, W., Rongen, L., 2015. Passive houses for different climate zones. Energy Build. 105, 71–87. Available from: https://doi.org/10.1016/j.enbuild.2015.07.032.
Shen, B., Ghatikar, G., Lei, Z., Li, J., Wikler, G., Martin, P., 2014. The role of regulatory reforms, market changes, and technology development to make demand response a viable resource in meeting energy challenges. Appl. Energy 130, 814–823. Available from: https://doi.org/10.1016/j.apenergy.2013.12.069.
Stazi, F., Tomassoni, E., Bonfigli, C., Di Perna, C., 2014. Energy, comfort and environmental assessment of different building envelope techniques in a Mediterranean climate with a hot dry summer. Appl. Energy 134, 176–196. Available from: https://doi.org/10.1016/j.apenergy.2014.08.023.
Stein, L., Diefenbach, M., Arcipowska, Z., 2016. Monitor Progress Towards Climate Targets in European Housing Stocks - Main Results of the EPISCOPE Project. Institut Wohnen und Umwelt GmbH, Darmstadt.
Taha, H., 1997. Urban climates and heat islands: albedo, evapotranspiration, and anthropogenic heat. Energy Build. 25, 99–103. Available from: https://doi.org/10.1016/S0378-7788(96)00999-1.
United Nations, Department of Economic and Social Affairs, 2018. World Urbanization Prospects 2018. United Nations, New York.
United Nations, Department of Economic and Social Affairs, 2019. World Population Prospects 2019. United Nations, New York.
Van Raaij W. and Verhallen Th., 1983. Patterns of residential energy behavior, J. Econ. Psychol. 4, 85–106.
van Ruijven, B.J., De Cian, E., Sue Wing, I., 2019. Amplification of future energy demand growth due to climate change. Nat. Commun. 10, 2762. Available from: https://doi.org/10.1038/s41467-019-10399-3.
Vigna, I., Pernetti, R., Pasut, W., Lollini, R., 2018. New domain for promoting energy efficiency: energy flexible building cluster. Sustain. Cities Soc. 38, 526–533. Available from: https://doi.org/10.1016/j.scs.2018.01.038.
Vivian, J., Chiodarelli, U., Emmi, G., Zarrella, A., 2020. A sensitivity analysis on the heating and cooling energy flexibility of residential buildings. Sustain. Cities Soc. 52, 101815. Available from: https://doi.org/10.1016/j.scs.2019.101815.
Wagener, P., 2017. IEA HPT Annex 42 - Heat Pumps in Smart Grids. Available at: https://heatpumpingtechnologies.org/publications/heat-pumps-in-smart-grids-final-report/. (accessed 7/5/2020)
Winslow, C.-E.A., Herrington, L.P., Gagge, A.P., 1937. Physiological reaction of the human body to varying environmental temperatures. Am. J. Physiol. Legacy Content 120, 1–22. Available from: https://doi.org/10.1152/ajplegacy.1937.120.1.1.
World Bank, n.d. The World Bank Data [online]. Washington, DC.
Yang, X., Li, Y., Luo, Z., Chan, P.W., 2017. The urban cool island phenomenon in a high-rise high-density city and its mechanisms. Int. J. Climatol. 37, 890–904. Available from: https://doi.org/10.1002/joc.4747.

Renewable energy integration as an alternative to the traditional ground-source heat pump system

Cristina Sáez Blázquez[1], David Borge-Diez[2], Ignacio Martín Nieto[1], Arturo Farfán Martín[1] and Diego González-Aguilera[1]
[1]Department of Cartographic and Land Engineering, Higher Polytechnic School of Ávila, University of Salamanca, Ávila, Spain, [2]Department of Electric, System and Automatic Engineering, University of León, León, Spain

Chapter Outline

Nomenclature   109
5.1 Introduction   110
5.2 Methodology   111
    5.2.1 Description of the proposed solution   111
    5.2.2 Test procedure   115
5.3 Technical calculation   116
    5.3.1 Thermal module   116
    5.3.2 Power module   120
    5.3.3 Contribution of the suggested installation   121
5.4 Economic and environmental analysis   122
    5.4.1 Economic analysis   122
    5.4.2 Environmental evaluation   123
5.5 Discussion   123
    5.5.1 Sensitivity analysis   126
5.6 Conclusions   128
Acknowledgments   128
References   129

Nomenclature

Symbols
C_T Total thermal contribution (kJ)
C_1 Thermal contribution (kJ)
C_2 Electrical contribution (kJ)
E_g Enthalpy gained in the geothermal borehole (kJ)
E_s Enthalpy gained in the solar thermal panel (kJ)
E_f Enthalpy provided by the photovoltaic panel (kJ)
E_s Enthalpy provided by the wind generator (kJ)
E_n Enthalpy provided by electricity network (kJ)

Energy Services Fundamentals and Financing. DOI: https://doi.org/10.1016/B978-0-12-820592-1.00005-1
© 2021 Elsevier Inc. All rights reserved.

E_n Enthalpy provided by the batteries (kJ)
N_b Energy needs of the building (kJ)
f f-Chart method value
D_1 f-Chart method factor
D_2 f-Chart method factor
E_a Energy absorbed by the thermal solar panel (kJ)
S_c Panel surface (m^2)
$F'_r(\tau\alpha)$ Factor calculated from the efficiency of the panel and characteristic values of each panel
R_1 Daily monthly radiation by unit of surface (kJ/m^2)
N Number of days in the month
H_1 Monthly heat load (kJ)
E_p Energy lost by the thermal solar panel (kJ)
$F'_r U_L$ Characteristic factor of the thermal panel losses
t_a Monthly ambient temperature (°C)
Δt Period of time considered (s)
K_1, K_2 Correction factors

Acronyms
GSHP Ground-source heat pump
GHG Greenhouse gas
EED Earth energy designer

5.1 Introduction

The increasing global energy demand closely related to the pivotal climate change has led to a notable growing of greenhouse gas (GHG) emissions. On this matter, strategies that aimed at maximizing the use of renewable energies are numerous nowadays (Bartolucci et al., 2018; Fikiin and Stankov, 2015; Bassetti et al., 2018; Jacobson et al., 2018; Guo et al., 2018). However, these energies are still far from being used across the board. Most of them suffer from relevant variations due to its dependence on climatic and weather conditions (Nanaki and Xydis, 2018). The economic component is another factor that negatively affects the expansion of these energy sources.

Focusing on the geothermal energy and in particular the very low-enthalpy geothermal systems, they are commonly used to produce domestic hot water or to warm/cool a certain space. Different from other renewable energies, they are not affected by environmental conditions, given that the earth temperature is constant from a particular depth. The main problem that complicates a generalized use of these installations is the economic issue. The high initial investments they require mean an important impediment for a large number of users. In this regard, the main economic contributions are constituted by the set of borehole heat exchangers and the geothermal heat pump. The number of boreholes and hence the surface of heat exchange depend on the particular geological characteristics of area in question. Therefore, the first item is mostly inevitable although it could be reduced. Geothermal heat pumps (also required in these systems) involve a high initial

Table 5.1 Principal characteristics of a traditional GSHP system and the model proposed in this work.

Traditional GSHP system	Proposed model
High initial investment	Moderate initial investment
Quite long amortization periods	Medium amortization periods
External power supply dependence	Reduction of the external power supply dependence (it could be autonomous under ideal conditions)
Monthly costs associated with electrical power consumption and system maintenance	Reduction of the monthly costs derived from the system operation and maintenance

payment in addition to the monthly costs associated with the electrical consumption. They are efficient devices whose function is increasing the temperature of the heat carrier fluid coming from the thermal exchange with the earth (Fraga et al., 2018; Jeong et al., 2018; Zarrella et al., 2018; Zhang et al., 2018; Luo et al., 2018). However, their high prices thoroughly raise the initial investment of the general geothermal system. The second item could be avoided by implementing an equivalent but more economical solution. Reducing the initial investment, and hence the amortization period of the installation, would make the implementation of these systems more attractive (Sorknæs, 2018; Sarbu and Sebarchievici, 2016).

Together with the economic field, the other weakness of these systems is related to the constant dependence of external power supply for the heat pumps operation. Since these devices are electricity consumers, a certain emission of GHGs is associated with their use.

Given the reasons previously described, the present research mainly focuses on finding an appropriate alternative to reduce the total budget commonly associated with a vertical geothermal installation of very low enthalpy. To that end, the possibility of substituting the geothermal heat pumps by other renewable solutions is considered along this work.

Although the mentioned system will be thoroughly described later, Table 5.1 presents the main characteristics of it in relation to a conventional geothermal installation.

5.2 Methodology

5.2.1 Description of the proposed solution

As already mentioned, the aim of the present work is to provide an alternative to the common configuration of a very low-enthalpy geothermal system. The proposed design is based on substituting the energy source of these installations (the geothermal heat pumps) by a combination of different renewable energies. Basically, the difference between a traditional geothermal system and the one presented in this chapter is the way of increasing the temperature of the heat carrier fluid after coming from the thermal exchange in the boreholes. While in a conventional plant, the temperature of the fluid is increased in the geothermal heat pump, in the system suggested in this research, the fluid raises its temperature thanks to a series of renewable modules. Thus Fig. 5.1 graphically shows the standard workflow of a geothermal installation, indicating the additional process contemplated in the suggested solution.

The renewable module is designed to make the function of the geothermal heat pump. This system is constituted by the following items:

- thermal solar energy
- wind energy
- thermal buffer tank constituted by an auxiliary electric heater
- computerized system of data acquisition and decision-making

The operation of the mentioned module is briefly explained in Fig. 5.2. As described in this figure, the first step is the thermal exchange of a heat carrier fluid (mixture of water–glycol) in the corresponding geothermal borehole. Up to here, it is the common procedure in a geothermal installation. When the heat carrier fluid leaves the borehole, it would get in the geothermal heat pump; however, in this research, this device is replaced by the renewable module. Thus the heat carrier fluid enters in a thermal solar panel where its temperature is increased again. In the following stage, the fluid gets in a buffer tank constituted by an electrical resistance and a set of sensors that belong to the intelligent system of data acquisition and decision-making. In this tank, the automated system analyzes if the fluid has got the required temperature or not. If the temperature is enough, the heat carrier fluid is conducted inside the building to carry out the thermal exchange by the underfloor heating. In the event that the temperature of the fluid was not the required, the electrical resistance of the buffer tank would increase it until the established temperature. Then, the fluid would also get into the building in question. The last component of this module is made up of a solar photovoltaic panel and a wind

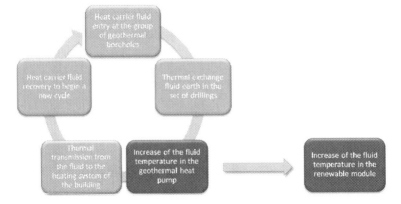

Figure 5.1 Workflow of a common geothermal installation showing the different steps in the system under consideration.

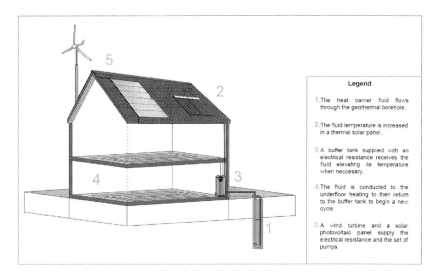

Figure 5.2 Description of the renewable module proposed in the present research.

turbine, which constitute the electrical contribution. These elements provide the electrical energy required by the set of pumps responsible for the fluid movement and the electrical resistance placed in the buffer tank. If the meteorological conditions were not favorable, an external electricity contribution would be considered.

The computerized control system is the key of the whole installation. The response capacity of the proposed plant is quite slow; hence, its limited flexibility is compensated by a suitable processing of the installation and external information. The brain of the computerized system is responsible for collecting a large amount of information coming from:

- sensors from each energetic module (geothermal borehole, thermal solar panel, buffer tank),
- sensors placed inside and outside the building,
- weather forecasts,
- data about the expected consumption provided by the user (personal prediction of anomalous consumptions, higher or lower, during a certain period of time),
- data from similar installations (solar and wind systems)
- data about the electrical supply rate with the aim of detecting the most appropriate hours if external energetic contribution is required.

Whit this set of information, the mentioned automated system will be capable of controlling the operation and storage functions, selecting which modules get involved to efficiently cover the energy needs for an established period of time. Based on the reliability of the meteorological information, this period could be of 24, 48, or 72 h. Fig. 5.3 describes how the mentioned system is connected with each of the components of the installation, taking information from the large number of sensors distributed over the whole renewable plant. As Fig. 5.3 shows, there are two groups of sensors: one of them provides the temperature of the heat carrier fluid in each module as well as the temperature inside the building. The remaining set of sensors collects relevant information from the whole installation: the electrical contribution of the renewable modules (solar and wind energy), the electric rate in a certain moment, information from the user, and the weather forecast. Due to the computerized system, the plant operates in the most efficient conditions, guaranteeing a constant energy supply to the user at the same time.

According to Fig. 5.3, the principal equations of the system are presented.

The total contribution of the whole installation can be expressed as follows:

$$C_T = C_1 + C_2 \tag{5.1}$$

where C_1 is the thermal contribution and C_2 is the electrical contribution.

$$C_1 = E_g + E_s \tag{5.2}$$

where E_g is the enthalpy gained in the geothermal borehole and E_s is the enthalpy gained in the solar thermal panel.

$$C_2 = E_f + E_e + E_n + E_b \tag{5.3}$$

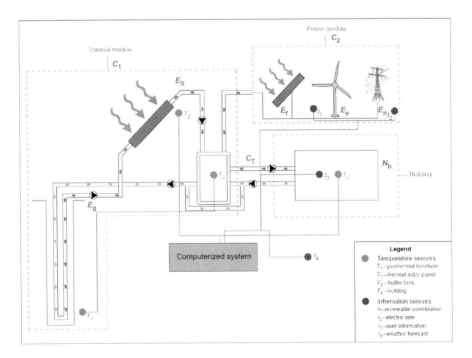

Figure 5.3 Schema of the computerized system operation.

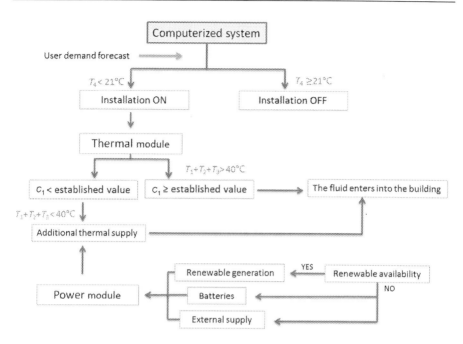

Figure 5.4 Process of decision-making in the computerized system.

where E_f is the enthalpy provided by the photovoltaic panel, E_e is the enthalpy provided by the wind generator, E_n is the enthalpy provided by electricity network, and E_b is the enthalpy provided by the batteries.

Finally, the energy needs of the building determine the total contribution required:

$$C_T = N_b \tag{5.4}$$

where N_b is the energy needs of the building.

In addition, the process of decision-making carried out by the computerized system is graphically described in Fig. 5.4.

It must be clarified from Fig. 5.4 that when the fluid leaves the building, the computerized system decides if its temperature is enough (in this case it returns to the borehole) or not (the fluid needs to go across the buffer tank). The established value mentioned in Fig. 5.4 will depend on the specific user needs. It usually takes a value around 40°C–60°C (Instituto Para la Diversificación Y Ahorro de la Energía, 2010).

5.2.2 Test procedure

With the aim of evaluating the solution previously described, it has been applied to a certain real case. Thus the procedure followed in the present research is constituted by the following steps:

- selection of a study case where the suggested system will be checked and establishment of a series of initial conditions;
- dimensioning of each of the components that constitute the plant, the geothermal borehole, and the renewable module;
- economic calculation and environmental analysis of the whole renewable installation;
- economic and environmental comparison of the proposed solution and a conventional geothermal system.

5.3 Technical calculation

As already mentioned, the first step is selecting the initial conditions of the real case where the system will be implemented. The selected building is a single family home of 200 m^2 placed in the province of Ávila (Spain). The system is considered to only cover the demand of the building in heating mode (cooling needs are not usual in this area). Henceforth, each of the renewable components was dimensioned.

5.3.1 Thermal module
5.3.1.1 Geothermal energy

In the first place, the geothermal module was calculated as a common ground-source heat pump (GSHP) system. With that aim, earth energy designer (EED) software developed by "Blocon Software" (Blázquez et al., 2017; Sarbu et al., 2017; Sliwa et al., 2016) was used to define the most relevant parameters of this kind of systems. It constitutes an useful tool for the design of vertical heat exchangers providing (among other things) the required borehole size and layout. Thus entering in the software with the characteristics of the building (energy demand), ground thermal and geological properties, and a series of parameters of the heat exchanger design (piping system, distance, material, and so on), it automatically defines the geothermal plant. Table 5.2 collects the main data introduced in EED software and the resultant parameters of its calculation.

Table 5.2 Design of the geothermal system by EED software.

Parameters used in EED software	
Location of the building	Province of Ávila (Spain)
Ground geology	Granitic materials
Energy demand in heating mode (kWh)	28,000
Piping design	Polyethylene double-U pipes of 32 mm in diameter
Heat carrier fluid	Mixture glycol/water
Results calculated by EED	
Number of boreholes	2
Total drilling length (m)	180
Nominal heat pump power (kW)	20.72

The aforementioned information corresponds to the usual dimensioning of a geothermal GSHP system. However, for the plant designed in the present research, some aspects must be reconsidered. The required heat pump power (20.72 kW) will be provided by the renewable module explained in the previous section. Regarding the number of boreholes and the drilling length, they will vary for the conditions of the suggested plant. In the case of the conventional geothermal installation, as Table 5.2 shows, two boreholes of 90 m are needed to supply the heat pump. The function of these boreholes is to increase the temperature of the heat carrier fluid before entering into the heat pump as well as dissipating the low temperature of the fluid when it leaves the heat pump. According to heat pump manufacturer data, the temperature of the fluid when it enters into the evaporator is 0°C and −3°C when it leaves the heat pump and returns to the set of boreholes. In addition, EED software simulates the temperature of the fluid during the lifetime of the installation as shown in Fig. 5.5.

On the contrary, in the installation proposed in this research, the fluid that returns to the boreholes comes from the buffer tank to a higher temperature than in the conventional plant (−3°C). Thus in this second case, the ground is only required to elevate the temperature of the heat carrier fluid to a certain value before getting in the solar panel. Since the conditions of the fluid are not as extreme as using a heat pump (the ground does not have to dissipate the cold of the entering fluid), the drilling length can be significantly reduced. For these reasons, only one borehole of 90 m was initially considered as the constituent of the geothermal section.

Regarding the temperature of the heat carrier fluid, for this assumption it is not possible to use the numerical simulation of EED since this software presupposes a geothermal installation constituted by a heat pump that notably decreases the outlet fluid temperature. Based on previous researches in a borehole of the study area

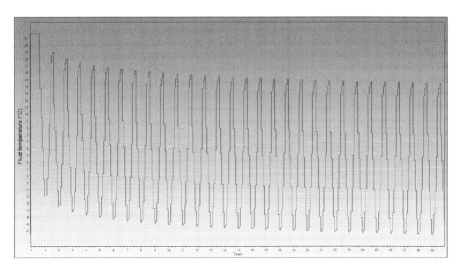

Figure 5.5 Evolution of the fluid temperature in the conventional geothermal installation.

(Blázquez et al., 2016), the temperature of the ground keeps constant to 14.6°C in the depth of 20 m. Considering the standard thermal gradient of 1°C each 33 m of depth, in the designed borehole of 90 m, a temperature of around 16°C will be easily achieved. Thus given the conditions of the system, the heat carrier fluid will return to the borehole with a temperature of around 10°C −12°C and will reach the temperature of the ground in a reasonable period of time.

5.3.1.2 Thermal solar energy

The following stage is the increase of the heat carrier fluid temperature by using the thermal solar energy. The corresponding selection of the solar module was carried out considering a series of data of the area where the system is planned to be implemented and the fact that the working fluid enters in the thermal solar panel with a temperature of 16°C (since it comes from the geothermal borehole). Table 5.3 shows the main geographical and climatic information belonging to the study area.

As already mentioned, the use of the suggested installation is limited to the cold months from September to May. Observing Fig. 5.6, it is easily perceptible how the elevation angle of the sun gets the lowest value during the winter months (priority use of the system). A habitual practice is considering the inclination of the solar panels as the sum of 10 degrees to the latitude of the place if the installation is especially designed to be working in the cold months (Instituto para la Diversificación y Ahorro de la Energía, 2006). Since it is 40.50 degrees in the study area, the solar panel will be installed with an inclination angle of 50.50 degrees. In Fig. 5.6, it is also possible to observe in the solar mask that the solar radiation is not affected by surrounding constructions due to the location of the area.

Fig. 5.7 shows how the inclination angle of the panel improves the global amount of energy received in a year in comparison with the gain of a solar panel installed in a horizontal position. In addition, this figure describes the annual evolution of the ambient temperature in the study area and the temperature of the fluid in the solar panel (set in 16°C since it comes from the borehole).

The characteristics of the thermal solar panel selected in function of the information presented earlier can be found in Table 5.4. This selection is based on the

Table 5.3 Geographical and climatic data in the location of the installation (Código Técnico de la Edificación, 2017a,b; Agencia Estatal de Meteorología, 2017).

Province	Ávila
Climatic zone	IV
Global solar radiation (MJ/m^2)	$16.6 \leq H \leq 18.0$
Latitude (degree)	40.50
Altitude (m)	2279
Medium relative humidity (%)	41.00
Mean wind velocity (km/h)	21.00
Maximum temperature in summer (°C)	30.00
Minimum temperature in winter (°C)	−6.00

Renewable energy integration as an alternative to the traditional ground-source heat pump system 119

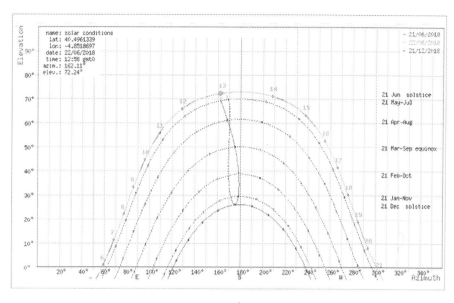

Figure 5.6 Solar chart/mask of the province of Ávila from the "Sun Earth Tool" application.

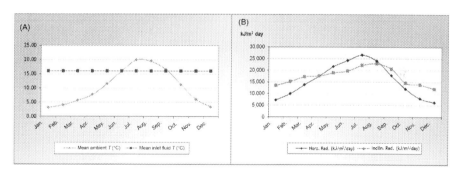

Figure 5.7 (A) Monthly ambient temperature and inlet fluid temperature. (B) Monthly solar radiation values (Instituto para la Diversificación y Ahorro de la Energía, 2006).

technical and economical availability of these panels in a large number of countries. In addition, the simplicity of installation and probed reliability motivated this decision. The design of the collector (plane panel) and the internal fluid circuits facilitates the inclusion in the geothermal fluid loop.

Finally, applying f-chart method (Okafor and Akubue, 2012), the heat contribution of the thermal solar system was determined for the month with the highest demand. The expression used by this method can be observed in Eq. (5.5).

$$f = 1.029D_1 - 0.065D_2 - 0.245D_1^2 + 0.0018D_2^2 + 0.0215D_1^3 \tag{5.5}$$

Table 5.4 Main information of the solar panel selected in this research.

Design	Plane panel
Model	SOLARIS CP1
Surface (m^2)	2.02
Efficiency factor	0.799
Global losses coefficient (W/(m$^2 \cdot$ °C))	3.4
Primary circuit flow ((l/h)/m^2)	50
Primary circuit specific heat (kcal/(kg $\cdot$ °C))	1
Panel efficiency	0.9

Substituting in Eq. (5.5) with the values of D_1 and D_2 whose calculation is presented in Appendix A, a factor f of 0.36 was obtained. It means that in the least favorable month, the geothermal–thermal module supply 36% of the energy demand. Since the annual demand is 28,000 kWh, 10,088 kWh is covered with the combination of these energies.

5.3.2 Power module

5.3.2.1 Photovoltaic solar energy

Based on data provided by meteorological observatories, the hours of sunlight in the study area for the months of the installation expected use was determined (Instituto para la Diversificación y Ahorro de la Energía, 2014). Then, by selecting a commercial solar kit constituted by nine photovoltaic panels, a contribution of 2090 W/h can be obtained for the period considered in the present research (January–May, September–December). Thus considering 1819 sunlight hours (applying the sun peak hour's method including information of the solar mask for the place) for the nine months of installation operation, the annual supply of this section would be of 3.80 MWh. It must be mentioned that the inverters, which these systems usually include, are not needed in this case since the electricity is directly conducted to the buffer tank. This fact will result in a significant economical saving. Table 5.5 shows the technical information of the solar kit selected as a part of the global suggested installation.

The selection of the model was made according to the availability and technical reliability of the equipment chosen. It is a standard device usually used by the common local solar installer.

5.3.2.2 Wind energy

Finally, the wind module was defined by the use of a free application developed by the company "Enair Energy." This tool allows knowing the contribution of a certain wind generator in the place where the system will be located. It is based on a method that uses the Weibull index and a series of parameters of the area in question (wind medium velocity, altitude, wind rose, and so on). Thus entering in the

Renewable energy integration as an alternative to the traditional ground-source heat pump system   121

Table 5.5 Technical data of the photovoltaic solar kit.

Design	Solar package
Model	Victron
Dimensions (mm)	1650 × 992 × 40
Power (Wp)	2250
Panel solar voltage (V)	31.4
Technology	Polycrystalline
Number of panels	9
Total kit weight (kg)	18.5

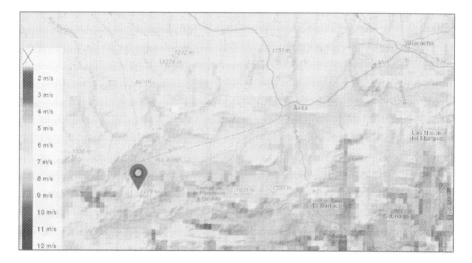

Figure 5.8 Velocity distribution in the area contemplated in this research.

application with the altitude and latitude and longitude coordinates of the area, it calculates (for a particular commercial model) the daily supplying of the wind resource providing at the same time the wind velocity. Fig. 5.8 shows the software simulation of the velocities in the area where the system will be placed.

From the output data of this tool, the annual contribution was calculated, for the model of generator described in Table 5.6. This parameter takes the value of 8.53 MWh/year for a wind velocity of 5.9 m/s.

5.3.3 Contribution of the suggested installation

By the implementation of the previously described modules, a fraction of the total demand would be covered. Thus considering the contribution of each of these systems, the suggested installation would provide:

Table 5.6 Technical description of the wind generator.

Design	Wind generator
Model	Enair
Number of blades	3
Nominal power (W)	6000
Voltage (V)	24/48/220
Diameter (m)	4.4
Swept area (m^2)	14.5
Minimum wind speed (m/s)	2

$$\text{Total contribution}$$
$$= \text{geothermal/solar module} + \text{photovoltaic module} + \text{wind module}$$
$$= 10.09 \text{ MWh} + 3.80 \text{ MWh} + 8.53 \text{ MWh} = 22.42 \text{ MWh}$$

Since the global energy demand reaches 28.00 MWh, the installation described in this research would be covering 80.07% of the total needs. This fact means that the remaining demand (19.93%) requires external energy supply.

5.4 Economic and environmental analysis

In order to justify the system described throughout the present research, an economic and environmental evaluation of the conventional GSHP system and the proposed installation was carried out.

5.4.1 Economic analysis

The first item of the technical calculation section was the geothermal module. The parameters of the traditional GSHP system calculated by EED software were presented in Table 5.2. Based on this information, this common option was economically evaluated. Thus Table 5.7 shows the initial investment for the conventional geothermal installation according to the needs of the building that were previously described.

In order to analyze the economic viability of the system proposed in this research, its initial investment was also calculated (collected in Table 5.8) to compare it with one of the conventional plant that was previously presented.

In addition to the initial investment of both systems, the annual costs derived from the installation use must be considered. Regarding the traditional geothermal plant, these costs correspond to the heat pump electricity consumption. If the nominal power of this device was of 20.72 kW and its coefficient of performance is 4.5, the energy consumption required by the heat pump would be around 4.6 kW, meaning 11,050.67 kWh per year (considering 2400 h of annual operation). In the

Renewable energy integration as an alternative to the traditional ground-source heat pump system   123

Table 5.7 Initial investment for the conventional geothermal installation.

Conventional GSHP system				
Element	**Quantity**	**Unit**	**Unit price (€)**	**Total price (€)**
Drilling	180.00	m	44.00	7,920.00
Geothermal heat exchanger	2.00	Unit	511.20	1,022.40
Injection tube	2.00	Unit	146.00	292.00
Guide	2.00	Unit	52.80	105.60
Assembly pack	2.00	Unit	27.20	54.40
Spacers	18.00	Unit	6.40	115.20
Grouting material	5.00	kg	0.70	3.50
Manifold	2.00	Unit	409.50	819.00
Geothermal heat pump	1.00	Unit	19,300.00	19,300.00
Heat pump accessories	—	—	—	6,553.57
Heat carrier fluid	1,050.00	L	4.00	4,200
Auxiliary piping	35.00	m	7.36	257.60
Thermal response test	1.00	Unit	4,118.36	4,118.36
Boost pump group	1.00	Unit	3,219.66	3,219.66
Total investment				**47,981.29**

*Prices were taken from the commercial catalogs of Enertres and ALB systems.

suggested system, the demand not covered by the renewable modules (19.93% of the total demand) needs an external electricity contribution, which also results in annual expenses. As already seen, the total demand was 28,000 kWh, therefore, 5,580.40 kWh must be externally provided. An additional characteristic of this system is that the mentioned electricity contribution can be made during the night taking advantage of more affordable rates. Table 5.9 presents the annual costs associated with both systems.

5.4.2 Environmental evaluation

As described in the previous section, the electricity supply required by the suggested installation is lower in comparison to the traditional GSHP system. This fact results in a reduction of the global CO_2 emissions since they are the product of the electricity use. Thus an analysis of the CO_2 emissions associated with each scenario must be made. Such evaluation can be found in Table 5.10.

5.5 Discussion

An alternative to the common GSHP system has been proposed and described throughout this study. Based on the technical calculations of both solutions (the common geothermal system and the one suggested) on the same case of study, an economic and environmental comparison was established. In Tables 5.7 and 5.8, the initial investment of each solution was thoroughly justified. In addition, in

Table 5.8 Initial investment for the suggested system.

Suggested renewable installation				
Element	**Quantity**	**Unit**	**Unit price (€)**	**Total price (€)**
Thermal module				
Drilling	90.00	m	44.00	3,960.00
Geothermal heat exchanger	1.00	Unit	511.20	511.20
Injection tube	1.00	Unit	146.00	146.00
Guide	1.00	Unit	52.80	52.80
Assembly pack	1.00	Unit	27.20	27.20
Spacers	9.00	Unit	6.40	57.60
Grouting material	2.50	kg	0.70	1.75
Manifold	1.00	Unit	409.50	409.50
Boost pump group	1.00	Unit	3,219.66	3,219.66
Heat carrier fluid	525.00	L	4.00	2,100.00
Auxiliary piping	50.00	m	7.36	368.00
Thermal response test	1.00	Unit	4,118.36	4,118.36
Buffer tank	1.00	Unit	1,088.67	1,088.67
Thermal solar panel	10.00	Unit	287.14	2,871.40
Power module				
Photovoltaic package	1.00	Unit	4,674.00	4,674.00
Wind generator	1.00	Unit	9,015.00	9,015.00
Storage battery	10.00	Unit	293.90	2,939.00
Computerized system				
Core unit	1.00	Unit	1,180.25	1,180.25
Sensor	8.00	Unit	3.25	26.00
Communication system	1.00	Unit	255.10	255.10
Total Investment				**37,021.49**

*Prices were taken from the commercial catalogues of Enertres and ALB systems.

Table 5.9 Annual expense of the traditional geothermal system and the one presented in this research.

	Electricity consumption (kWh)	Rate (€/kWh)	Annual expense (€)
Conventional plant	11,050.67	0.12779[a]	1,412.17
Suggested system	5,580.40	0.06580[b]	317.19

[a]Price of kWh in the common rate by EDP Company.
[b]Price of kWh in the nocturnal rate by EDP Company.

Renewable energy integration as an alternative to the traditional ground-source heat pump system

Table 5.10 Annual CO_2 emissions of the traditional geothermal system and the one presented in this research.

	Electricity consumption (kWh)	Conversion factor (kg CO_2/kWh)[a]	Annual CO_2 emissions (kg)
Conventional plant	11,050.67	0.399	4,409.22
Suggested system	5,580.40	0.399	2,226.58

[a]Normalized values provided by the Spanish Institute of the Diversification and Energy save (Instituto para la Diversificación y Ahorro de la Energía, 2014).

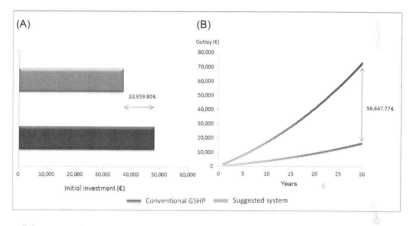

Figure 5.9 Economic comparison of the common GSHP system and the one presented in this research. (A) Initial investment. (B) Annual costs.

Table 5.9, the annual expense associated with the use of each of the systems was also calculated. Thus by observing the mentioned data, the economic balance is clearly more favorable for the new solution presented in this research. In Fig. 5.9, it is easily observable the economic differences between both solutions in terms of initial investment and annual expenses. It can be seen that the initial investment of the common GSHP system is €10,959.80 higher than the investment required for the suggested solution.

Regarding the annual costs, from data of Table 5.9, the evolution of the costs of each system in the period of useful life was calculated. With the aim of updating the costs of each year to the real value of the moment, data are expressed in terms of the net present value using a discount rate of 1.8%. In this way, in the last year of operation (year 30), the cumulative cost for the common geothermal plant is considerably higher (€56,647.77) than the cost associated with the suggested system.

In relation to the environmental dimension, as shown in Table 5.10, the CO_2 emissions of the alternative presented here are half of the emissions associated with the traditional GSHP system. The difference in CO_2 emissions grows during the operative phase of both systems reaching the value of 65,479.13 kg in the last year (year 30).

In connection with all the aforementioned discussions, a series of considerations about the system suggested in this work can be addressed:

- The installation proposed as an alternative of the common GSHP system means an economic saving of around 22.84% regarding the initial investment of both solutions. The savings are possible thanks to the substitution of the geothermal heat pump and the reduction of the total drilling length. This fact will make the geothermal energy more attractive and affordable for the final user, promoting at the same time a more extensive use of renewable energies.
- The annual electricity costs of the conventional geothermal systems are considerably low in comparison with other technologies. However, the new solution reduces even more costs, in a percentage of 77.54%, which involves an economic saving of €56,647.77 for the global operative phase of the installation.
- The environmental side also influences the selection of the system suggested in this research. The reason of this choice derives from the reduction of the CO_2 emissions in this solution, being 49.50% lower.
- The new system still depends on an external electricity supply. However, increasing the contribution of the thermal and photovoltaic modules and wind energy, the total demand could be covered avoiding the external dependence. It would cause an increase in the initial investment but the annual expenses would disappear. The solution presented here finds equilibrium between both parameters (initial investment and annual costs) so that from the economic and environmental point of view, it is more recommendable than the common GSHP.

5.5.1 Sensitivity analysis

In this section, a sensitivity analysis is presented to evaluate how the different values of a set of independent variables affect the viability of the project suggested in this research. With that aim, three different variables were modified keeping the remaining parameters as constant in a pessimistic, optimistic, and an expected scenario.

5.5.1.1 Electricity price

The first variable considered in this analysis is the electricity price. In the results presented in Fig. 5.9, a discount rate of 1.8% was applied. This rate belongs to the expected scenario. In a hypothetical optimistic scenario the discount rate estimated is 1.5% and is 2.1% for a pessimistic situation. As shown in Fig. 5.10, the conventional system starts being sensitive from year 15, and this sensitivity progressively grows in the remaining considered period. Regarding the suggested system, it is less sensitive to the variation of the electricity price than the conventional one. In both systems, the sensitivity is the same in any of the scenarios analyzed, optimistic and pessimistic.

5.5.1.2 Electric rate

In this case, the electric rates showed in Table 5.9 belong to the expected scenario for both the conventional and the suggested systems. These rates have been reduced

Renewable energy integration as an alternative to the traditional ground-source heat pump system   127

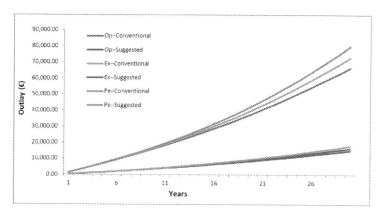

Figure 5.10 Sensitivity analysis to the variation of the electricity price parameter. *Op*, optimistic scenario; *Ex*, expected scenario; *Pe*, pessimistic scenario.

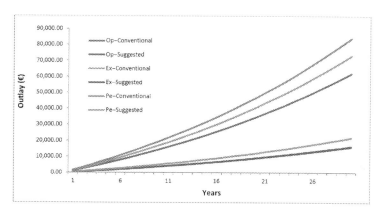

Figure 5.11 Sensitivity analysis to the variation of the electric rate parameter. *Op*, optimistic scenario; *Ex*, expected scenario; *Pe*, pessimistic scenario.

and increased in a percentage of 15% to represent the optimistic and pessimistic scenarios, respectively. As presented in Fig. 5.11, the sensitivity to the variation of the electric rate is also higher in the conventional system for which the sensitivity is the same for the optimistic and pessimistic scenario. In the case of the suggested system, it is quite sensitive to the increase of the electric rate (pessimistic scenario). However, its sensitivity to the reduction of the electric rate is practically null.

5.5.1.3 CO_2 emission factor

The last variable considered in the sensitivity analysis is the CO_2 emission factor. Observing Table 5.10, the expected emission factor is 0.399. As in the previous cases, optimistic and pessimistic scenarios have also been analyzed. Thus this factor has been reduced and increased in a 15%, obtaining for the optimistic situation an

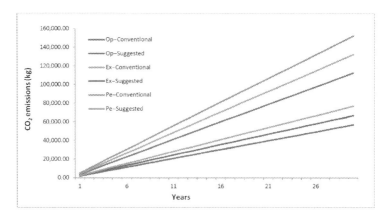

Figure 5.12 Sensitivity analysis to the variation of the CO_2 emission factor. *Op*, optimistic scenario; *Ex*, expected scenario; *Pe*, pessimistic scenario.

emission factor of 0.339 and 0.459 for the pessimistic one. Fig. 5.12 shows how the sensitivity to the increase or reduction of the emission factor is approximately the same in both systems. However, as in the previous cases, the sensitivity of the conventional system is higher compared to the suggested one.

5.6 Conclusions

The mitigation of the climatic change includes the constant search of new solutions that tend to reduce the electricity dependence and the use of renewable energies. The current schema of the geothermal system is frequently discarded by the general user due to the high initial outlay they demand. In this regard, it is highly recommendable to find new alternatives that promote this kind of energy. On this matter, a new schema of the common GSHP system was proposed. A case study was conducted for evaluating the viability of the new solution which considers the substitution of the geothermal heat pump by a combination of renewable modules. By the implementation of the mentioned system, an important economic save would be achieved as well as the environmental side would be improved due to the reduction of the CO_2 emissions. The external electricity contribution is considerably reduced (not eliminated) and, in addition, the configuration of the system allows using the most economic rate of electricity supply. The model could be readily extended to alternative scenarios and areas where the contribution of each module will not be the same. Regardless, the system would still present important advantages even in the least favorable cases where the renewable contribution was considerably lower.

Acknowledgments

Authors would like to thank the Department of Cartographic and Land Engineering of the Higher Polytechnic School of Ávila, University of Salamanca, for allowing us to use their facilities and their collaboration during the experimental phase of this research. Authors also want to thank the Ministry of Education, Culture and Sport for providing a FPU Grant (Training of University Teachers Grant) to the corresponding author of this chapter, which has made possible the realization of the present work.

References

Agencia Estatal de Meteorología (AEMET), 2017. Servicios climáticos. Ministerio de Agricultura y Pesca, Alimentación y Medio Ambiente, Madrid.

Bartolucci, L., Cordiner, S., Mulone, V., Rocco, V., Rossi, J.L., 2018. Hybrid renewable energy systems for renewable integration in microgrids: influence of sizing on performance. Energy 152, 744−758.

Bassetti, M.C., Consoli, D., Manente, G., Lazzaretto, A., 2018. Design and off-design models of a hybrid geothermal-solar power plant enhanced by a thermal storage. Renew. Energy 128 (Part B), 460−472.

Blázquez, C., Martín, A.F., García, P.C., Pérez, L.S.S., del Caso, S.J., 2016. Analysis of the process of design of a geothermal installation. Renew. Energy 89, 1−12.

Blázquez, C.S., Martín, A.F., Nieto, I.M., González-Aguilera, D., 2017. Measuring of thermal conductivities of soils and rocks to be used in the calculation of a geothermal installation. Energies 10, 795−813.

Código Técnico de la Edificación (CTE), 2017a. DB-HE-1: Limitación de la demanda energética. Ministerio de Economía y Competitividad, Madrid.

Código Técnico de la Edificación (CTE), 2017b. DB-HE-2: Rendimiento de las instalaciones térmicas. Ministerio de Economía y Competitividad, Madrid.

Fikiin, K., Stankov, B., 2015. Integration of renewable energy in refrigerated warehouses, Handbook of Research on Advances and Applications in Refrigeration Systems and Technologies, 28. IGI Global, Hershey, PA, pp. 803−853.

Fraga, C., Hollmuller, P., Schneider, S., Lachal, B., 2018. Heat pump systems for multifamily buildings: Potential and constraints of several heat sources for diverse building demands. Appl. Energy 225, 1033−1053.

Guo, S., Liu, Q., Sun, J., Jin, H., 2018. A review on the utilization of hybrid renewable energy (Review). Renew. Sustain. Energy Rev. 91, 1121−1147.

Instituto para la Diversificación y Ahorro de la Energía (IDAE), 2006. Energía solar térmica. Ministerio de Industria, Turismo y Comercio, Madrid.

Instituto Para la Diversificación Y Ahorro de la Energía (IDAE), 2010. Agua Caliente Sanitaria Central. Ministerio de Industria, Turismo y Comercio, Madrid.

Instituto para la Diversificación y Ahorro de la Energía (IDAE), 2014. Factores de emisión de CO2 y coeficientes de paso a energía primaria de diferentes fuentes de energía final consumidas en el sector edificios en España. Ministerio de Industria, Energía y Turismo, Madrid.

Jacobson, M.Z., Cameron, M.A., Hennessy, E.M., Petkov, I., Meyer, C.B., Gambhir, T.K., et al., 2018. 100% Clean and renewable wind, water, and sunlight (WWS) all-sector energy roadmaps for 53 towns and cities in North America. Sustain. Cities Soc. 42, 22−37.

Jeong, J., Hong, T., Kim, J., Chae, M., Ji, C., 2018. Multi-criteria analysis of a self-consumption strategy for building sectors focused on ground source heat pump systems. J. Clean. Prod. 186, 68−80.

Luo, J., Luo, Z., Xie, J., Xia, D., Huang, W., Shao, H., et al., 2018. Investigation of shallow geothermal potentials for different types of ground source heat pump systems (GSHP) of Wuhan city in China. Renew. Energy 118, 230−244.

Nanaki, E.A., Xydis, G.A., 2018. Deployment of renewable energy systems: barriers, challenges, and opportunities. Adv. Renew. Energies Power Technol. 2, 207−229.

Okafor, I.F., Akubue, G., 2012. F-chart method for designing solar thermal water heating systems. Int. J. Sci. Eng. Res. 3 (9).

Sarbu, I., Sebarchievici, C., 2016. Solar-assisted heat pump systems. Renew. Energy Sources Appl. Emerg. Technol. 1, 79−130.

Sarbu, I., Sebarchievici, C., Dorca, A., 2017. Simulation of ground thermo-physical capacity for a vertical closed-loop ground-coupled heat pump system. Int. Multidiscip. Sci. GeoConf. Surv. Geol. Min. Ecol. Manag. 17 (42), 557−565.

Sliwa, T., Nowosiad, T., Vytyaz, O., Sapinska-Sliwa, A., 2016. Study on the efficiency of deep borehole heat exchangers. SOCAR Proc. 2, 29−42.

Sorknæs, P., 2018. Investigation of shallow geothermal potentials for different types of ground source heat pump systems (GSHP) of Wuhan city in China. Energy 152, 533−538.

Zarrella, A., Zecchin, R., Pasquier, P., Guzzon, D., De Carli, M., Emmi, G., et al., 2018. A comparison of numerical simulation methods analyzing the performance of a ground-coupled heat pump system. Sci. Technol. Built Environ. 24 (5), 502−512.

Zhang, X., Li, H., Liu, L., Bai, C., Wang, S., Song, Q., et al., 2018. Optimization analysis of a novel combined heating and power system based on biomass partial gasification and ground source heat pump. Energy Conver. Manag. 163, 355−370.

Energy-saving strategies on university campus buildings: Covenant University as case study

Sunday O. Oyedepo[1], Emmanuel G. Anifowose[1], Elizabeth O. Obembe[1] and Shoaib Khanmohamadi[2]

[1]Department of Mechanical Engineering, Covenant University, Ota, Nigeria, [2]Department of Mechanical Engineering, Kermanshah University of Technology, Kermanshah, Iran

Chapter Outline

6.1 Introduction 131
    6.1.1 Energy modeling software for buildings 134
    6.1.2 Energy conservation measures in buildings 136
6.2 Materials and methods 136
    6.2.1 Study location 136
    6.2.2 Procedure for data collection 137
    6.2.3 Instrumentation and procedure for data analysis 137
    6.2.4 Economic analysis 138
    6.2.5 Assessment of environmental impacts 139
6.3 Results and discussions 139
    6.3.1 Result of energy audit in cafeterias1 and 2 139
    6.3.2 Result of energy audit in Mechanical Engineering building 141
    6.3.3 Result of energy audit in university library 145
    6.3.4 Result of energy audit in health center 147
    6.3.5 Result of energy audit in the students' halls of residence 148
    6.3.6 Qualitative recommendation analysis 150
6.4 Conclusion 152
References 153

6.1 Introduction

The study of building energy demand has become a topic of great importance, because of the significant increase of interest in energy sustainability, which has grown up after the emanation of the energy performance building (EPB) European Directive (Sretenovic, 2013). Considering the constant increase of fuel prices, threats of global warming, and implications of carbon emissions from traditional fuels, there is a growing interest in improving energy efficiency. Notable improvements have been made in obtaining and realizing energy efficiency in new buildings since the oil embargo of 1973 (Coakley et al., 2014). Seeking to reduce energy consumption brings about a positive cash flow as expenses on energy bills would be reduced. To achieve the goal of the Paris agreement (COP21), which is to hold the increase of global temperature below 2°C above preindustrial levels, efforts to reduce the energy consumption needs to be intensified (United Nations, 2015). In the past time, different industries and organizations have established energy management programs to check and regulate energy use in buildings through energy audit systems.

Campus energy potential studies involve an energy auditing process that provides an opinion of the availability of energy efficiency resources on a campus and allows the development of cost and savings' strategies. On university campuses, the vast majority of the energy consumption takes place within buildings, and the environmental consequences of this consumption are considerable (Petersen et al., 2007). One of the major environmental issues resulting from energy consumption is the emission of carbon dioxide (CO_2), which contributes to the global warming (Tang, 2012).

On university campuses, there is a considerable population, including students, academic and administrative staff, researchers, and others who work or study in universities. Thus large amount of energy needed for operations, including teaching and research, provision of support services, and in residential and hostel areas, it is almost comparable to "mini city" (Choong et al., 2012). Hence, energy consumption in higher institutions should be effectively managed to incur the minimum cost and to reduce the environmental impacts (Alshuwaikhat and Abubakar, 2008).

Since universities involve the large number of building users and facilities, the environmental degradation caused by the huge amount of energy consumption by universities is getting to be a major concern. Thus it is imperative for universities' authority to establish energy management programs to integrate sustainability into campus operations in order to act more responsibly in practice for a sustainable future (Pike et al., 2003). Moreover, with the help of several measures such as organizational, technological, and energy optimization, the energy waste on university campus can be considerably reduced (Kolokotsa et al., 2016).

Due to the existence of old buildings and other sources of energy wastage on university campuses, the potential of effective and efficient energy utilization on such campuses are usually very low. Therefore, it is important to assess energy consumption patterns on university campus to determine the ways of improving the energy efficiency (Chung and Rhee, 2014). Energy consumption patterns on the university campus have been investigated and documented by researchers: Escobedo et al. (2014) evaluated the energy consumption pattern and the resulting greenhouse gas (GHG) emission in the main university campus of National Autonomous University of Mexico with the aim of identifying energy-saving strategies on the campus. The study suggested that the energy use in the university buildings could not be compared with that of most other universities because most of the other universities are in regions where space heating and cooling are more relevant. The energy audit carried out in the study identified several forms of energy wastage through inefficient lighting systems, refrigerators, and water heating. Ishak et al. (2016) evaluated the energy consumption patterns among students in the four selected universities in Malaysia. The mean daily energy consumed per student was estimated as 6.1 kWh. The study classified energy users into four groups: high, medium, low, and conserve. A centrographic approach was used to analyze the behavioral factors affecting the energy consumption. A prediction model was developed from the study, and a potential energy saving of about 55 kWh was reported. Wen and Palanichamy (2018) evaluated the energy consumption profile of Curtin University Malaysia with the aim of suggesting ways to reduce and control the university's energy expenses. An energy audit, which was performed on the campus, showed that heating, ventilation, and air-conditioning (HVAC) systems consumed the highest amount of energy (>70%) followed by the office equipment (>13%). The load profile of the university showed that the maximum power demand is at noon just before lunchtime, while the average load per day is 942.10 kW. Minimum demand was observed to be 174 kW at night (9:00 p.m.) and then remains constant till the next morning.

As regard to energy-saving strategies on the campus, few researches have been carried out to identify possible energy-saving strategies on the university campuses. Saleh et al. (2015) investigated the efforts taken by several universities toward sustainability and energy management. The study concluded that the efforts of these universities toward sustainability has not been very productive; hence, they identified 5 clusters of 23 critical success factors that would help the university tailor her efforts toward sustainability. Deshko and Shevchenko (2013) affirmed that energy certification is one of the major ways to improve the energy efficiency of buildings on university campuses. The study emphasized that university campuses are in different classes; hence, they require different approaches to energy certification. The methodology chosen for the energy efficiency assessment includes selection of the determining factors; collecting and verifying information; distribution of university campuses by types; adaptation and normalization of the data on energy consumption; development of an energy consumption and efficiency assessment scale; and finally, choosing the best optimal variant. Faghihi et al. (2015) studied the relationship between the energy savings on the university campus and the funds required to provide this savings. It was observed that funding is the major challenge that campuses face in designing and operating sustainability improvement programs. Two main categories of improving sustainability were identified: the energy efficiency and energy conservation. The study discovered that both energy efficiency and conservation save significant amounts of money; however, the latter requires maintenance to extend the energy-saving practices since it deals with human behavior, which is not constant. Finally, they developed a dynamics model that helps to improve the understanding of sustainability programs that lead to an increase in the energy and monetary savings. In other study, Zhou et al. (2013) discovered that the energy consumption in private universities per student or building area is higher than that of public universities because of better conditions of teaching and research. Hence, there is a lot of potential of energy saving in these universities. The study also identified five major energy conservation measures on the university campus, which includes electricity submetering, utilization of renewable energy, installation of energy-saving appliances, and so on. Odunfa et al. (2015) investigated the effect of different building orientations of three lecture theaters in the

University of Ibadan on their respective energy efficiencies. The required cooling load of each building based on their orientation was used as the yardstick of energy efficiency. The study affirmed that North/South building orientation gives the best energy efficiency. Spirovski et al. (2012) identified ways in which the South East European University could implement a climate action plan to help reduce the university's GHG emissions and carbon footprint. A GHG inventory was taken; then possible measures to reduce GHG emissions were identified. The study identified 13 methods of reducing GHG emissions, which includes the use of solar thermal, the replacement of lamps, the use of solar photovoltaic, and so on. It further classified these methods based on the corresponding payback time. Last, the study suggested the use of several educational strategies such as teaching, climate change seminars, study programs, research, and so on to solve the challenge of GHG emissions. In their study, Allab et al. (2017) developed an energy audit protocol, which addressed both the challenges of energy efficiency and that of thermal comfort in a building. A standard energy audit of the building was carried out alongside a thermal comfort survey; then a building energy model was developed to assess the impact of the recommendations of the building retrofitting that were made. The study affirmed that the primary energy utilized in higher educational institutions is about double the energy used in the primary and secondary schools. A higher energy consumption was observed during the winter than during the summer signifying that the heating systems utilize more energy than the cooling systems. The most significant observation during the energy audit was the poor energy management system in the higher institution buildings. It was discovered that the energy-saving mode was not used in the buildings even during periods of low occupancy. It was recommended that during periods of high occupancy, the energy management should be able to provide adequate comfort, while energy-saving mode is used during low occupancy periods such as nights, weekends, holidays, and so on to achieve the double goal of thermal comfort and energy efficiency.

The prime objectives of the current study are to:

1. determine the energy usage pattern in Covenant University,
2. identify areas of energy wastage on the campus buildings,
3. identify existing techniques of energy savings in Covenant University and determine their effectiveness,
4. demonstrate the use of eQUEST to model and determine the thermal envelope of selected buildings on the campus.

6.1.1 Energy modeling software for buildings

Building energy modeling software is a physics-based software simulation of a building's energy use. The software takes building geometries, lighting density, construction materials, HVAC system design, and so on as input parameter to build the energy model. It also takes the usage schedules of the buildings as input.

The simulation usually produces hourly reports of the building energy use. In addition, the simulation usually considers the effect of interrelated systems such as lighting and heating in a building. This sophisticated approach to energy modeling makes the software very attractive to building energy professionals. The use of building energy modeling software includes:

- Architectural design
- HVAC design and operation
- Building performance rating
- Building stock analysis

There are different building energy modeling softwares available with each having its own strength. These softwares include Building Loads Analysis and System Thermodynamics; BSim; Designer's Simulation Toolkits; DOE-2.1E; ECOTECT; Ener-Win; Energy Express; EnergyPlus; ESP-r; Hourly analysis program; HEED; TRACE 700; TRNSYS; eQUEST. The considered software in this study is eQuest. This is because it is one of the most appealing softwares. This software has the workflow within the GUI, from high-level information about the building to the more detailed modifications of each object in the building. Moreover, it carries out energy performance, conceptual design, performance analysis, simulation, energy efficiency calculation, and other applications. It is an easy to use software compared to other softwares such as EnergyPlus, TRNSYS, and so on, and it is a free software courtesy of the State of California's Energy Design Resources Program.

Based on the aforementioned features, several researchers have carried out simulation and optimization of energy consumption on buildings using eQuest software: Lou et al. (2017) analyzed the possibility of achieving zero-energy consumption in a secondary school building. A model of a six-story secondary school building was developed in eQuest using input parameters such as window-to-wall ratio, air-conditioning settings, occupancy, and lighting schedules, and then the energy consumption of the building was simulated to determine the annual energy consumption of the building. The study affirmed that zero-energy school buildings are achievable in a hot and humid climate. Ma et al. (2017) applied eQuest to investigate the energy consumption status of 119 public buildings in northern China with the aim of proposing measures to reduce energy consumption in these buildings. The buildings investigated included hospitals, offices, and schools. Of these three, the highest energy consumer building was the hospital followed by the offices and then the schools. This observation was justified by the fact that the usage time of air conditioners and heating systems in the hospital is longer than in the offices and schools. In addition, it was shown that generally, the energy consumption of older buildings was more than that of newer buildings because the old buildings do not have a properly insulated building envelope. Ke et al. (2013) examined the energy-saving performance contract of an office building by ensuring the measurement and verification of energy savings. In the study, an eQuest model of a 21-story office building in Taiwan was developed; then sensitization was carried out on several energy consumption parameters to determine the possibility of energy savings. Some of these parameters include building walls, window glass, air-conditioning system, lighting power density, use schedules, occupancy, and equipment power. Dehghani et al. (2018) investigated measures through which energy savings could be achieved in an office building. The eQuest model of a four-story building in

Ohio was used for this study. The results of the eQuest model for annual electricity demands of the building corresponded with the actual utility bills with a difference less than 3%, hence validating the model.

6.1.2 Energy conservation measures in buildings

Energy conservation in buildings is the reduction in building energy consumption without reducing thermal comfort. It usually results in better indoor air quality and occupant's productivity. Energy conservation measures do not always yield instant financial incentives; however, they increase the national energy security, reduce environmental pollution, reduce dependence on fossil fuel, and so on.

Before any energy conservation measure can be recommended, an understanding of the current energy consumption pattern and utility rates need to be gotten. The results of the data analysis delineated earlier should provide the understanding after which energy conservation measures can be suggested. Some of the common energy conservation measures in buildings can be grouped into building envelope, electrical appliances, HVAC systems, energy management control, behavioral controls, and novel technologies.

6.2 Materials and methods

In this study, the employed methods in data acquisition include physical observations and identification of the various electrical appliances, lighting fixtures, and HVAC systems in the selected buildings in Covenant University campus. Various measurements were carried out to determine their energy consumption; then the observations and analysis of the data has been done; and energy conservation measures were suggested. Utility cost details of the whole campus were gotten from Canaanland Power plant—the utility vendor of Covenant University over a period of 5 years; then this data were then used to analyze the monthly variation in energy consumption of the university.

Furthermore, an energy modeling software, eQuest, was used to build energy models of each selected building, and then the energy efficiency measure wizard in the software was used to simulate and predict the impact of the variations of key energy parameters such as lighting density and window properties on the energy consumption of the building.

The selected buildings in this study comprises of 10 students' halls of residence, university library, health center, the engineering buildings, and two cafeterias.

6.2.1 Study location

Covenant University is a modern campus that includes several types of buildings of different functionalities and purposes. Some of the major buildings are four college buildings (College of Business Studies, College of Engineering, College of Science and Technology, and the College of Development Studies), four engineering buildings that tailor to the different fields of engineering (Chemical/Petroleum,

Mechanical, Civil, and Electrical and Information Engineering), one university library built to match the amount of student's daily use, two lecture theaters, and a university chapel. Other buildings include 10 halls of residence, two cafeterias, staff quarters, and the university guesthouse.

Energy demand of Covenant University is increasing with developmental activities and progress as more buildings are being built, and more modern equipment are being introduced into the university, to carry out different functions that are relevant to enable the university attain its maximum potential and commitment and to deliver a world class teaching environment for the benefits of its students. It is therefore necessary to understand the status of energy consumption in Covenant University, to ensure the sustainable use of energy usage, and to reduce the cost and environmental impacts associated with its use.

6.2.2 Procedure for data collection

This study used the detailed energy audit method to analyze the energy-saving opportunities in the campus. The data collection for this study was in three stages, which include the walk-through survey stage, utility bill stage, and energy model parameters stage.

6.2.3 Instrumentation and procedure for data analysis

The systems audited include electrical consumption, lighting, and HVAC systems. The parameters to be measured are:

- light, which was measured with a lux meter;
- temperature, which was measured with an infrared thermometer;
- power consumption, which was measured with a wattmeter.

The collected data include the electricity consumption of the selected buildings in the university, utility bills of the university for 5 years, and the energy model parameters of the selected buildings.

The utility bills were investigated to determine recurring patterns in energy use and how weather variations or seasonal changes affected energy consumption in the university. The electricity consumption data collected from each building were analyzed using Microsoft Excel; then individual contributions of each electrical appliance to the energy mix were determined. In addition, the energy usage of the different classes of buildings were compared.

Finally, the results of simulations in eQUEST were obtained from the software; then an energy baseline model was developed from which energy-saving strategies were evaluated. The energy consumption data obtained from eQUEST simulations cannot be validated since buildings in the university do not have meters, which can be used to calibrate the energy models. However, reasonable values of energy consumption produced by the simulation runs suggest that the models are accurate. Furthermore, a minimalist approach was used while gathering data regarding energy consumption in the buildings. This ensures that the model does not yield exaggerated values of energy consumption, which could hamper the economic analysis. The summaries of the data analysis are presented in figures and graphs to ensure ease of comparison.

6.2.4 Economic analysis

One of the major problems that led to this study is the exorbitant amounts of money spent on purchasing energy on university campuses; hence, an economic analysis of the suggested energy conservation measures need to be undertaken before these suggestions can turn to recommendations. Furthermore, since most energy conservation measures have a delayed reward, one must be sure that the savings resulting from the retrofit would exceed its initial outlay over its lifetime.

An economic evaluation involves the comparison between competing alternatives from which the best option is selected. The alternatives are judged based on the various financial performance indicators such as payback time, net present worth, rate of return, benefit–cost ratio, life cycle cost analysis, and so on. This study makes use of the payback time because it is simple to calculate, and it is suitable for situations in which the energy market is relatively stable.

The parameters needed in calculating the payback time include:

- investment cost of purchasing the energy retrofits
- unit cost of energy
- amount of energy saved by using the retrofit

Relevant equations needed for the economic analysis can be found below:

$$\text{Electrical energy}, E(\text{kWh}) = \frac{(\text{Power rating in W}) \times \text{Usage factor} \times \text{time(h)}}{1000} \tag{6.1}$$

$$\text{Energy saving} = \text{Present energy consumption} - \text{energy-efficient solution} \tag{6.2}$$

$$\text{Simple payback period} = \frac{\text{Net project cost}}{\text{Regular annual cost saving}} \tag{6.3}$$

$$\text{Percentage saving} = \frac{\text{Net energy saving}}{\text{Resent energy consumption}} \times 100\% \tag{6.4}$$

$$\text{Cost of replacement} = \text{Unit cost of LED lamp} \times \text{number of fixtures} \tag{6.5}$$

$$\text{Annual energy saving} = \text{Daily energy saving} \times \text{active school days in year} \tag{6.6}$$

$$\text{Annual cost saving} = \text{Annual energy saving(kWh)} \times \text{unit cost of electricty} \quad (6.7)$$

$$\text{Estimated project cost} = \text{Cost of panels/W} \times \text{power rating in W} \quad (6.8)$$

$$\text{Annual energy saving} = \frac{\text{Power rating in W} \times \text{sun hours} \times \text{active school days in year}}{1000} \quad (6.9)$$

6.2.5 Assessment of environmental impacts

A concomitant problem associated with the energy consumption is environmental degradation as a result of GHG emissions. A conversion factor of 0.523 kg of CO_2 per kWh of electrical energy consumed is used to calculate the associated environmental pollution. This factor is used to calculate the GHG savings associated with the energy savings.

The equation used to evaluate the GHG savings is given as follows:

$$\text{CO2 emissions(tons)} = \frac{\text{Annual energy saving} \times \text{conversion factor}}{1000} \quad (6.10)$$

6.3 Results and discussions

This section presents the results of energy audit carried out in selected buildings in Covenant University. The results were analyzed using Excel and eQuest software.

6.3.1 Result of energy audit in cafeterias1 and 2

Since the energy model of both cafeteria 1 and cafeteria 2 are similar, results of energy audit of cafeteria 1 are presented as follows.

Fig. 6.1 shows the building envelope of cafeteria 1. The building envelope is primarily made up of windows and walls. The windows occupy a large area toward the entrance and to the sides thereby making the building suitable for daylighting recommendations. In addition, there is a courtyard in the middle of the building to allow for natural ventilation toward the kitchen side.

In Fig. 6.2, energy end use pattern of cafeteria 1 is presented. These results were generated from eQuest based on the energy consumption data and occupancy profile provided. The software was then able to predict what the energy consumption of the building for a year would be. The results showed that refrigeration consumes the highest amount of energy.

140   Energy Services Fundamentals and Financing

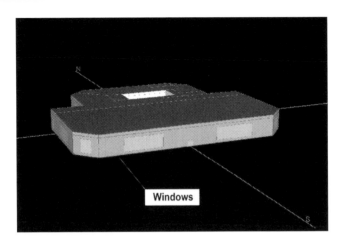

Figure 6.1 Building envelope of cafeteria 1.

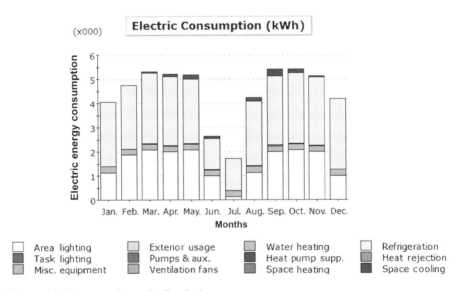

Figure 6.2 Energy end use of cafeteria 1.

Fig. 6.3 and Table 6.1 show the results of the parametric analysis carried out on the energy model of cafeteria 1. Daylighting with dimming was applied to the building to determine the impact it would have on the energy consumption of the building.

Fig. 6.4 shows the results of the walk-through energy audit carried out in cafeteria 1. From the energy audit data and the aforementioned simulation, it can be seen that the highest consumer of energy in cafeteria is refrigeration system. This is because, a lot of perishable goods and drinks need to be kept fresh, and this could only be done with a refrigerator. In addition, the cold rooms in the cafeteria take a

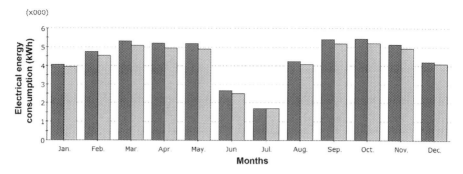

Figure 6.3 Daylighting control for cafeteria 1: dimming up to 30%.

lot of power, hence making them the highest consumers of energy. This is followed by lighting systems. Since the dining halls have a large area, they require a lot of lighting fixtures to be properly illuminated. Furthermore, the lighting fixtures used in the cafeterias—incandescent bulbs and fluorescent tubes—are not energy efficient; hence, the high amount of electric energy is consumed. A daylighting control of 30% dimming resulted in a 2.5% decrease in energy consumption. This shows that utilization of daylighting could help to reduce energy consumption in both cafeteria 1 and cafeteria 2.

Based on the earlier results, the following recommendations are hereby made as measures for energy conservation strategies in cafeteria:

- The thermal insulation around the cold rooms can be increased to prevent escape of thermal energy thereby resulting in lower energy consumption.
- The inefficient lighting fixtures in the cafeterias can be changed to more energy-efficient fixtures such as the light-emitting diode (LED) bulbs.
- Since the cafeteria has a large window area, daylighting control could be used to reduce energy consumption in the day.

6.3.2 Result of energy audit in Mechanical Engineering building

Fig. 6.5 presents building envelope of the Mechanical Engineering Department building. The building envelope is also primarily made up of windows and walls. However, since the building is more enclosed than cafeteria, daylighting is not considered as an effective recommendation. The building has a large corridor with a lot of open space; hence, sensors are recommended to control lighting on the corridors.

Fig. 6.6 shows energy end use pattern in Mechanical Engineering Department building. These results were generated from eQuest software based on the energy consumption data and occupancy profile provided. The software was then able to predict what the energy consumption of the building for a year would be. The results showed that space cooling systems consume the highest amount of energy.

Fig. 6.7 shows the results of walk-through energy audit carried out in Mechanical Engineering Department building. Mechanical Engineering Department building is chosen to represent academic buildings since energy consumption

Table 6.1 Daylighting control for cafeteria 1: dimming up to 30%.

	January	February	March	April	May	June	July	August	September	October	November	December	Total
Baseline ($\times 10^3$ kWh)	4.05	4.74	5.31	5.2	5.18	2.64	1.71	4.22	5.41	5.42	5.12	4.19	53.2
Daylighting ($\times 10^3$ kWh)	3.94	4.54	5.06	4.93	4.9	2.5	1.69	4.07	5.17	5.19	4.92	4.1	51.03

Energy-saving strategies on university campus buildings: Covenant University as case study

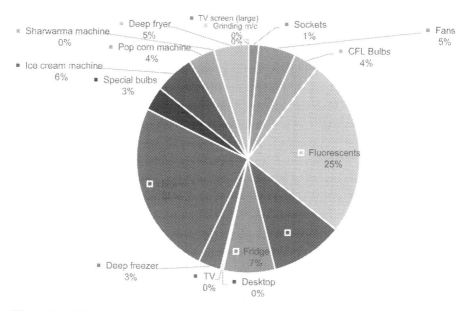

Figure 6.4 Cafeteria 1: energy audit data.

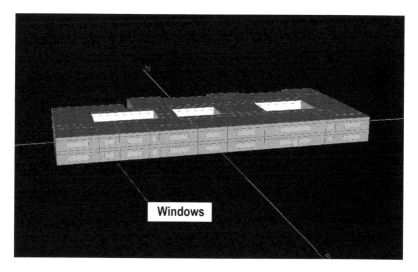

Figure 6.5 Building envelope of Mechanical Engineering Department building.

pattern is similar in all the academic buildings. From Fig. 6.7, it can be seen that air conditioners take the highest amount of energy in the academic buildings. The reason for this is because of the number of offices in the buildings (over 30 offices in Mechanical Engineering Department) in which all have air conditioners installed. In addition, the large number of fluorescents and negative behavior with regards to the turning off the corridor lights has led to this high energy usage.

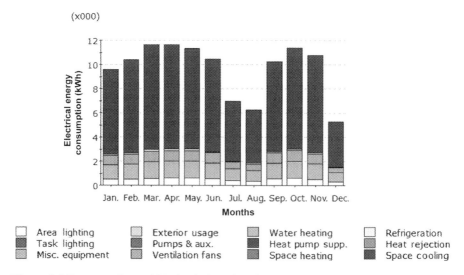

Figure 6.6 Energy end use of Mechanical Engineering Department building.

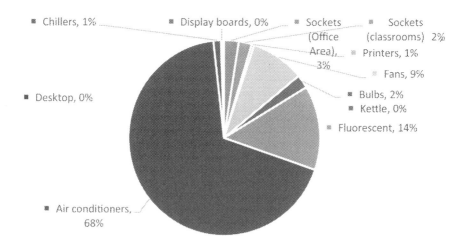

Figure 6.7 Energy audit data of Mechanical Engineering Department building.

Based on the aforementioned results, the following recommendations are proposed as measures of energy conservation strategies in academic buildings:

- The large roof area in each of the academic buildings could have photovoltaic solar panels installed on them since the climate in Covenant University favors that. This cheap renewable energy source would help to offset some of the peak energy demand during the day.
- The windows in each of the offices should be replaced with double glazed ones to reduce the energy required for cooling.
- In addition, automated control systems such as proximity sensors could be used to turn off corridor lights when no one is there.

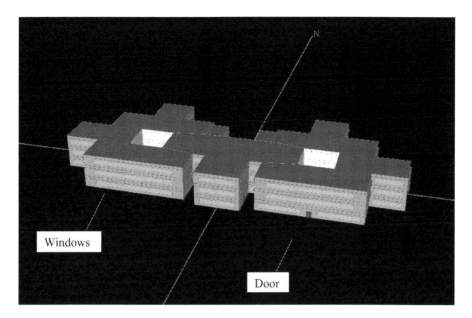

Figure 6.8 Building envelope of the university library.

6.3.3 Result of energy audit in university library

Fig. 6.8 presents the building envelope of the university library. The building envelope is primarily made up of windows and walls. The building is an open space; hence, lighting and space cooling consume the highest amounts of energy in the building.

The energy end use pattern of the university library is shown in Fig. 6.9. These results were generated from eQuest based on the energy consumption data and occupancy profile provided. The software predicts what the energy consumption of the building for a year would be. The results showed that space cooling and task lighting consume the highest amount of energy. In addition, ventilation fans and miscellaneous equipment such as plug loads also consume a reasonable amount of energy.

Results of the walk-through energy audit carried out in the university library building are presented in Fig. 6.10. From Fig. 6.10, it is observed that the highest energy consumer in the library is clearly the air conditioners (71%). This is due to the large open space in the library requires a lot of energy to cool it. In addition, since adequate illumination is required for users of the library and inefficient fluorescent tubes are still being used, the lighting fixtures are the second largest consumer (24%) of electric energy.

The following recommendations to minimize energy consumption in the university library are made:

- The large roof could also have solar photovoltaic panels installed on them to reduce the energy taken from the grid.
- In addition, the fluorescent tubes should be changed to LED tubes to reduce energy consumption.

146   Energy Services Fundamentals and Financing

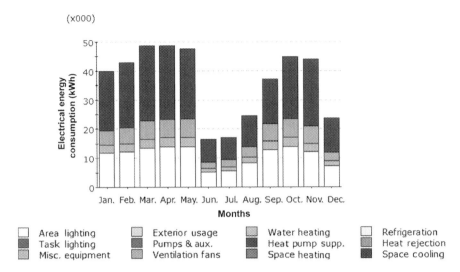

Figure 6.9 Energy end use of the university library.

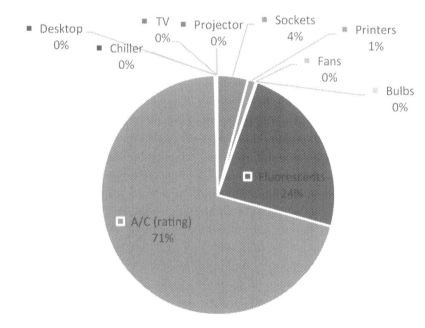

Figure 6.10 Energy audit data of the university library.

Energy-saving strategies on university campus buildings: Covenant University as case study   147

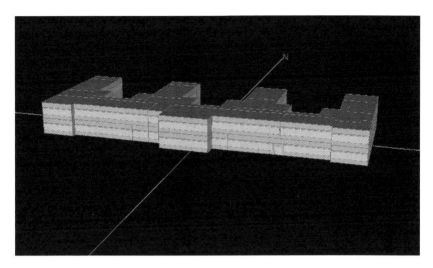

Figure 6.11 Building envelope of the health center.

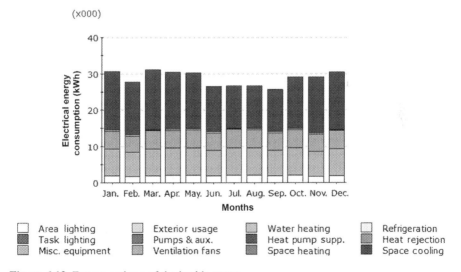

Figure 6.12 Energy end use of the health center.

6.3.4 Result of energy audit in health center

The building envelope of Covenant University health center is presented in Fig. 6.11. The building envelope is primarily made up of windows and walls. The building is highly compartmentalized, and it does not have a lot of open space; hence, daylighting would not be suitable here.

Energy end use pattern of the health center is presented in Fig. 6.12. These results were generated from eQuest based on the energy consumption data and

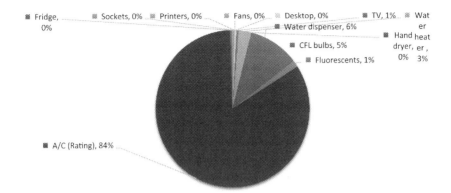

Figure 6.13 Energy audit data of the university health center.

occupancy profile provided. The software was used to predict what the energy consumption of the building for a year would be. The results showed that space cooling and miscellaneous equipment also known as plug loads consume the highest amount of energy in the building.

The result of walk-through energy audit carried out in health center is shown in Fig. 6.13. From Fig. 6.13, it seen that in the health center, the highest energy consumer is air conditioners (84%). This is due to the fact that the whole building is fully air-conditioned and the air conditioner runs 24 hours per day for in-house patients.

Based on the results of energy audit of the university health center, the following recommendations are made as suggestions to conserve energy in the building:

- The large roof could also have solar photovoltaic panels installed on them to reduce the energy taken from the grid.
- In addition, the fluorescent tubes should be changed to LED tubes to reduce energy consumption by lighting systems.

6.3.5 Result of energy audit in the students' halls of residence

The building structures of the students' halls of residence are similar. Hence, one male hall (Daniel hall) is taken as sample for analysis. Fig. 6.14 shows the building envelope of Daniel hall. The building envelope is primarily made up of windows and walls. The building has a lot of open spaces and corridor spaces in the middle, but it is surrounded by the walls; hence, it is susceptible to daylighting recommendations. The population density in the hall is also very high with about 400 rooms and 1000 students.

In Fig. 6.15, the energy end use pattern of Daniel hall is presented. These results were generated from eQuest based on the energy consumption data and occupancy profile provided. The software predicts what the energy consumption of the building for a year would be. The results showed that miscellaneous equipment also known as plug loads and ventilation fans consume the highest amount of energy in

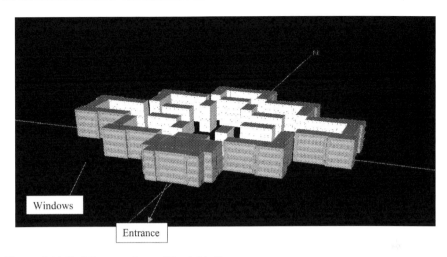

Figure 6.14 Building envelope of Daniel hall.

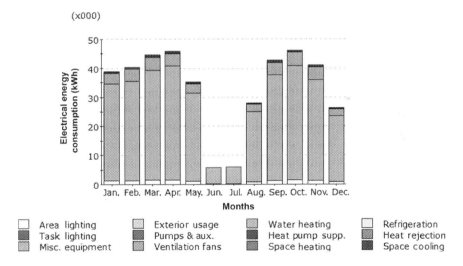

Figure 6.15 Energy end use of Daniel hall.

the building. In addition, since there is no air conditioner in the students' room apart from the hall officers' rooms, space cooling does not contribute substantially to the energy consumption in the building.

The result of walk-through energy audit carried out in Daniel hall is presented in Fig. 6.16. Daniel hall was chosen to represent the halls of residence. The population of students in Daniel hall is about 1000 with each of them owning a laptop, iron, electric kettle, and so on. These plug loads are responsible for majority of the energy consumption in the hall. In addition, the type of lighting bulbs in the hall is inefficient fluorescent tubes thereby leading to excessive energy consumption.

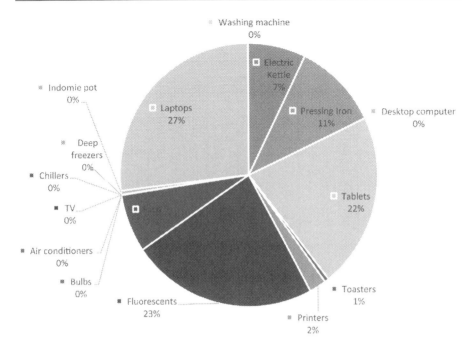

Figure 6.16 Energy audit data of the Daniel hall.

Based on the result of this study, laptop is the major energy consumer (27%) in the students' halls of residence, and this is followed by fluorescent bulbs (23%).

Based on the walk-through energy audit carried in the halls of residence, the following recommendations are made to conserve energy in the halls:

- There should be an increased energy efficiency and conservation awareness among the students. The best way of doing this is to organize competitions among the halls for the hall using the least amount of energy to motivate students to manage their energy use.
- An association of energy enthusiasts should be created, which would see them participate in energy hackathons with cash prizes and industrial placement.
- Hebron Energy Club is the proposed name for this association, which bears similitude to Stanford Energy Club and MIT Energy Club, which are already in existence. The energy club should be inaugurated and empowered by the management to coordinate energy efficiency and conservation awareness on campus.
- Furthermore, inefficient lighting fixtures should be changed to more energy-efficient LED bulbs with automated control in place to control corridor lights.

6.3.6 Qualitative recommendation analysis

There have been several recommendations for energy conservation strategies based on the results of the energy audit in different buildings audited; however, a lot of these recommendations cannot be evaluated economically because of their nature. An example is the increase in awareness of energy efficiency in the students' halls

Table 6.2 Economic analysis of lighting fixtures' replacement with LED bulbs.

Fixtures	No. of fixtures	Daily energy saved (kWh)	Cost of replacement (naira)	Cost of replacement (dollars)	Cost saved annually (naira)	Cost saved annually (dollars)
Fluorescents	11,266	951.42	₦6,759,600.00	$18,269.19	₦6,850,190.26	$18,514.03
CFL bulbs	472	37.31	₦283,200.00	$765.41	₦268,620.69	$726.00
Special bulbs	22	5.19	₦13,200.00	$35.68	₦37,346.40	$100.94
	11,760	993.91	₦7,056,000.00	$19,070.27	₦7,156,157.35	$19,340.97

Table 6.3 Environmental analysis of lighting fixtures' replacement with LED bulbs.

Fixtures	No. of fixtures	Daily energy saved (kWh)	CO_2 savings (tons)
Fluorescents	11,266	951.42	119.42
CFL bulbs	472	37.31	4.68
Special bulbs	22	5.19	0.65
	11,760	993.91	124.76

of residence, which would take a lot of time and analysis to evaluate. However, two of the suitable recommendations would be evaluated for their economic justification: replacement of lighting fixtures with LED bulbs and installation of solar panels on the roofs of selected buildings.

6.3.6.1 Replacement of lighting fixtures with light-emitting diode bulbs

Table 6.2 presents the economic analysis of lighting fixtures' replacement with LED bulbs. Results in Table 6.2 were obtained by using Eqs. (6.1)–(6.10).

The basic assumptions made are as follows:

- Number of active school days in a year = 240 days.
- The cost of a 10 W LED bulb = ₦600.
- The cost of electricity in Covenant University = ₦30/kWh.
- The exchange rate is $1 to ₦370.

Based on the assumptions and Table 6.2, replacing the lighting fixtures in the selected facilities would yield annual savings of over ₦7 million (over $19,000) with a payback period of 0.99 years.

In Table 6.3, the environmental analysis of lighting fixtures' replacement with LED bulbs is presented.

Basic assumption made includes the following:

- CO_2 conversion factor is 0.523 kg of CO_2/kWh.

Table 6.4 Economic and environmental analysis of solar panels installation.

Power (W)	Cost of solar panels (naira)	Cost of solar panels (dollars)	Energy savings (kWh)	Cost savings (naira)	Cost savings (dollars)	CO_2 savings (tons)
530,505.21	₦132,626,303.46	$358,449.47	763,927.51	₦22,917,825.24	$61,940.07	406.41

From Table 6.3, it is evident that replacing inefficient lighting fixtures with more efficient ones will save the environment from pollution to the order of 125 tons of CO_2 emissions annually.

6.3.6.2 Installation of solar panels on the roofs of selected buildings

Electrical energy from solar panels could be fed directly into the grid system of the university campus during peak demand, which usually occurs in the afternoon thereby eliminating the need for batteries in the solar system. The large amounts of electrical energy consumed in the halls of residence would not make rooftop solar panels an economically feasible; hence, the other facilities apart from students' halls of residence are evaluated.

Table 6.4 shows the economic and environmental analysis of solar panels installation.

The following basic assumptions are made in the analysis:

- The cost of solar panels/W of energy = ₦250/W.
- Average daily sun hours = 6 hours.
- School days in a year = 240 days.
- Cost of energy (₦/kWh) = ₦30.
- The exchange rate is $1 to ₦370.

The aforementioned analysis shows that installing rooftop solar panels would save the cost about 23 million naira annually with a payback period of 5.79 years and eliminate about 400 tons of CO_2 annually.

6.4 Conclusion

From this study, it was deduced through the walk-through energy audit method that there is potential for energy saving on the selected buildings in Covenant University campus. There are a lot of cases where the loading of equipment and appliances on building was relatively high such that reconstructions were carried out causing the buildings to have more offices and laboratories than they were initially designed to accommodate, thus increasing the load and energy demand on the building. It was also observed from the walk-through energy audit that there were places where energy was not efficiently utilized and being wasted due to the:

- use of low energy-efficient appliances,

- habit of not switching off an appliance or equipment when not in use,
- multiple use of inefficient heating equipment,
- use of high-power rating appliances and equipment when lower power rated equipment or appliances could be used.

Results of economic and environmental analyses revealed that over ₦30 million (over $81,000) can be saved annually with a payback time of less than 6 years if the recommendations regarding the replacement of inefficient bulbs with LED bulbs and installation of solar panels were implemented. In addition, about 500 tons of CO_2 emissions can be eliminated annually based on adoption of the recommended strategies.

References

Allab, Y., Pellegrino, M., Guo, X., Nefzaoui, E., Kindinis, A., 2017. Energy and comfort assessment in educational building: case study in a French university campus. Energy Build. 143, 202−219.

Alshuwaikhat, H.M., Abubakar, I., 2008. An integrated approach to achieving campus sustainability: assessment of the current campus environmental management practices. J. Clean. Prod. 16 (16), 1777−1785.

Choong, W.W., Chong, Y.F., Low, S.T., Mohammed, A.H., 2012. Implementation of energy management key practices in Malaysian universities. Int. J. Emerg. Sci. 2 (3), 455−477.

Chung, M.H., Rhee, E.K., 2014. Potential opportunities for energy conservation in existing buildings on university campus: a field survey in Korea. Energy Build. 78, 176−182.

Coakley, D., Raftery, P., Keane, M., 2014. A review of methods to match building energy simulation models to measured data. Renew. Sustain. energy Rev. 37, 123−141.

Dehghani, M.J., McManamon, P., Ataei, A., 2018. Toward building energy reduction through solar energy systems retrofit options: an eQUEST model. J. Appl. Eng. Sci. 8 (1), 53−60.

Deshko, V.I., Shevchenko, O.M., 2013. University campuses energy performance estimation in Ukraine based on measurable approach. Energy Build. 66, 582−590.

Escobedo, A., Briceño, S., Juárez, H., Castillo, D., Imaz, M., Sheinbaum, C., 2014. Energy consumption and GHG emission scenarios of a university campus in Mexico. Energy Sustain. Dev. 18, 49−57.

Faghihi, V., Hessami, A.R., Ford, D.N., 2015. Sustainable campus improvement program design using energy efficiency and conservation. J. Clean. Prod. 107, 400−409.

Ishak, M.H., Sipan, I., Sapri, M., Iman, A.H.M., Martin, D., 2016. Estimating potential saving with energy consumption behaviour model in higher education institutions. Sustain. Environ. Res. 26 (6), 268−273.

Ke, M.T., Yeh, C.H., Jian, J.T., 2013. Analysis of building energy consumption parameters and energy savings measurement and verification by applying eQUEST software. Energy Build. 61, 100−107.

Kolokotsa, D., Gobakis, K., Papantoniou, S., Georgatou, C., Kampelis, N., Kalaitzakis, K., et al., 2016. Development of a web based energy management system for university campuses: the CAMP-IT platform. Energy Build. 123, 119−135.

Lou, S., Tsang, E.K., Li, D.H., Lee, E.W., Lam, J.C., 2017. Towards zero energy school building designs in Hong Kong. Energy Procedia 105, 182−187.

Ma, H., Du, N., Yu, S., Lu, W., Zhang, Z., Deng, N., et al., 2017. Analysis of typical public building energy consumption in northern China. Energy Build. 136, 139–150.

Odunfa, K.M., Ojo, T.O., Odunfa, V.O., Ohunakin, O.S., 2015. Energy Efficiency in Building: Case of Buildings at the University of Ibadan, Nigeria. J. Build. Construct. Plan. Res 3, 18–26.

Petersen, J.E., Shunturov, V., Janda, K., Platt, G., Weinberger, K., 2007. Dormitory residents reduce electricity consumption when exposed to real-time visual feedback and incentives. Int. J. Sustain. High. Educ. 8 (1), 16–33.

Pike, L., Shannon, T., Lawrimore, K., McGee, A., Tylor, M., Lamoreaaux, G., 2003. Science education and sustainability initiatives: a campus recycling case study shows the importance of opportunity. Int. J. Sustainability High. Educ. 4 (3), 218–229.

Saleh, A.A., Mohammed, A.H., Abdullah, M.N., 2015. Critical success factors for sustainable university: a framework from the energy management view, Procedia-Soci. Behav. Sci. 172, 503–510.

Spirovski, D., Abazi, A., Iljazi, I., Ismaili, M., Cassulo, G., Venturin, A., 2012. Realization of a low emission university campus through the implementation of a climate action plan, Procedia-Soc. Behav. Sci. 46, 4695–4702.

Sretenovic, A., 2013. Analysis of Energy Use at University Campus (Master's thesis). Norwegian University of Science and Technology, Trondheim, pp. 8–83.

Tang, F.E., 2012. An energy consumption study for a Malaysian university, World Academy of Science, Engineering and Technology. Int. J. Environ., Earth Sci. Eng. 6 (8), 99–105.

United Nations, 2015. Framework Convention on Climate Change, Adoption of The Paris Agreement. Agreement 1–27.

Wen, T.W., Palanichamy, C., 2018. Energy and environmental sustainability of Malaysian Universities Through Energy Conservation Measures. Int. J. Energy Econom. Policy 8 (6), 186–195.

Zhou, X., Yan, J., Zhu, J. and Cai, P., Survey of energy consumption and energy conservation measures for colleges and universities in Guangdong province, Energy Build. 66, 112–118.

Energy conversion systems and Energy storage systems

Jian Zhang[1], Heejin Cho[2] and Pedro J. Mago[3]
[1]Department of Mechanical Engineering, University of Wisconsin Green Bay, Green Bay, WI, United States, [2]Department of Mechanical Engineering, Mississippi State University, Mississippi State, MS, United States, [3]Department of Mechanical and Aerospace Engineering, West Virginia University, Morgantown, WV, United States

Chapter Outline

7.1 Introduction 155
7.2 Energy systems in buildings 156
    7.2.1 Energy generation systems   156
    7.2.2 Energy conversion systems   168
    7.2.3 Energy storage systems   171
7.3 Conclusion   175
References   175

7.1 Introduction

According to the report "energy efficiency: buildings" made by the International Energy Agency (IEA) (International Energy Agency IEA, n.d.), the building and building construction sectors take up about 36% of the global final energy consumption (as shown in Fig. 7.1) and almost 40% of the total CO_2 emissions. Up-to-date, electricity is mostly provided by fossil-fuel-based central power plants and distribution systems, which have low overall efficiency and high air pollution. In recent years, more renewable or distributed generation has been encouraged and adopted worldwide. Based on the latest report, "global energy perspective 2019: reference case" (McKinsey, 2019), made by McKinsey company, renewable generation is projected to make up over 50% by 2035. Therefore, carbon emissions are projected to decline due to the decreasing coal demand.

Under this circumstance, the importance of energy services has been widely recognized to achieve energy- and environment-saving goals. Understanding of the fundamentals and concepts of energy systems used in buildings is critical to provide/receive energy services in an impactful, yet cost-effective manner. This chapter presents an introduction to common energy systems used in building applications. These energy systems can generally be classified into three types: energy generation systems (discussed in Section 7.2.1), energy conversion systems (discussed in Section 7.2.2), and energy storage systems (discussed in Section 7.2.3). The focus of this chapter is to review and survey the application of

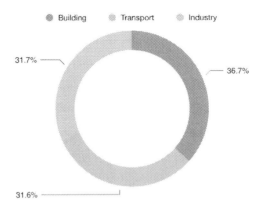

Figure 7.1 Global final energy consumption by sector.
Source: Data from International Energy Agency (IEA), n.d. Energy efficiency: buildings. https://www.iea.org/topics/energyefficiency/buildings/ (accessed 06.08.19).

the aforementioned building energy systems, with the aims of providing fundamental information on energy systems in buildings for providers and clients of energy services.

7.2 Energy systems in buildings

In this section, a series of conventional energy systems used in building applications are introduced. These energy systems are classified into three groups: energy generation systems, energy conversion systems, and energy storage systems.

7.2.1 Energy generation systems
7.2.1.1 Combined heat and power system

Combined heat and power (CHP) systems are designed to utilize the waste heat energy from an on-site power generation unit (PGU) so that it can satisfy both the electric and thermal load at the same time in an effective manner (Cho et al., 2010; Liu et al., 2014; Zhang et al., 2016). In addition, CHP systems provide alternative solutions to reduce electricity grid dependence, save energy costs, and other energy and emission problems (Cho et al., 2009; Wu and Wang, 2006; Cho et al., 2014). Normally, a CHP system consists of a PGU, which generates the electricity and heat energy at the same time; energy recovery components, which are used to recover useful heat; and energy management system, which regulates the operation of the system. Fig. 7.2 illustrates a schematic of a CHP system for building applications. In addition, the network flow model of a typical CHP system is illustrated in Fig. 7.3.

Typically, PGU is the combination of a prime mover and a generator, and the CHP system is identified by the type of prime mover (Mago and Luck, 2011; U.S.

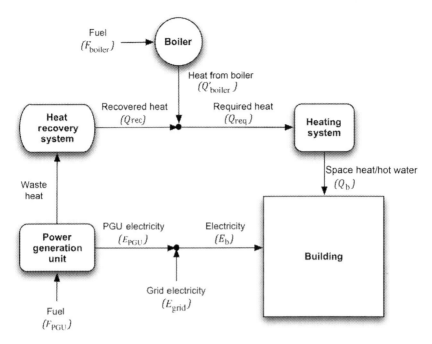

Figure 7.2 Schematic of a CHP system (Mago and Smith, 2012).

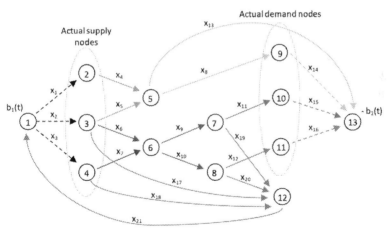

Node 1: Total energy supply, b₁(t)
Node 2: Electric Grid (EG)
Node 3: Cogeneration Systems (CS)
Node 4: Supplementary Heating Devices (SHD)
Node 5: Electric energy provided from CS and EG
Node 6: Thermal energy provided from CS and SHD
Node 7: CHP cooling components
Node 8: CHP heating components
Node 9: Electric energy demand
Node 10: Cooling energy demand
Node 11: Heating energy demand
Node 12: Total energy loss
Node 13: Total energy demand, b₂(t)

x_1: Electric energy supply
x_2, x_3: Fuel energy supply into CS and SHD
x_4, x_5: Electric energy flow from EG and CS
x_6, x_7: Thermal energy flow from CS and SHD
x_8: Electric energy flow into CHP building
x_9, x_{10}: Thermal energy flow into CHP components
x_{11}: Cooling energy flow into CHP building
x_{12}: Heating energy flow into CHP building
x_{13}: Excess electric energy flow export to EG
x_{14}, x_{15}, x_{16}: Electric, cooling, and heating energy demands
$x_{17}, x_{18}, x_{19}, x_{20}$: Energy loss
x_{21}: Total energy loss

Figure 7.3 Network flow model of a typical CHP system (Cho et al., 2009).

Environmental Protection Agency Combined Heat and Power Partnership, 2017). Typical prime movers include reciprocating internal combustion engines, combustion turbines, steam turbines, microturbine, and fuel cells.

Reciprocating engines, also called piston engines, are the devices that convert pressure to kinetic energy (rotating) (Wikipedia, 2019b). Reciprocating engines currently used by CHP systems are spark-ignition (SI) engines and compression-ignition (CI) engines. The main difference between these two engines is the way the fuel is ignited. SI engines usually use natural gas as fuel, and the fuel is initiated by the spark plug, whereas CI engine operates on diesel or heavy oil, and the fuel is ignited by the compressed and heated air (Liu et al., 2014).

Combustion turbines, that is, gas turbines, are operated based of the Brayton cycle with air as the working fluid (Wikipedia, 2019a). In a combustion turbine, fuel is sprayed into the air and combusted to produce high-temperature and high-pressure gas, which then enters the turbine to expand and generate shaft work. Combustion turbines have a wide range of sizes to match various CHP systems (500 kW–300 MW (U.S. Environmental Protection Agency Combined Heat and Power Partnership, 2017).

Steam turbines are devices that convert thermal energy (extracted from pressurized steam) to mechanical shaft work (Wikipedia, 2019c). Steam turbines are widely used in power generation industries, including large-scale CHP systems due to the merits of high efficiency and low cost. However, for small-scale power generation applications, steam turbines are not good candidates because of the low electric efficiency at partial load.

Microturbines, in essence, belong to combustion turbines. They are designed in small sizes and fueled with natural gas, gasoline, diesel, and other similar fuels. Compared to other heat engines, the construction of microturbines is simple and compact, which make them quite suitable for distributed generation such as CHP systems.

Fuel cells use electrochemical processes to convert chemical energy into electricity and provide heat energy at the same time in the form of hot water or steam. Fuel cells are considered as a clean and reliable method for cogeneration because of low emission (only water) and few moving parts contained (Kordesch and Simader, 1996).

According to the report made by U.S. Environmental Protection Agency (U.S. Environmental Protection Agency Combined Heat and Power Partnership, 2017), reciprocating engines make up the largest share of CHP systems over the US region, whereas the combustion turbines make up the largest share of CHP capacity. Fig. 7.4 shows the US installed CHP facilities and capacity by different technologies (U.S. Environmental Protection Agency Combined Heat and Power Partnership, 2017). The performance, characteristics, and cost for CHP systems with different prime movers are summarized in Table 7.1.

7.2.1.2 Solar photovoltaic system

Solar photovoltaic (PV) systems are considered as promising renewable technology, and they are expanding rapidly in recent years. Based on the report given by Inter

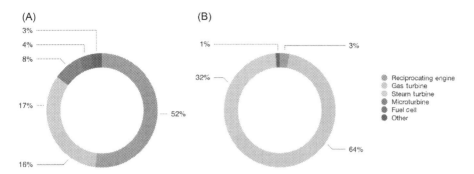

Figure 7.4 US installed CHP facilities (A) and capacity (B) by different technologies.
Source: Data from U.S. Environmental Protection Agency, 2017. Combined Heat and Power Partnership. Catalog of CHP Technologies. Washington, DC.

Solar (Intersolar, 2019), the solar PV installation in 2017 (99.1 GW) is as much as the cumulative installation in 2012 (100.9 GW), which resulted in a total global solar PV capacity of over 400 GW in 2017. Solar PV systems convert solar energy into direct current electricity using solar cells, which are normally made by semiconducting materials. Common technologies used in PV industries include crystalline silicon, thin-film, compound semiconductor, and nanotechnology (El Chaar et al., 2011).

Silicon is the most commonly used material in PV cells till now, making up around 90% of the PV modules (U.S. Department of Energy DOE, 2013). In a crystalline silicon cell, crystal lattice is formed by silicon atoms connected to each other. The crystal lattice is the unit that converts the light into electricity. In the PV industry, the cells that use crystalline silicon include monocrystalline PV cells, polycrystalline PV cells, and emitter-wrap-through PV cells. Among these types of PV cells, the monocrystalline cells are the most widely used due to a good balance between efficiency and cost. Polycrystalline cells once were the most popular when the cost of Si was high. These cells cost less while they are less efficient than monocrystalline cells. Emitter-wrap-through cells refers to a cell design concept that uses small laser-drilled holes to connect the contacts on the back surface to the emitter on the front surface (Gee et al., n.d.). The emitter-wrap-through cells provides an alternative way to increase the cell efficiency other than material improvements.

Thin-film PV cells are the second-generation solar cells that are created by depositing one or multiple thin films of PV material on a plastic, glass, or metal backing material. Compared to crystalline silicon cells, thin-film cells have the potential to reduce the cost of PV modules since thin films largely reduce the requirement of PV material for each cell. As a consequence, the efficiency of the thin-film cell is lower than that of crystalline silicon cell because of the same reason mentioned earlier, reaching efficiencies of 12%–20% (Barnett et al., 2001).

Compound semiconductor PV cells are designed to utilize most of the solar radiation by joining up a stack of semiconductor material layers with different band

Table 7.1 Performance, characteristics, and cost for CHP systems with different prime movers.

Prime mover	Reciprocating engine	Gas turbine	Steam turbine	Microturbine	Fuel cell
Efficiency	27%–41%	24%–36%	5%–40%	22%–28%	30%–63%
Overall efficiency	77%–80%	66%–71%	80%	63%–70%	55%–80%
Typical capacity (MWe)	0.005–10	0.5–300	0.5 to several hundred MW	0.03–1	0.005–2
CHP installed costs ($/kW)	1500–2900	1200–3300	670–1100	2500–4300	5000–6500
O&M costs ($/kWh)	0.009–0.025	0.009–0.013	0.006–0.01	0.009–0.013	0.032–0.038

Source: Data from U.S. Environmental Protection Agency, 2017. Combined Heat and Power Partnership. Catalog of CHP Technologies. Washington, DC.

(energy) gaps, so that different portions of the solar spectrum could be absorbed by different layers. As for the nanotechnology, it has been applied to the PV cell production to reduce the manufacturing and installation costs even though nanotechnology PV cells are less efficient than traditional ones.

7.2.1.3 Solar thermal system

In building applications, solar thermal systems are typically designed to convert solar energy to thermal (heat) energy. In a solar thermal system, working fluid (water, air, oil, and so on) is heated by the solar collector, which absorbs solar irradiation as heat, and then the heat energy carried by the working fluid can be either used directly or stored in the thermal energy storage (TES) system for later use. This section focuses on the review of solar collectors, whereas the review of TES systems is presented in Section 7.2.3.

Solar collectors are generally classified into nonconcentrating collectors and concentrating collectors on the basis of concentration ratios (de Winter, 1991). Nonconcentrating collectors have the same aperture area as their absorber area and are usually used for building space or hot water heating. Flat-plate collectors, evacuated tube collectors, and hybrid PV/thermal (PVT) collectors are common nonconcentrating collectors. Concentrating collectors have a much larger aperture area than the receiver area and are typically used in solar thermal power generation plants, such as parabolic trough collectors and parabolic dish collectors.

Flat-plate collectors consist of a transparent cover, an absorber plate with working fluid passages, insulation layers, and other auxiliaries. The transparent cover is designed to decrease the convection losses from the absorber plate and the radiation losses from the collector. Typically, it is made of glass or other substance with high transmissivity of high-frequency radiation but low transmissivity of low-frequency radiation. Low-iron glass is considered as a good choice in high-performance solar collector designs due to its transmittance characteristics (Khoukhi and Maruyama, 2005; Tian and Zhao, 2013). The absorber plate normally has a dark-colored coating (Tripanagnostopoulos et al., 2000) with the aim of absorbing as much solar energy as possible. The working fluid circulates through the passages on the absorber to remove the heat from the collector. The working fluid is typically water; however, if other fluids (e.g., oil) are used, then a heat exchanger may be needed to transfer the heat from the working fluid to a water tank.

Evacuated tube collectors perform better than flat-plate solar collectors for high-temperature applications and are the most popular solar thermal technology in China (International Energy Agency IEA, 2019). In an evacuated tube solar collector, the absorber is placed in a vacuum glass tube so that the convection heat loss is effectively reduced, and therefore the conversion efficiency is increased. The absorber can be made of either metal or glass. Several ways are commonly used to extract heat from evacuated tubes including heat pipe, flow through absorber, all-glass tubes, and storage absorber (Morrison et al., 2004). Compared to flat-plate collectors, evacuated tube collectors are more efficient but less cost-effective.

Hybrid PVT collectors are designed to utilize solar energy for both electricity and heat generation. A PVT collector combines a PV cell, which is used to produce electricity, with a solar thermal absorber, which acts as a heat removal device. On one hand, the solar thermal absorber utilizes the waste heat from PV module to produce hot water; on the other hand, the PV model is cooled down for more efficient performance. Therefore, the PVT collectors can achieve a higher energy conversion efficiency than solar thermal or solar PV alone (Guarracino et al., 2019).

7.2.1.4 Organic Rankine cycle system

Organic Rankine cycle (ORC) systems are similar to conventional steam power generation systems but use organic fluids instead of water as working fluid. Potential working fluid for ORC systems include hydrocarbons, hydrofluorocarbons, chlorofluorocarbons, hydrofluoroolefins (HFOs), among others (Tchanche et al., 2011; Mago et al., 2008; Wang et al., 2011). However, on the view of safety and environmental influence, HFOs are the better selection since they are nonflammable, nontoxic, and environment friendly with zero-ozone depletion potential values and a low global warming potential value (Yang et al., 2017; Fang et al., 2019). In recent years, ORC has been recognized as a promising technology to convert low-grade heat energy into power. Common heat resources include waste heat energy from other thermal systems, solar energy, geothermal energy, biomass products, and so on (Tchanche et al., 2011; Rahbar et al., 2017; Yang et al., 2014). Fig. 7.5 shows a schematic of a basic ORC system. The system consists of four key components: a pump, an evaporator, an expander (a turbine), and a condenser. The working fluid at low pressure, saturated liquid state is pressurized by the pump and then heated in the evaporator to high pressure, saturated or superheated vapor state. Then the vapor expands in the expander and generates shaft work, which is converted to electricity if connected with a generator. Finally, the working fluid is condensed to saturated liquid in the condenser.

In order to improve the energy conversion efficiency, different configurations of ORC system have been developed, including ORC systems with recuperator, reheat ORC systems, and regenerative ORC system. Fig. 7.6 displays an ORC system with a recuperator (an internal heat exchanger). The recuperator allows heat transfer between the superheated vapor at the turbine exit and the subcooled liquid at the pump outlet. The schematic of a reheat ORC system is shown in Fig. 7.7. In a reheat ORC system, the superheated vapor first expands in high-pressure first-stage turbine and then is delivered back to the evaporator, where it is reheated to a certain temperature before expanding in low-pressure second stage turbine. Fig. 7.8 illustrates the schematic of a regenerative ORC system. In a regenerative ORC system, a regenerative tank (a heat exchanger in essence) is employed (Zhang et al., 2014). During the expansion process, part of the working fluid vapor is extracted at an intermediate pressure and routed to the regenerative tank, where it mixes and exchanges heat with the subcooled liquid pressurized by Pump 1. Subsequently, the mixture is pressurized by Pump 2 and delivered to evaporator.

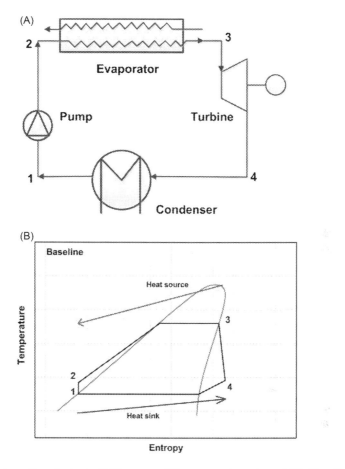

Figure 7.5 Schematics of a basic ORC system: (A) system schematic and (B) $T-s$ diagram (Li, 2016a).

7.2.1.5 Geothermal system

Geothermal energy is a clean and sustainable energy generated by the earth. Compared to the conventional fossil-fueled systems, geothermal energy systems are usually environment friendly with much lower emissions (Wu, 2009; Neves et al., 2020). Generally, geothermal energy is utilized via three technologies: electricity production, direct heating, and geothermal heat pumps, providing indirect heating and cooling for buildings (Wu, 2009). According to a worldwide review on geothermal energy made by Lund et al. Lund and Boyd (2016), geothermal heat pumps make up the largest share of energy usage and capacity installation, accounting for 55.15% of the energy usage and 70.90% of the installed capacity.

Geothermal heat pumps, also known as ground-source heat pumps (GSHPs), are considered as one of the most energy-efficient technology for space heating and

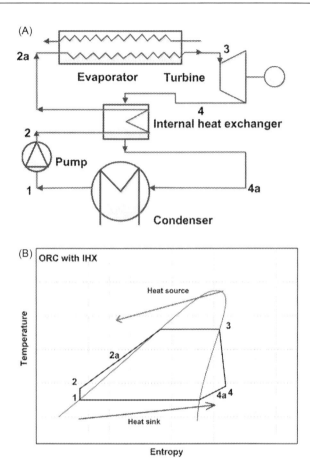

Figure 7.6 Schematics of an ORC with a recuperator: (A) system schematic and (B) $T-s$ diagram (Li, 2016a).

cooling in buildings. Typically, GSHP systems consist of three core subsystems (U.S. Department of Energy DOE, 2019; Self et al., 2013):

1. Earth connection: circulating the working fluid (water or antifreeze solution) to absorb heat from or reject heat to the ground via a ground heat exchanger (GHE) loop.
2. Heat pump: transfer heat between the building and the earth connection.
3. Heat distribution: distribute heated or cooled air throughout the building space.

The way that a GSHP system works is based on the fact that the temperature beneath a certain distance of the earth's surface remains at a nearly constant temperature throughout the year, warmer in the winter and cooler in the summer than the air. In the winter time, a GSHP system extracts heat from the ground and distributes it to the building space, whereas in the summer time, it removes heat from the building and transfers it to the ground for cooling. Based on the geothermal

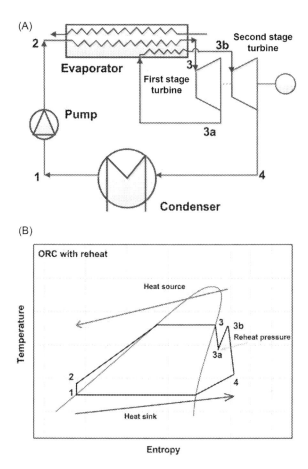

Figure 7.7 Schematics of a reheat ORC system: (A) system schematic and (B) $T-s$ diagram (Li, 2016a).

sources, that is, groundwater, surface water, and ground, GSHP systems have been classified into three categories by ASHRAE (2011): groundwater heat pump (GWHP) systems, which transfer heat with ground water; surface water heat pump (SWHP) systems, which use the surface water such as pond, reservoir, lake, and so on as heat source or sink; and ground-coupled heat pump (GCHP) systems. The schematics of each system are displayed in Fig. 7.9. Among different types of GSHP systems, the GCHP is the most attractive one. In a GCHP system, a GHE is employed to combine the refrigerant loop and the water loop as a closed loop. Based on the arrangement of the pipes connected to GHE, the closed loop systems are classified as vertical closed loop and horizontal closed loop. Literally, the pipes in the ground run vertically in a vertical closed loop, whereas they run horizontally in a horizontal closed loop, as shown in Fig. 7.10.

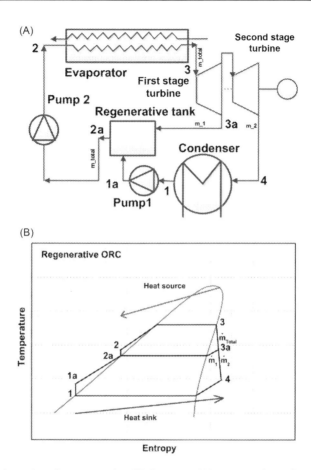

Figure 7.8 Schematics of a regenerative ORC system: (A) system schematic and (B) $T-s$ diagram (Li, 2016a).

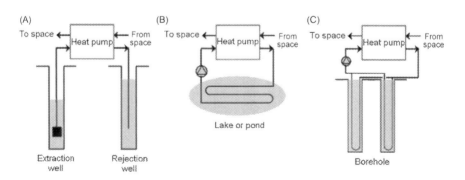

Figure 7.9 Schematic of different GSHP systems: (A) GWHP, (B) SWHP, and (C) GCHP (Sarbu and Sebarchievici, 2014).

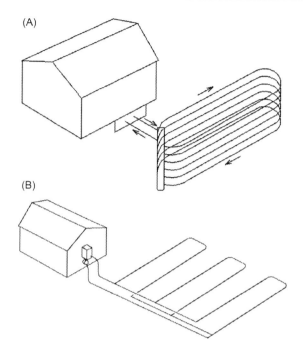

Figure 7.10 (A) Vertical closed loop and (B) horizontal closed loop (Sarbu and Sebarchievici, 2014).

7.2.1.6 Wind turbine system

Wind turbines, also called wind energy converters, are used to convert the air kinetic energy to mechanical energy for electricity generation. Despite various configurations, wind turbines are usually classified as two types: horizontal-axis wind turbine (HAWT) and vertical-axis wind turbine (VAWT).

HAWT is the most commonly used wind turbine due to the merits of simple configuration, high efficiency, and low cost (Tawfiq et al., 2016). Most HAWTs have three blades and are operated "upwind," with the rotation axis of the blades placed in a horizontal position at the top of the tower. When the wind blows through blades, the kinetic energy of the wind converts into rotational shaft energy. An electric generator is placed at the top of the tower, and a gear box is used to increase the rotation speed in order to drive the electric generator. One of the drawbacks of HAWTs is that they are sensitive to the wind direction and turbulence, which means that an additional control system is required to control the direction of the turbine blades (Polinder et al., 2006). For VAWTs, the axis of rotation is arranged vertically to the ground. Compared to the HAWTs, the wind is always perpendicular to the blades of VAWTs. Thus there is no need to control the blades' direction. In addition, the gearbox and generator in VAWTs are placed on the ground, which makes the installation and maintenance processes much easier. Table 7.2 shows a comparison between HAWTs and VAWTs.

Table 7.2 Comparison between HAWTs and VAWTs.

Feature	HAWTs	VAWTs
Efficiency	Around 70%	Less than 60%
Rotation speed	High	Low
Height	Large	Small
Direction of wind	Dependent	Independent
Noise	5–60 dB	0–10 dB
Effect on birds	Large	Small
Application	Offshore and Onshore	Onshore

Source: Data from Tawfiq, K.B., Mansour, A.S., Ramadan, H.S., Becherif, M., El-Kholy, E.E., 2019. Wind energy conversion system topologies and converters: comparative review. Energy Procedia 162, 38–47. https://doi.org/10.1016/j.egypro.2019.04.005.

7.2.2 Energy conversion systems

The energy conversion systems, mentioned in this section, mainly refer to the heating and cooling energy conversion systems in buildings, that is, heating, ventilation, and air-conditioning (HVAC) systems. HVAC systems are designed to keep desired or comfortable conditions in a space, and they are an important part of building structures.

7.2.2.1 Heating systems

Heating systems are used to provide heat for buildings. Typically, boilers, furnaces, or heat pumps are adopted to heat water, steam, or air. The heat energy is usually transferred by convection or radiation.

Boilers are enclosed pressure vessels in which heat generated by combustion or electrical resistance elements is transferred to a fluid. In most cases, the fluid is liquid water or steam. Boiler systems give out the heat energy as the hot water or steam passes through radiators or other type of heat exchangers in the space. Then the cool water is routed to the boiler to be reheated. Normally, boilers can be classified into different types according to working pressure and temperature, fuel, construction material, and whether they are condensing or noncondensing (ASHRAE, 2016). Based on the working temperature and pressure, boilers are grouped as low-pressure boilers, which are limited to 160 psi (1.103 MPa) and 250°F (121°C), and high-pressure boilers, which are operated above 160 psi and 250°F. Boilers can be designed to burn fuels such as coal, oil, wood, and fuel gas or use electricity as electric boilers. As for the construction material, the most common construction material for boilers is cast-iron, with other materials include copper, steel, stainless steel, and aluminum as alternatives. Traditional boilers are operated without condensing the fuel gas in order to prevent cast-iron, copper, or steel parts being corroded. However, condensing boilers have higher efficiency and are now available from many manufacturers.

Furnaces are combustion and heating appliances in which heat is transferred from burned gas or oil, or electric resistance elements to air. Then the warm air is delivered to the conditioned space through ductwork. Typically, furnaces consist of several components: casing or cabinet, heat exchangers, burners and controls, venting devices, air blower, air filter, and other accessories. Furnaces can be used for both commercial and residential applications and are classified based on different characteristics (ASHRAE, 2016):

1. Fuel types: natural gas, oil, liquefied petroleum gas, or electricity.
2. Airflow direction: horizontal, upflow, or downflow.
3. Installation location: within conditioned space (indoors) or outside conditioned space.
4. Combustion air source: direct vent (outdoor air) or indoor air.

Heat pumps extract heat from a low-temperature source and reject it to a high-temperature source. Thus heat pumps can be used for both heating and cooling purpose. In summer, heat pumps are used to move heat from cool indoor space to warm outside, while in the winter, heat pumps reverse the trick, extracting heat from the cool outside space and delivering it to warm inside. Heat pumps contain all the main components of air conditioners, that is, compressor, evaporator, condenser, fan, filters, tube, and controls. In the building applications, two common types of heat pumps are available: air-source heat pumps and GSHPs. Air-source heat pumps use the outside air as the high-temperature source during winter and low-temperature source during summer. GSHPs, as discussed in Section 7.2.1.5, obtain heat from underground. Despite being less efficient compared to GSHPs, air-source heat pumps are much more popular because of the low cost and easy installation.

7.2.2.2 Cooling systems

Compared to heating systems, cooling systems, or air-conditioning systems, are more complicated since they are designed to remove heat and moisture from the indoor space instead of generating heat. The air-conditioning is commonly achieved through a vapor compressor cycle, and the term "air-conditioning," in a more general sense, can refer to many different technologies such as heat pumps, absorption chillers, and solar cooling.

Air conditioners are used to cool and dehumidify the air when air flows over the cooling coil surface. The cooling coil is typically an air-to-liquid heat exchanger with rows of tubes through which the liquid cooling medium passes the coil. Based on the liquid used in the cooling coil, air conditioners can be classified as direct-expansion (DX) systems and chilled-water systems. In a DX equipment, the refrigerant is the cooling medium that passes through the cooling coil, and the air is cooled directly by the refrigerant without a secondary medium. DX systems are widely used in many applications, while limitations on capacity, application, and performance reduce their use in larger and more complex HVAC applications (Stanford, 2016). In these applications, chilled water systems are usually employed. In a chilled water system, a chiller system is located outside the building to cool the water, then the chilled water is pumped to cooling coil. In the cooling coil, the chilled water is used as cooling medium instead of refrigerant to cool the air that passes over the cooling coil. Chilled water systems are more suitable for large building applications though they are normally less efficient than DX systems.

The chillers used in chilled water systems are normally grouped as two categories: mechanical and absorption chillers. Mechanical chiller systems, which use the vapor compression cycle to chill the water, consist of a mechanical compressor, an evaporator, a condenser, and control devices. Absorption chillers use heat to generate cooling energy. Instead of using a mechanical compressor and the vapor compression cycle, absorption chillers are composed of an absorber, a generator, a condenser, an evaporator and control devices and use absorption cooling cycle to chill water. Lithium bromide or ammonia/water is usually used as absorbent.

Another type of cooling system is the solar cooling system, which provides an alternative idea to utilize solar energy. Typically, solar cooling can be achieved through three different approaches: absorption cycles, desiccant cycles, and solar mechanical cycles (Duffie and Beckman, 2013). Solar absorption cooling systems consist of solar collectors, hot water storage tank, and cooling tower in addition to an absorption chiller system. Solar collectors are used to heat the water. Then the hot water is routed to the generator of the absorption chiller system to heat the absorbent solution. In this way, the solar system is integrated into the absorption chiller system to produce cooling. Desiccant cooling cycles use hygroscopic substances to dehumidify the air, and the hygroscopic substances can be regenerated when applying solar heat during the cycle. Solar mechanical cooling cycles use the power generated by the solar power systems (such as a Rankine cycle) to drive conventional air-conditioning systems to produce cooling.

7.2.2.3 Ventilation systems

Ventilation systems are used to introduce outdoor air into the air-conditioned space for the purposes of controlling the quality of indoor air and maintaining thermal comfort. In this section, primary components/equipments in ventilation systems (fans and air-handling units [AHUs]) are discussed. In addition, heat recovery ventilation systems are also introduced.

Fans work as the prime mover in ventilation systems to move the air and provide continuous airflow so that the space air, conditioned air, exhaust air, and the fresh air can be transported through air ducts or other air passages (Wang, 2000). Fans are usually grouped based on the direction of airflow through the impeller as axial, centrifugal, mixed, or cross-flow fans. For all fans, air is moved due to the pressure difference between the fan inlet and outlet, and pressure is generated by changing the airflow's velocity. More in detail, axial fans produce pressure mainly by changing the air velocity when air passes through the impeller blades. On the other hand, centrifugal fans produce pressure from the centrifugal force due to the rotation of air and the kinetic energy conveyed to the air.

AHUs, also known as air handler, are primary devices used to regulate and condition the air and distributes it through ductwork ventilation systems to each conditioned space. Generally, AHUs are large metal boxes consisting of fans, heating or cooling coils, filters, humidifiers, controls, and other accessories. According to the structure, that is, whether the supply fan is installed at the same level as the filters and coils or not, AHUs can be classified as horizontal or vertical units. Based on the location of the supply fan, AHUs are classified as draw-through units and blow-through units. As for draw-through units, the air is drawn through the coil section since the supply fan is placed downstream from the coils, whereas for blow-through units, air is blown through the coils as the supply fan is placed upstream from the coils. According to other characteristics, AHUs can also be grouped as (Wang, 2000) outdoor air AHU or mixing AHU, single-zone or multizone AHU, and rooftop or indoor AHU, and so on.

Heat recovery ventilation systems take advantage of additional heat exchangers installed between the supply and exhaust airflow to reduce the energy requirement in preconditioning the fresh air. Heat recovery ventilation is considered as a promising technology and is a popular design for ventilation systems. Nowadays, several different types of heat recovery ventilation systems are available:

1. Fixed-plate heat exchangers are commonly used for heat recovery ventilation. Fixed-plate heat exchangers are commonly used for heat recovery ventilation. The supply and exhaust airflows pass through the heat exchanger and transfer heat from one airflow to the other.
2. Rotary thermal wheels utilize a rotating porous metal wheel to transfer heat from one airflow to the other one by passing through each airstream alternatively.
3. Heat pipes use a multiphase process to transfer heat between exhaust and supply airflow.

7.2.3 Energy storage systems

Energy storage systems are used to capture and store the energy generated at one time for later use. Energy storage systems usually convert energy from forms that are hard to store in more convenient ways. In the building applications, battery energy storage (BES) and TES are two commonly used technologies.

7.2.3.1 Battery energy storage system

Batteries store electricity in the form of electrochemical energy. BES systems allow electricity generated at times of low demand and cost or from intermittent energy sources (PV, wind turbine, and so on) to be used at times of high demand and cost or when no other generation is available (McLarnon and Cairns, 1989; Baker and Collinson, 1999; Chen et al., 2009). Thus BES systems can provide considerable benefits to load shifting and peak shaving as well as increase the renewable energy source penetration (Zhang et al., 2017; Zhang et al., 2018; Shivashankar et al., 2016). Batteries currently used in BES systems include lead acid battery, nickel battery, sodium sulfur battery, and lithium-ion battery. A comparison of different BES technologies is summarized in Table 7.3.

Lead acid batteries are the oldest and most popular batteries. A lead acid battery consists of positive and negative electrodes made of lead plates immersed in a sulfuric acid solution. Despite the toxic and potential of environmental pollution, lead—acid batteries are still the most widely used batteries because of low cost, large power-to-weight ratio, and high reliability and efficiency. Nickel batteries use

Table 7.3 Comparison of different BES technologies.

Type	Power Rating	Cycling capability	Lifetime (years)	Energy efficiency (%)	Self-discharge per day (%)	Storage period	Capacity cost ($/kW)	Power cost ($/kWh)
Lead acid	0–20 MW	200–500	5–15	70–80	0.1–0.3	Minutes to days	300–600	200–400
NiCd	0–40 MW	3500	13–16	72	0.2–0.6	Minutes to days	500–1500	800–1500
NaS	50 kW–8 MW	2500	12–20	75–85	20	Seconds to hours	1000–3000	300–500
Li-ion	0–100 kW	1500	14–16	90	0.1–0.3	Minutes to days	1200–4000	500–2500

Sources: Data from Chen, H., Cong, T.N., Yang, W., Tan, C., Li, Y., Ding Y., 2009. Progress in electrical energy storage system: a critical review. Prog. Nat. Sci. 19, 291–312. doi:10.1016/J.PNSC.2008.07.014; Mahlia, T.M.I., Saktisahdan, T.J., Jannifar, A., Hasan, M.H., Matseelar, H.S.C., 2014. A review of available methods and development on energy storage; technology update. Renew. Sustain. Energy Rev. 33, 532–545. doi:10.1016/J.RSER.2014.01.068; Díaz-González, F., Sumper, A., Gomis-Bellmunt, O., Villafáfila-Robles, R., 2012. A review of energy storage technologies for wind power applications. Renew. Sustain. Energy Rev. 16, 2154–2171. doi:10.1016/J.RSER.2012.01.029.

nickel oxide hydroxide as positive electrode. Among all nickel batteries, two types, that is, nickel−cadmium (NiCd) and nickel−metal hydride (NiMH) batteries are the most popular. NiCd batteries use metallic cadmium as the negative electrode and come with two types: sealed and vented. The advantages of NiCd are long cycle life, robust reliability, and able to tolerate deep discharge. However, the toxicity of cadmium and the susceptibility to memory effect are the two major drawbacks of NiCd batteries. NiMH batteries use hydrogen absorbing alloy instead of cadmium as the negative electrode. Compared to NiCd batteries, NiMH batteries have high energy density and are environment friendly but suffer from high self-discharge rate.

Sodium sulfur (NaS) batteries are molten-salt batteries and consist of liquid sodium and liquid sulfur as active materials at the positive and negative electrodes, respectively. A solid beta alumina ceramic electrolyte also acts as the separator for active materials. NaS batteries are attractive for the applications in large-scale BES systems due to the merits of high charge and discharge efficiency, high energy density, long cycle life, and are fabricated from inexpensive materials (Mahlia et al., 2014; Hameer and van Niekerk, 2015). However, a high operating temperature (300°C−350°C) is required for NaS batteries, and thus energy input is needed to maintain the proper temperature. Lithium batteries use lithium compounds and graphitic carbon as the positive and negative electrodes. They are commonly used for portable electronics and growing rapidly for other applications. Advantages such as high energy density (twice that of NiCd batteries), high efficiency (almost 100%), and low maintenance required make lithium batteries a promising technology, but the high cost due to the special packing and internal overcharge protection circuits limit their applications in large-scale storage right now.

7.2.3.2 Thermal energy storage system

TES systems are designed to store excess thermal energy and use it hours, days, or even months later. They can help to exploit various renewable energy sources (solar thermal, wind, and so on), balance energy demand and supply, increase the overall system efficiency, and reduce the greenhouse gas emissions (International Renewable Energy Agency IRENA, 2013). In building applications, ongoing researches focus on three types of TES technologies: sensible heat storage, latent heat storage, and thermochemical heat storage. Typical limitations on parameters including capacity (kilowatt-hour per ton), power, efficiency, and storage period for TES systems are presented in Table 7.4.

Sensible heat storage is a simple and mature technology and is usually used to provide hot water in building applications. Sensible heat storage systems store the thermal energy by heating or cooling storage materials that are not subject to a phase change during the storage process. The amount of energy stored in the process depends on the temperature change, the mass of storage material, and the specific heat of the material. Materials with the following characteristics can be good candidates for sensible heat storage: high specific heat and density, high thermal conductivity, good thermal diffusivity, long-term stability, low cost, low

Table 7.4 Typical limitations for TES systems.

	Capacity (kWh/ton)	Power (MW)	Efficiency (%)	Storage period
Sensible (hot water)	10−50	0.001−10.0	50−90	days/months
Phase change material (PCM)	50−150	0.001−1.0	75−90	hours/months
Chemical reactions	120−250	0.01−1.0	75−100	hours/days

Source: Data from Hauer, A., 2011. Storage technology issues and opportunities. In: Proceedings of the Strategic and Cross-Cutting Workshop on Energy Storage: Issues and Opportunities, Paris, France.

corrosivity, and low environmental impacts (Li, 2016b). Sensible heat storage materials can be either liquid or solid mediums. Water is the most common liquid material for sensible heat storage because of its low price and high specific heat. Other common liquid materials are thermal oil and organic liquids such as ethanol, propane, butane, and so on (Tatsidjodoung et al., 2013), whereas the common solids used for sensible heat storage include brick, sand, rock, concrete, aluminum, cast-iron, among others.

Latent heat storage, usually refer to phase change storage, is based on the characteristic of some materials that release or absorb heat during the phase change process. These materials are called PCMs. Compared to the sensible heat storage, the advantage of latent heat storage is that it offers higher energy storage density and almost no temperature change during the phase change process. This allows the latent heat storage system to store more energy by using the same amount of materials (Guelpa and Verda, 2019). In addition, the constant temperature during the phase change process in latent heat storage helps to reduce the heat losses substantially compared to an equivalent sensible energy storage (Mazman et al., 2009). PCMs are usually classified into organic, inorganic, and eutectic types. In the real applications, the most widely used materials for latent heat storage are hydrated salts, paraffins, fatty acids, and ice water (for cold storage).

The third technology is thermochemical heat storage. Thermochemical heat storage systems use thermochemical materials to store and release heat in a reversible endothermic/exothermic reaction process (Sarbu and Sebarchievici, 2018). During the charging process, the thermochemical material absorbs heat and generates two products, which will be stored separately. Thermal energy is converted and stored as chemical energy. When heat is required, the reaction products in the charging process are mixed and undergo a reverse reaction; at the same time, heat is released. The advantages of thermochemical heat storage are higher energy density and lower heat losses compared to sensible heat storage and latent heat energy storage. Higher energy density makes thermochemical heat storage systems more compact, whereas lower heat losses make it more suitable for long-term energy storage. Magnesium sulfate, iron carbonate, and iron hydroxide are considered as promising thermochemical materials (Mahlia et al., 2014).

7.3 Conclusion

This chapter presented an introduction to conventional energy systems used in building applications. The energy systems presented in this chapter include energy generation systems, energy conversion systems, and energy storage systems. This chapter reviewed and surveyed the use of the building energy systems mentioned earlier to provide essential and valuable information on these systems for providers and clients of energy services.

References

ASHRAE, 2011. Handbook AF. American Society of Heating, Refrigerating and Air-Conditioning Engineers, Atlanta, GA.

ASHRAE, 2016. ASHRAE Handbook HVAC Systems and Equipment. ASHRAE, Atlanta, GA.

Baker, J.N., Collinson, A., 1999. Electrical energy storage at the turn of the Millennium. Power Eng. J. 13, 107–112. Available from: https://doi.org/10.1049/pe:19990301.

Barnett, A., Rand, J., Hall, R., Bisaillon, J., DelleDonne, E., Feyock, B., et al., 2001. High current, thin silicon-on-ceramic solar cell. Sol. Energy Mater. Sol Cell 66, 45–50. Available from: https://doi.org/10.1016/S0927-0248(00)00157-4.

Chen, H., Cong, T.N., Yang, W., Tan, C., Li, Y., Ding, Y., 2009. Progress in electrical energy storage system: a critical review. Prog. Nat. Sci. 19, 291–312. Available from: https://doi.org/10.1016/J.PNSC.2008.07.014.

Cho, H., Luck, R., Chamra, L.M., 2010. Supervisory feed-forward control for real-time topping cycle CHP operation. J. Energy Resour. Technol. 132. Available from: https://doi.org/10.1115/1.4000920.

Cho, H., Luck, R., Eksioglu, S.D., Chamra, L.M., 2009. Cost-optimized real-time operation of CHP systems. Energy Build. 41, 445–451. Available from: https://doi.org/10.1016/J.ENBUILD.2008.11.011.

Cho, H., Mago, P.J., Luck, R., Chamra, L.M., 2009. Evaluation of CCHP systems performance based on operational cost, primary energy consumption, and carbon dioxide emission by utilizing an optimal operation scheme. Appl. Energy 86, 2540–2549. Available from: https://doi.org/10.1016/J.APENERGY.2009.04.012.

Cho, H., Smith, A.D., Mago, P., 2014. Combined cooling, heating and power: a review of performance improvement and optimization. Appl. Energy 136, 168–185. Available from: https://doi.org/10.1016/J.APENERGY.2014.08.107.

de Winter, F., 1991. Solar Collectors, Energy Storage, and Materials. The MIT Press, Cambridge, MA.

Díaz-González, F., Sumper, A., Gomis-Bellmunt, O., Villafáfila-Robles, R., 2012. A review of energy storage technologies for wind power applications. Renew. Sustain. Energy Rev. 16, 2154–2171. Available from: https://doi.org/10.1016/J.RSER.2012.01.029.

Duffie, J.A., Beckman, W.A., 2013. Solar Engineering of Thermal Processes, fourth ed. John Wiley & Sons;, Hoboken, NJ.

El Chaar, L., lamont, L.A., El Zein, N., 2011. Review of photovoltaic technologies. Renew. Sustain. Energy Rev. 15, 2165–2175. Available from: https://doi.org/10.1016/J.RSER.2011.01.004.

Fang, Y., Yang, F., Zhang, H., 2019. Comparative analysis and multi-objective optimization of organic Rankine cycle (ORC) using pure working fluids and their zeotropic mixtures for diesel engine waste heat recovery. Appl. Therm. Eng. 157, 113704. Available from: https://doi.org/10.1016/J.APPLTHERMALENG.2019.04.114.

Gee, J.M., Schubert, W.K., Basore, P.A. n.d., Emitter wrap-through solar cell. In: Conference Record of the Twenty Third IEEE Photovoltaic Specialists Conference - 1993 (Cat. No.93CH3283-9). IEEE, Louisville, KY, pp. 265−270. doi:10.1109/PVSC.1993.347173.

Guarracino, I., Freeman, J., Ramos, A., Kalogirou, S.A., Ekins-Daukes, N.J., Markides, C.N., 2019. Systematic testing of hybrid PV-thermal (PVT) solar collectors in steady-state and dynamic outdoor conditions. Appl. Energy 240, 1014−1030. Available from: https://doi.org/10.1016/J.APENERGY.2018.12.049.

Guelpa, E., Verda, V., 2019. Thermal energy storage in district heating and cooling systems: a review. Appl. Energy 252, 113474. Available from: https://doi.org/10.1016/J.APENERGY.2019.113474.

Hameer, S., van Niekerk, J.L., 2015. A review of large-scale electrical energy storage. Int. J. Energy Res. 39, 1179−1195. Available from: https://doi.org/10.1002/er.3294.

Hauer, A., 2011. Storage technology issues and opportunities. In: Proceedings of the Strategic and Cross-Cutting Workshop on Energy Storage: Issues and Opportunities, Paris, France.

International Energy Agency (IEA), n.d. Energy efficiency: buildings. https://www.iea.org/topics/energyefficiency/buildings/ (accessed 06.08.19).

International Energy Agency (IEA), 2019. Sloar heat worldwide. https://www.iea-shc.org/solar-heat-worldwide.

International Renewable Energy Agency (IRENA), 2013. Thermal energy storage technology brief. https://www.irena.org/publications/2013/Jan/Thermal-energy-storage (accessed 13.05.2020).

Intersolar, 2019. Global market outlook. https://www.intersolarglobal.com/en/news-app/download-resources/global-market-outlook (accessed 13.05.2020).

Khoukhi, M., Maruyama, S., 2005. Theoretical approach of a flat plate solar collector with clear and low-iron glass covers taking into account the spectral absorption and emission within glass covers layer. Renew. Energy 30, 1177−1194. Available from: https://doi.org/10.1016/J.RENENE.2004.09.014.

Kordesch, K., Simader, G., 1996. Fuel Cells and their Applications. VCH, Weinheim:.

Li, G., 2016a. Organic Rankine cycle performance evaluation and thermoeconomic assessment with various applications, part I: energy and exergy performance evaluation. Renew. Sustain. Energy Rev. 53, 477−499. Available from: https://doi.org/10.1016/j.rser.2015.08.066.

Li, G., 2016b. Sensible heat thermal storage energy and exergy performance evaluations. Renew. Sustain. Energy Rev. 53, 897−923. Available from: https://doi.org/10.1016/J.RSER.2015.09.006.

Liu, M., Shi, Y., Fang, F., 2014. Combined cooling, heating and power systems: a survey. Renew. Sustain. Energy Rev. 35, 1−22. Available from: https://doi.org/10.1016/j.rser.2014.03.054.

Lund, J.W., Boyd, T.L., 2016. Direct utilization of geothermal energy 2015 worldwide review. Geothermics 60, 66−93. Available from: https://doi.org/10.1016/J.GEOTHERMICS.2015.11.004.

Mago, P.J., Chamra, L.M., Srinivasan, K., Somayaji, C., 2008. An examination of regenerative organic Rankine cycles using dry fluids. Appl. Therm. Eng. 28, 998−1007. Available from: https://doi.org/10.1016/J.APPLTHERMALENG.2007.06.025.

Mago, P.J., Luck, R., 2011. Prime mover sizing for base-loaded combined heating and power systems. Proc. Inst. Mech. Eng. A J. Power Energy 226, 17–27. Available from: https://doi.org/10.1177/0957650911406055.

Mago, P.J., Smith, A.D., 2012. Evaluation of the potential emissions reductions from the use of CHP systems in different commercial buildings. Build. Env. 53, 74–82. Available from: https://doi.org/10.1016/J.BUILDENV.2012.01.006.

Mahlia, T.M.I., Saktisahdan, T.J., Jannifar, A., Hasan, M.H., Matseelar, H.S.C., 2014. A review of available methods and development on energy storage; technology update. Renew. Sustain. Energy Rev. 33, 532–545. Available from: https://doi.org/10.1016/J.RSER.2014.01.068.

Mazman, M., Cabeza, L.F., Mehling, H., Nogues, M., Evliya, H., Paksoy, H.Ö., 2009. Utilization of phase change materials in solar domestic hot water systems. Renew. Energy 34, 1639–1643. Available from: https://doi.org/10.1016/J.RENENE.2008.10.016.

McKinsey, 2019. Global energy perspective 2019: reference case. https://www.mckinsey.com/~/media/McKinsey/Industries/OilandGas/OurInsights/GlobalEnergyPerspective2019/McKinsey-Energy-Insights-Global-Energy-Perspective-2019_Reference-Case-Summary.ashx%0A.

McLarnon, F.R., Cairns, E.J., 1989. Energy storage. Annu. Rev. Energy 14, 241–271. Available from: https://doi.org/10.1146/annurev.eg.14.110189.001325.

Morrison, G.L., Budihardjo, I., Behnia, M., 2004. Water-in-glass evacuated tube solar water heaters. Sol. Energy 76, 135–140. Available from: https://doi.org/10.1016/J.SOLENER.2003.07.024.

Neves, R., Cho, H., Zhang, J., 2020. Techno-economic analysis of geothermal system in residential building in Memphis, Tennessee. J. Build. Eng. 27, 100993. Available from: https://doi.org/10.1016/J.JOBE.2019.100993.

Polinder, H., Van Der Pijl, F.F.A., De Vilder, G.-J., Tavner, P.J., 2006. Comparison of direct-drive and geared generator concepts for wind turbines. IEEE Trans. Energy Convers. 21, 725–733. Available from: https://doi.org/10.1109/TEC.2006.875476.

Rahbar, K., Mahmoud, S., Al-Dadah, R.K., Moazami, N., Mirhadizadeh, S.A., 2017. Review of organic Rankine cycle for small-scale applications. Energy Convers. Manag. 134, 135–155. Available from: https://doi.org/10.1016/J.ENCONMAN.2016.12.023.

Sarbu, I., Sebarchievici, C., 2014. General review of ground-source heat pump systems for heating and cooling of buildings. Energy Build. 70, 441–454. Available from: https://doi.org/10.1016/J.ENBUILD.2013.11.068.

Sarbu, I., Sebarchievici, C., 2018. A comprehensive review of thermal energy storage. Sustainability 10. Available from: https://doi.org/10.3390/su10010191.

Self, S.J., Reddy, B.V., Rosen, M.A., 2013. Geothermal heat pump systems: status review and comparison with other heating options. Appl. Energy 101, 341–348. Available from: https://doi.org/10.1016/J.APENERGY.2012.01.048.

Shivashankar, S., Mekhilef, S., Mokhlis, H., Karimi, M., 2016. Mitigating methods of power fluctuation of photovoltaic (PV) sources - a review. Renew. Sustain. Energy Rev. 59, 1170–1184. Available from: https://doi.org/10.1016/j.rser.2016.01.059.

Stanford H.W. III HVAC Water Chillers and Cooling Towers. 2016, CRC Press, Boca Raton, Florida. doi:10.1201/b11510.

Tatsidjodoung, P., Le Pierrès, N., Luo, L., 2013. A review of potential materials for thermal energy storage in building applications. Renew. Sustain. Energy Rev. 18, 327–349. Available from: https://doi.org/10.1016/J.RSER.2012.10.025.

Tawfiq, K.B., Abdou, A.F., EL-Kholy, E.E., Shokrall, S.S., 2016. Application of matrix converter connected to wind energy system. In: Eighteenth International Middle East Power Systems Conference, Cairo, Egypt. IEEE,Piscataway, NJ, pp. 604−609. doi:10.1109/MEPCON.2016.7836954.

Tawfiq, K.B., Mansour, A.S., Ramadan, H.S., Becherif, M., El-Kholy, E.E., 2019. Wind energy conversion system topologies and converters: comparative review. Energy Procedia 162, 38−47. Available from: https://doi.org/10.1016/j.egypro.2019.04.005.

Tchanche, B.F., Lambrinos, G., Frangoudakis, A., Papadakis, G., 2011. Low-grade heat conversion into power using organic Rankine cycles − a review of various applications. Renew. Sustain. Energy Rev. 15, 3963−3979. Available from: https://doi.org/10.1016/J.RSER.2011.07.024.

Tian, Y., Zhao, C.Y., 2013. A review of solar collectors and thermal energy storage in solar thermal applications. Appl. Energy 104, 538−553. Available from: https://doi.org/10.1016/J.APENERGY.2012.11.051.

Tripanagnostopoulos, Y., Souliotis, M., Nousia, T., 2000. Solar collectors with colored absorbers. Sol. Energy 68, 343−356. Available from: https://doi.org/10.1016/S0038-092X(00)00031-1.

U.S. Department of Energy (DOE), 2013. Solar photovoltaic cell basics. https://www.energy.gov/eere/solar/articles/solar-photovoltaic-cell-basics (accessed 05.07.19).

U.S. Department of Energy (DOE), 2019. Geothermal heat pumps. https://www.energy.gov/eere/geothermal/geothermal-heat-pumps (accessed 15.07.19).

U.S. Environmental Protection Agency, 2017. Combined Heat and Power Partnership. Catalog of CHP Technologies. Washington, DC.

Wang, E.H., Zhang, H.G., Fan, B.Y., Ouyang, M.G., Zhao, Y., Mu, Q.H., 2011. Study of working fluid selection of organic Rankine cycle (ORC) for engine waste heat recovery. Energy 36, 3406−3418. Available from: https://doi.org/10.1016/J.ENERGY.2011.03.041.

Wang, S.K., 2000. Handbook of Air Conditioning and Refrigeration, second ed. McGraw-Hill, New York.

Wikipedia, 2019a. Gas turbine. https://en.wikipedia.org/wiki/Gas_turbine (accessed 03.07.19).

Wikipedia, 2019b. Reciprocating engine. https://en.wikipedia.org/wiki/Reciprocating_engine (accessed 01.07.19).

Wikipedia, 2019c. Steam turbine. https://en.wikipedia.org/wiki/Steam_turbine (accessed 03.07.19).

Wu, D.W., Wang, R.Z., 2006. Combined cooling, heating and power: a review. Prog. Energy Combust. Sci. 32, 459−495. Available from: https://doi.org/10.1016/j.pecs.2006.02.001.

Wu, R., 2009. Energy efficiency technologies−air source heat pump vs. ground source heat pump. J. Sustain. Dev. 2, 14−23.

Yang, F., Cho, H., Zhang, H., Zhang, J., 2017. Thermoeconomic multi-objective optimization of a dual loop organic Rankine cycle (ORC) for CNG engine waste heat recovery. Appl. Energy 205, 1100−1118. Available from: https://doi.org/10.1016/J.APENERGY.2017.08.127.

Yang, F., Dong, X., Zhang, H., Wang, Z., Yang, K., Zhang, J., et al., 2014. Performance analysis of waste heat recovery with a dual loop organic Rankine cycle (ORC) system for diesel engine under various operating conditions. Energy Convers. Manag. 80, 243−255. Available from: https://doi.org/10.1016/J.ENCONMAN.2014.01.036.

Zhang, J., Cho, H., Knizley, A., 2016. Evaluation of financial incentives for combined heat and power (CHP) systems in U.S. regions. Renew. Sustain. Energy Rev. 59, 738−762. Available from: https://doi.org/10.1016/j.rser.2016.01.012.

Zhang, J., Cho, H., Luck, R., Mago, P.J., 2018. Integrated photovoltaic and battery energy storage (PV-BES) systems: an analysis of existing financial incentive policies in the US. Appl. Energy 212, 895−908. Available from: https://doi.org/10.1016/j.apenergy.2017.12.091.

Zhang, J., Knizley, A., Cho, H., 2017. Investigation of existing financial incentive policies for solar photovoltaic systems in U.S. regions. AIMS Energy 5, 974−996. Available from: https://doi.org/10.3934/energy.2017.6.974.

Zhang, J., Zhang, H., Yang, K., Yang, F., Wang, Z., Zhao, G., et al., 2014. Performance analysis of regenerative organic Rankine cycle (RORC) using the pure working fluid and the zeotropic mixture over the whole operating range of a diesel engine. Energy Convers. Manag. 84, 282−294. Available from: https://doi.org/10.1016/j.enconman.2014.04.036.

Energy systems in buildings

Getu Hailu
Department of Mechanical Engineering, University of Alaska Anchorage, Anchorage, AK, United States

Chapter Outline

8.1 Introduction 181
8.2 Energy-efficient building envelopes 182
    8.2.1 Increasing thermal resistance of the building envelope 182
    8.2.2 Climate-specific design of energy-efficient envelopes 183
8.3 Renewable energy sources for building energy application 183
    8.3.1 Analyzing electrical/thermal loads of a building 184
    8.3.2 Consideration of local codes and requirements for renewable energy systems 184
    8.3.3 Solar energy systems 184
    8.3.4 Building-integrated photovoltaic systems 188
8.4 Solar thermal energy storage 191
    8.4.1 Types of thermal energy storage technologies 192
8.5 Wind energy 196
    8.5.1 Brief introduction 196
    8.5.2 Wind resource assessment 197
    8.5.3 Building-integrated/mounted wind turbine 197
    8.5.4 Optimizing building-integrated/mounted wind turbine devices 198
    8.5.5 Small/micro wind turbines for building application 199
8.6 Heat pumps 199
    8.6.1 Air-source heat pumps 200
    8.6.2 Ground-source heat pumps 201
    8.6.3 Working principles of heat pumps 202
    8.6.4 Performance measures 204
8.7 Biomass 204
8.8 Summary 205
References 205

8.1 Introduction

According to the 2017 global status report, building sectors consumed nearly 125 EJ[1] in 2016, or 30% of total final energy use (Dean et al., 2016). Building construction, including the manufacturing of materials for building such as steel and cement,

---

[1] EJ is equal to $1.0E + 18$ J.

Energy Services Fundamentals and Financing. DOI: https://doi.org/10.1016/B978-0-12-820592-1.00008-7
© 2021 Elsevier Inc. All rights reserved.

accounted for an additional 26 EJ (nearly 6%) in estimated global final energy use (Dean et al., 2016). As buildings consume considerable amount of energy, they are widely regarded as major contributors to the greenhouse gas (GHG) emissions. Reducing building energy consumption from fossil fuels, consequently, leads to reducing GHG emission. Many countries around the world have set target dates for substantially reducing the use of fossil fuel-based energy resources in the building sector. They aim to meet substantial part of building energy consumption from renewable sources. The approaches and technologies that help meet these goals include:

- the use of energy-efficient building envelopes,
- appropriate building orientation,
- daylighting,
- passive solar heating,
- efficient heating, refrigerating, ventilating, and air-conditioning (HVAC) and lighting,
- renewable energy utilization such as solar thermal, solar electric [photovoltaics (PVs)], wind, and geothermal energy.

In the following sections, use of these technologies in the building sector is discussed. The focus of this chapter is to give an overview of these technologies, their current status, their advantages, and disadvantages.

8.2 Energy-efficient building envelopes

A building envelope is a physical barrier between the external environment and the internal conditioned space, keeping the residents comfortable. A building envelope consists of fenestration (doors and windows), roofs, walls, and insulations. Since a building envelope separates the unconditioned exterior environment from the conditioned interior space, it is one of the key factors that impact building energy consumption. Building envelopes of energy-efficient buildings are not simply barriers between interior and exterior; they are building systems that create comfortable spaces by actively responding to the building's external environment, and substantially reduce the buildings' energy consumption (Aksamija, 2015). Energy-efficient building envelopes:

- have high thermal resistant materials in the facade of the building,
- use vapor barriers and are effective in vapor control,
- have efficient window and door seals,
- have effective airflow control to minimize infiltration of outdoor air.

8.2.1 Increasing thermal resistance of the building envelope

Increasing the thickness of the insulating material is the simplest way to increase the thermal resistance of a facade. The method was very popular, especially in Northern Europe. This is evident from the fact that the thickness of insulation in building has increased since the early 1970s, almost doubling in Northern Europe (Papadopoulos, 2005). Increasing the thickness of insulation results in increased

Table 8.1 Climate-specific design of energy efficient envelopes.

Climate type	Design strategies for energy-efficient facades
Heating-dominated	• The building envelope receives solar heating. • Walls can be used as thermal masses for thermal storage. • Better insulation to minimize thermal losses. • Natural daylight is used. Facades have increased glazing areas to allow for natural light; light shelves that redirect light into interior spaces are used.
Cooling-dominated	• Appropriate shading techniques can be employed to protect from direct solar gain. • Use insulation to reduce solar heat gain. • Design to facilitate natural ventilation (wing walls). • Natural daylight should be used in such a way that solar heat gain is minimized.
Mixed climates	• Use shading devices to protect facade from direct solar radiation during warm days. • Use passive solar design for heating during cold seasons. • Use natural daylight and with increased glazed areas of walls with shading devices.

cost. Walls with thermal insulation also have a higher possibility of surface condensation when the relative humidity of ambient air is greater than 80%, provided the convective and radiative heat transfer coefficients of the exterior wall are small (Sadineni et al., 2011). This is undesirable because it promotes microbial growth raising health issues, as well as reducing the life of building envelope.

8.2.2 Climate-specific design of energy-efficient envelopes

Other methods of attaining high energy efficiency of building envelopes and thermal comfort level take into account the local climate (Oral et al., 2004). Strategies that work in dry and cold climates may not necessarily work in humid and hot regions. Researchers have developed design strategies for different locations. One of such strategies (Aksamija, 2015) is summarized in Table 8.1.

8.3 Renewable energy sources for building energy application

Commonly used renewable energy resources for building energy application are solar, wind, geothermal, and biomass. There are several factors that need to be considered in selecting the possible renewable energy source for building application. According to US DOE (United States Department of Energy), planning for a home renewable energy system is a process that includes analyzing the existing load (heating, cooling, and electricity), understanding the local codes and requirements, deciding if the systems should be on or off of the electric grid, and understanding the technology options that are available for the particular location of the building (US DOE).

8.3.1 Analyzing electrical/thermal loads of a building

The first step in any case is determining the load (electrical, heating, and cooling) of the building. For example, in case of electricity load, a thorough examination of electricity needs to be done. That includes conducting a load analysis such as recording the wattage and average daily use of all the electrical devices that are plugged into a central power source such as televisions, refrigerators, lights, and power tools. Some loads use electricity all the time. A refrigerator is one such load. A power tool uses electricity intermittently. Selectable loads are such loads that use electricity intermittently. If selectable loads are to be used only during the times of extra power availability, a smaller renewable energy system may be enough.

8.3.2 Consideration of local codes and requirements for renewable energy systems

Each country, state, local government, and municipality may have their own sets of regulations and codes that need to be followed. These regulations and codes can affect the type of renewable energy system to be installed and who installs it. The regulations and codes may also affect whether the intended renewable energy system needs to be grid tied or can be used as a stand-alone system. Some of the requirements include building codes, local agreements and regulations, and technology-specific requirements. Municipalities require that a desired system complies with the industry standard is safe and does not pose danger to the neighbors. In many cases, it may also be necessary to check for building codes that may require that the zoning board grants a conditional use permit or a variance from the existing code before a permit is issued. Furthermore, it may be necessary that the system to be installed is composed of certified equipment, and that it obeys local requirements and appropriate technical industry standards (US DOE). Some neighborhoods and communities have regulations specifying what homeowners can and cannot do with their properties. Sometimes these regulations forbid the use of renewable energy systems for esthetic or noise-control purposes. However, sometimes these regulations have provisions, supporting the renewable energy systems.

8.3.3 Solar energy systems

Solar energy systems for building applications include solar PV systems and solar thermal systems. Solar PV system is direct conversion of sunlight into electrical energy by solar PV panels. Solar PV systems can be applied to both small residential and large buildings such as offices. Solar thermal systems are used to produce heat from the sun's thermal energy for the purpose of heating and cooling a building. Solar thermal or heating systems can be air-based or water-based. Water-based solar thermal systems are preferred because of the high heat capacity of water.

8.3.3.1 Solar water heating

Solar water heating (SWH) accounts for a large portion of energy use at residential, commercial, and office buildings. SWH systems can be used effectively at facilities that have south-facing roof or nearby unshaded grounds for installation of a solar thermal collector. Almost any type of building can take advantage of SWH systems including swimming pools, residential buildings, hotels, laundries, hospitals, and restaurants. SWH systems can be active or passive. Passive SWH systems depend on heat-driven convection to circulate the heating fluid in the system. Active systems depend on the use of one or more pumps to circulate the working fluid in the system. Active SWH systems can broadly be classified into two categories: direct (open loop) circulation and indirect (closed loop) circulation systems. In direct circulation systems, water is circulated through the solar thermal collector and into the home for direct use (Fig. 8.1). These systems are good for locations without freezing temperatures. In places where freezing occurs, a nonfreezing fluid (such as glycol and water mixture) is circulated and the heat is transferred through heat exchangers to the usable water. An active SWH system consists of a solar thermal collector array, an energy transfer system (the fluid), and a storage tank. Thermal storage is generally required to couple the timing of the intermittent solar resource with the timing of the hot water load. In general, 1−2 gal of storage water per square foot of solar thermal collector area is adequate (Walker, 2016).

Solar thermal collector is the main component of an active SWH system. It absorbs the solar radiation and converts it to thermal energy, which heats the energy transfer fluid. The efficiency of an active SWH system is highly dependent on the

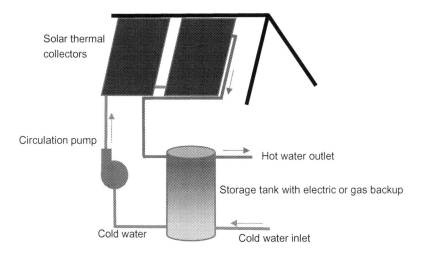

Figure 8.1 An active solar water heating system.

effectiveness of the solar thermal collector. There are two types of solar thermal collectors that are applicable to buildings: flat-plate and evacuated tube collectors.

8.3.3.1.1 Flat-plate collectors
A typical flat-plate solar thermal collector consists of glazing covers, absorber plates, insulation layers, and recuperating tubes (filled with heat transfer fluids). Single or multiple sheets of glass or other materials with high transmissivity of short-wave radiation and low transmissivity of long-wave radiation can be used as glazing material. The glazing is intended to minimize convection losses from the absorber plate and also reduce irradiation losses from the collector due to the greenhouse effect (Tian and Zhao, 2013). The absorber plate is usually coated black so that it can absorb as much heat as possible. Further thermal performance improvements have been achieved through honeycomb insertion, which is made of transparent material and is placed in the airspace between the glazing and the absorber, leading to heat loss reduction (Tian and Zhao, 2013). Other performance enhancements in flat-plate collectors that have been reported in the literature include replacing a solar water heater metal absorber plate with tar covering water tubes (Ammari and Nimir, 2003). It was found that the tar-based collector performed better than the conventional flat-plate collector in the late afternoon and evening hours.

8.3.3.1.2 Evacuated tube solar thermal collectors
Evacuated tube solar thermal collectors have excellent thermal performances and much more higher efficiencies than flat-plate collectors (Jamar et al., 2016; Morrison et al., 2004; Zubriski and Dick, 2012). They can collect both direct and diffuse radiations. Evacuated tube solar thermal collectors consist of a heat pipe filled with a liquid inside a glass enclosure (Fig. 8.2). The thermal energy from the sun is captured, and the heat is transferred to the working fluid while undergoing a phase change: evaporation and condensation cycles.

Zubriski and Dick (Zubriski and Dick, 2012) evaluated performance of evacuated tube solar collectors under various operating conditions. They examined different collector configurations, under Canadian climatic conditions, and found that the efficiency varied by approximately 5% in general. Vertical installations were found to be slightly more efficient than others during winter months, and slightly less during summer. Martínez-Rodríguez (Martínez-Rodríguez et al., 2018) developed preliminary design approach for a network of evacuated tube solar collector networks that can be used at early design stages for assessing the cost benefit of the integration of solar energy into low energy intensity processes. They concluded that the number of solar collectors required to achieve the thermal targets depends on the ambient conditions chosen for the design. Larger surface areas are required at lower solar radiation intensities and lower inlet temperatures. Innovative approaches to the design of evacuates tube solar collectors include those proposed by Alvarez et al. Alvarez et al. (2004), who developed and tested a single-glass air solar collector with an absorber plate made of recyclable aluminum cans. The aim was

Energy systems in buildings

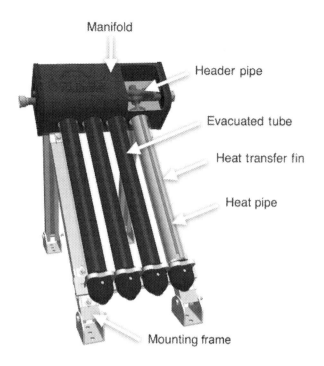

Figure 8.2 Evacuated tube solar thermal collectors (Evacuated Tube Solar Collectors, 2019).

production cost reduction. They reported that the maximum efficiency reached was 74% for an air solar collector with an absorber plate made of recyclable aluminum cans.

8.3.3.1.3 Choice of solar thermal collectors
The choice whether to install flat or evacuated tube solar collector depends on many factors: the major ones being cost, performance, and easiness in installation.

8.3.3.1.3.1 Cost In many literatures, it has been reported that flat-plate collectors are less costly (20%−40% less). Cost should be considered in terms of cost per kW (BTU) capacity and year-round performance. When such comparison is done, evacuated tube collectors will have a lower cost per kW (BTU) capacity in colder climates.

8.3.3.1.3.2 Performance It has been reported that generally, evacuated tubes perform better in colder and/or cloudier conditions than their fla-platet solar thermal collectors (Solar Panels Plus, 2014). This is because of the vacuum in the glass tube, which allows evacuated tube collectors to retain a high percentage of collected heat (Solar Panels Plus, 2014). Evacuated tube solar collectors do not corrode with the same level as flat-plate solar collectors do. There are contradicting reports regarding snow shading abilities of evacuated tube solar collectors, some reporting

Figure 8.3 Evacuated tube solar thermal collectors on rooftop. Notice the lack of snow on the evacuated tubes after this heavy March snowfall (Simple Solar).

that evacuated tube solar collectors are better in shading snow (Simple Solar) (Fig. 8.3) and others reporting not so (Vijayakumar et al., 2017).

The temperature range of flat-plate collectors is between 170°F and 180°F. On the other hand, evacuated tube solar collectors can work well over 250°F. This leads to a risk of overheating of the collectors in warmer climates. Hence, careful approach should be taken not to undersize storage tanks. Because of the curved shape of the tubes, evacuated solar collectors are less sensitive to sun angle and orientation than the flat-plate solar collectors. Their curved design allows sunlight to pass at an optimal angle throughout the day enhancing their performance.

8.3.3.1.3.3 Installation In terms of easiness for handling and installation, evacuated tubes are generally lighter and easy to handle. However, they are more fragile than flat-plate collectors. Flat-plate collectors on the other hand are heavier and require more space for installation.

8.3.4 Building-integrated photovoltaic systems

PV modules have been used for electricity generation since the 1960s. Nevertheless, the efficiency of PV modules is reduced by heat, resulting in poor electrical performance (Devarakonda and Mil'shtein, 2012; Zhang et al., 2012). Circulating a fluid behind PV modules improves their electrical efficiencies by removing the heat. A hybrid photovoltaic/thermal (PV/T), working in such a way not only improves the electrical performance of the PVs, but also provides thermal energy that can be used for space heating and other purposes. Because of this added functionality of producing heat, PV/T systems have been studied since the mid-1970s (Florschuetz, 1975; Wolf, 1976). The goal of early studies (Florschuetz, 1975, 1979; Wolf, 1976; Michael et al., 2015; Hendrie, 1979) was validating the concept of PV/T theoretically and experimentally. Over the last 40 years, PV/T systems have been significantly researched by many investigators, mainly to improve their design, performance, and cost-effectiveness (Zondag, 2008; Raghuraman, 2010; Skoplaki and Palyvos, 2009; Tripanagnostopoulos, 2007; Braunstein and Kornfeld, 1986; Lalović et al., 1986; O'Leary and Clements, 1980; Mbewe et al., 1985; Joshi et al., 2009; Chow, 2010). According to a literature review by Chow (2010), effective use of PV/T systems is highly dependent on geographical location. At locations with low ambient temperatures and solar radiation; space heating is required most of the year; hence, PV/T with air circulation can be useful and cost-effective (Chow, 2010). In conclusion, a PV/T system improves electrical performance of the PVs in addition to providing the possibility of extracting heat energy. As a result, a shorter payback time than the PV system alone is possible. The payback time can further be shortened by integrating a PV/T system to a building structure such as roof or facade. Such a system is known as building-integrated photovoltaic thermal (BIPV/T) system. BIPV/T systems are arrays of PV panels, which are integral parts of the building structures such as roofs and facades, which can produce both thermal and electrical energy (Chen et al., 2007, 2010a,b; Dembo, 2010). BIPV/T systems produce heat and electricity to further offset or eliminate fossil fuel demand in buildings. In most current design, PV/T systems are treated as separate and distinct systems from the building envelope showing a lack of system integration. On the other hand, BIPV/T systems are integrated into the building envelope so that the assembly replaces elements of the facade and/or the roof, and thereby reduce overall cost. Like PV/T system, heat recovery in BIPV/T systems is accomplished by fluid circulation behind the array of PV modules (Fig. 8.4). Integration of BIPV/T systems can be easily applied at the design stage of a new building (Chen et al., 2007, 2010a,b; Safa et al., 2015; Hailu et al., 2015; Kamel et al., 2015; Getu et al., 2014; Athienitis et al., 2014).

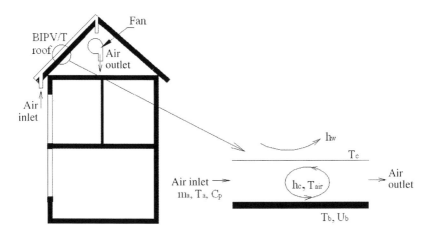

Figure 8.4 Schematic of the roof-integrated PV/T and the thermal energy recovery system.

Fig. 8.4 illustrates a BIPV/T system comprised of PV modules mounted on a building roof, with a gap in between the roof and the PV to create a channel. Air is circulated by a fan behind the PV modules, providing the PV cooling. The warm air is then directed to indoor for further use. A real example of BIPV/T system application is the ÉcoTerra house in Québec, Canada. The ÉcoTerra house, which incorporates a BIPV/T system, consumes only 26.8% of a typical Canadian home. The ÉcoTerra house (Fig. 8.5) couples a BIPV/T system with a geothermal heat pump (Doiron et al., 2011). To summarize, a BIPV/T system, circulating air behind the PVs has multiple advantages. It decreases the temperature of the PVs and improves their electrical efficiency, increases the life of the PV by minimizing the tendency of the modules to delaminate, and helps recover heat energy. This heat energy can directly be used for space and/or domestic water heating. Another use of the warm air collected from behind the PV modules is that it can be fed to an air-source heat pump (ASHP) to improve performance. As a result, coupling ASHPs with BIPV/T systems can further reduce heating and cooling costs of a building and dependence on nonrenewable heating sources.

BIPV/T systems can also meet all building envelope requirements, such as thermal insulation and mechanical resistance (Chen et al., 2010a). Because of this, BIPV/T systems have the potential of improving cost-effectiveness of residential buildings compared to add-on PV/T systems. The thermal coupling of BIPV/T system with a heat pump leads to further improvements in the thermal efficiency of the heat pump. Compared to ground-source heat pumps (GSHPs), which are popular in colder climates, ASHPs offer the alternative of being of low cost, with a 40% reduction in installation cost (Safa et al., 2015). However, their coefficient of performance decreases in colder outdoor temperatures (Bertsch and Groll, 2008). For this reason, a large capacity ASHP may be necessary during the heating season to

Figure 8.5 A photograph of ÉcoTerra house as seen from the southwest, Eastman, Québec, Canada (Doiron et al., 2011).

meet the required building heating demand. The compressor of such a large capacity ASHP generally operates at part loads to meet the building demand at milder winter temperatures. This on−off part load operation causes a reduction in efficiency. A BIPV/T system can provide heat inputs that would allow for the necessary "pretreatment" of the outdoor air so that efficient operation of the ASHP is possible even at very low outdoor winter temperatures. Two-stage variable capacity ASHPs (TS VC ASHPs) offer potential improvements in the efficiency and reliability of operation. These improvements result from enhanced performance at lower capacities and a reduction in cyclic operating time. The improved performance of TS VC ASHPs in mild winter temperatures can make it an attractive sustainable technology. In conclusion, by providing heating and electricity production, a BIPV/T system coupled with TS VC ASHP is expected to ensure thermal comfort and low environmental impact. The system can bring existing and new buildings closer to the goal of net-zero energy status in an integrated, cost-effective, and environment-friendly manner.

8.4 Solar thermal energy storage

Solar thermal energy storage is not an ancient concept. Caves have been used by early humans because they maintain almost uniform temperature throughout the year with almost negligible temperature variations with season. Currently, there is huge research initiative in the thermal energy storage area. Some of the thermal energy storage technologies have been fully developed and commercialized, some are under demonstration and deployment stage, and some are still being researched and developed. Fig. 8.6 shows stages of different thermal energy storage technologies (International Energy Agency, 2014).

There are different types of thermal storages, varying in size and type. Ranging from simple, small water tank storages to large borehole storages, the aim is to store

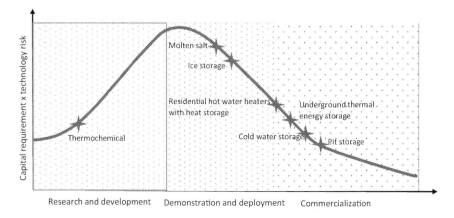

Figure 8.6 Maturity of thermal energy storage technologies.

thermal energy for short-term or long-term use. Large, borehole and aquifer thermal energy storages have been applied for both heating and cooling in Canada, Germany, the Netherlands, Norway, and Sweden. In the United States, an estimated 1 GW of ice storage has been deployed to reduce peak energy consumption in areas with high numbers of cooling degree days (Adamson and O'Donnell, 2012). The Drake Landing Solar Community in Okotoks, Canada is the first major implementation of borehole seasonal thermal energy storage in district heating in North America. It is also the first system of this type designed to supply more than 90% space heating with solar energy and the first operating system in such a cold climate (Gao et al., 2015).

8.4.1 Types of thermal energy storage technologies

In thermal energy storage technology, heat is transferred to a suitable medium and stored in it for later use. The use can be either heating, cooling applications and/or power generation. In seasonal thermal energy storage system, energy is stored for days, weeks, or months to compensate for a periodic variability of the source of heat, such as solar or wind. Solar and wind are intermittent, and storage is necessary to meet the demand without interruption, for example, storing heat in the summer for use in the winter via underground thermal energy storage systems (International Energy Agency, 2014). There are attractive advantages and reasons in storing thermal energy. There is a possibility of achieving increased overall efficiency, higher reliability, cost-effectiveness, and less GHG emission to the environment (Dincer, 2010). Consideration in selecting thermal energy storage includes required storage duration, that is, seasonal diurnal, cost, available energy source, and operating conditions. The questions that should be answered in selecting the appropriate thermal energy storage include (Duffie and Beckman, 2013):

1. What is the heat capacity of the thermal storage per unit volume?
2. What is the temperature range over which the thermal storage operates? What is the temperature at which heat is added to and removed from the thermal energy storage system?
3. How is heat added or removed from the thermal energy storage?
4. What are the temperature differences associated with removing and adding the heat?
5. How does the temperature stratification in the storage unit look like?
6. What is the power requirement for adding or removing heat?
7. What type of thermal storage tanks, containers, or other structural elements is needed?
8. What are the means of controlling thermal losses from the storage system?
9. How much does it cost?

8.4.1.1 Sensible heat storage system

The most common type of thermal energy storage is sensible heat storage (SHS) system. Liquids and solids can be used as a sensible thermal storage medium. Solids such as sand, rock, clay, earth, and liquids such as oil and water have been used as sensible thermal storage mediums. In SHS system, the temperature of the medium is either increased or decreased by adding heat or removing heat. Heat is removed from the storage medium whenever required to meet the load requirement, such as for space heating or for domestic hot water use. In all the cases, SHS medium consists of an insulated container, which is an enclosure to contain the storage medium, the heat storage medium and a pump or a fan for moving the heat transfer fluid for the purpose of adding and removing heat. In sensible hot storage mediums, heat is added for the purpose of increasing the temperature of the storage medium. In sensible cold storage mediums, heat is removed for the purpose of lowering the temperature of the storage medium. The amount of energy stored is proportional to the mass of the storage medium, the medium's heat capacity, and the difference between the storage medium's input and output temperatures (Dincer, 2002). The heat loss from sensible thermal storages is directly proportional to the temperature difference between the environment and the storage. It is important to consider the rate at which heat can be released and extracted, which depends on the thermal conductivity of the thermal storage material (Shah, 2018). An effective thermal energy storage has high energy density (i.e., high density and specific heat) and good thermal conductivity. The thermal conductivity of the thermal storage for residential applications is usually above 0.3 W/m K. The ability of storing heat in a given container depends on the value of the thermal capacity, ρC_p (Dincer, 1999). An estimated rule of thumb for sizing is to use 300–500 kg of rock per square meter of solar thermal collector area for space heating applications (Shah, 2018). Rock or pebble-bed storages can also be used for much higher temperatures, up to 1000°C (Shah, 2018).

8.4.1.1.1 Sensible solid heat storage system

There are several advantages of using solid mediums over liquids for SHS system. Solids have higher temperature changes as compared to liquids, solids do not melt, therefore do not flow, and there is no leakage from the container. Fig. 8.7 illustrates basic components and utilization of solid sensible thermal storage systems. In the system, a solid sensible thermal storage is used to seasonally store thermal energy. The thermal storage itself is situated underneath a garage floor, and it contains fine sand and pit-run gravel as a thermal storage medium. The sand bed was insulated underneath with a polystyrene foam, which has a thermal resistivity of RSI-5.64 (US R-32). The four sides of the sand thermal storage are also insulated with polystyrene foam board on both sides of a poured concrete foundation wall. The heat to be stored comes from evacuated tube solar thermal collectors. The evacuated solar thermal collectors heat a water–glycol mixture that, during normal operation, passes through a heat exchanger to heat domestic hot water. When the domestic hot water tank is not calling for heat, the excess heat is directed to the thermal storage under the garage floor for heating the space (Hailu et al., 2017). The system has dual purpose: heating the garage by radiation and convection and heating domestic water.

8.4.1.1.2 Sensible liquid heat storage system

One limitation in liquid thermal energy storage systems is that the temperature range that can be reached is limited by their boiling points. This determines what

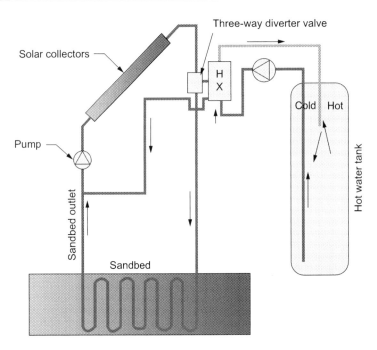

Figure 8.7 Schematic of the evacuated tube solar collectors and the thermal storage (Hailu, 2019).

the type of liquid to be used as a storage medium. Water, having high specific heat, is the most common liquid storage medium, if the temperature requirement is below 100°C. Confined underground water such as aquifers is cost-effective for large-scale thermal storage. In naturally occurring aquifers, the existing cold groundwater is displaced by pumping hot water for storage. This kind of approach can lower the cost of the thermal storage as the only investment required is the cost of drilling openings for injecting and collecting water. If there is a desire to use water for higher temperature applications, (temperatures above 100°C), it must be pressurized. It must be noted that this adds to the initial cost. For such cases, the limitation of water is the critical point, that is, 374°C (Dincer, 2002). Other liquid storage mediums include organic liquids and high molecular weight oils, which are also effective at higher temperature. Although oils such as Terminol can be used without pressurization in the range of −10°C to 320°C, they have the drawbacks of having low specific heat (2.3 kJ/kg K as compared to 4.19 kJ/kg K for water). Another drawback of oils is that they are liable to high-temperature cracking, polymerization, and formation of volatile products.

8.4.1.2 Sensible cold storage system

Heat is removed from the storage medium in sensible cold storage systems, lowering their temperature. Low operating costs can be achieved by using cheaper electricity rates during off-peak hours. Air-conditioning systems can take advantage of heat sinks that can be used as cold storages to which heat is rejected. Chillers can be coupled to cold storages for better performance and efficiency, although the initial investment cost can be higher as compared to conventional air-conditioning systems without cold storage (Hailu, 2019). When water is used as cold storage medium, a large amount is needed. This is because its useful temperature is limited as compared to when it is used as sensible hot storage medium.

8.4.1.3 Latent heat storage system

Latent heat storage (LHS) systems involve the storage of energy in phase change materials (PCMs). Thermal energy is stored and released when the storage material changes phase. LHS systems are advantageous in that they are usually compact. For a given amount of heat storage, the volume of PCM is significantly less compared to the volume of SHS material. This is important because it leads to the use of less insulation material and applicability in places where there is less space. Additional advantages of PCMs include ability to store large amount of heat under small temperature changes, ability to be applied under strict working temperature as they can work under isothermal conditions and having high storage density. LHS systems have about five to ten times higher storage densities as compared to SHS systems. In addition, the volume of PCM storage is two times lesser than that of water (Shah, 2018). Phase change can be from solid to liquid, from solid to gas, from solid to solid, from liquid to gas, and vice versa. In case of solid-to-solid phase change, heat is stored as the material transitions from one crystalline arrangement to another. This type of phase change, that is, solid-to-solid transition has low latent heat. Higher latent heat release and higher volume change are associated with solid-to-gas and liquid-to-gas transformations. However, the large change in volume can be an issue as a massive container is required. In some cases, this makes the system more complex and impractical. Because of the reasons mentioned earlier, the most desirable phase change is the transition from solid to liquid because it is associated with a small change in volume, even though solid-to-liquid transitions have low latent heat compared to liquid-to-gas. Solid-to-liquid PCMs are also cost-effective as thermal energy storage mediums. PCMs themselves cannot be used as heat transfer mediums. A separate heat transfer mediums must be used with a heat exchanger in between to transfer energy from the source to the PCM and from PCM to the load. Attention should be given to the heat exchanger design in view of the low thermal diffusivity of PCMs in general. The general requirement is that the PCM container must be compatible with the PCM and be able to handle changes in volume.

8.4.1.4 Thermochemical storage

In thermochemical storages, reversible chemical reactions occur. In principle, let us assume that heat is added to a material Z. Material Z absorbs energy and is converted into two materials X and Y. X and Y can be stored separately. The reverse reaction, that is, combining X and Y in order to form Z leads to the release of heat. This reversible reaction can be symbolically written as follows:

$$Z + heat \leftrightarrow X + Y$$

This released energy constitutes the recovered thermal energy from the thermochemical energy storage. Currently, there is significant research in the area of thermochemical energy storage due to the possibility to achieve energy storage densities of 5−20 times greater than sensible storage (International Energy Agency, 2014). However, thermochemical energy storage technology is still in its R&D stage, and more research and testing is required before it becomes available on the market to be implemented in the building sector.

8.5 Wind energy

8.5.1 Brief introduction

Another old technology that has seen revival due to the environmental concerns is technology associated with wind energy. Wind turbines convert the kinetic energy in the wind to mechanical energy and then to useful electricity. One of the major challenges associated with electricity production using wind is intermittency, that is, wind power plants generate electricity only when the wind is available within a range of wind velocity. This means that wind power plants need a backup, adding cost and complexity to system. Accurate wind speed and direction forecasting plays a major role in minimizing the impacts associated with intermittency of the wind on the electricity production. Another major challenge is integration f electricity produced by wind farms into the grid. This is because wind farms are located in remote areas, requiring, in some cases, construction of transmission lines. These challenges make installing wind turbines without costly incentives.

Wind turbines are classified broadly into horizontal-axis wind turbines (HAWTs) and vertical-axis wind turbines (VAWTs) based on their configurations. In HAWTs, the rotating axis of the wind turbine is horizontal, that is, parallel to the ground. HAWTs dominate most of the wind energy industry. Both, HAWTs and VAWTs have their own advantages and disadvantages. In general, HAWTs have the advantage of having variable pitch, tall tower exposed to higher wind speeds (high energy production) and high efficiency. On the other hand, HWATs have the disadvantage of being big and heavy, having tall towers and large blades thus difficult to transport, challenging to install. In case of HAWTs, main components are installed atop of tower, posing difficulty for maintenance and high visibility. In addition, a yaw control may be necessary. In VAWTs, the rotational axis of the turbine is vertical, that is, perpendicular to the ground. VAWTs are primarily used in small wind projects and residential applications. VAWTs have the advantage of being smaller, not needing yaw control, having their generator components located on the ground and having less noise compared to HAWTs. VAWTs can also take advantage of higher wind speeds produced by local structures and geography. The disadvantage of VAWTs is that they are generally exposed to lower wind speeds due to shorter structure.

8.5.2 Wind resource assessment

Wind resource maps are available that can help determine if an area of interest should be further explored or not. Care should be taken in doing so because wind resource at a microlevel can vary significantly. It is important to evaluate the specific area of interest before deciding to invest in wind systems. Before making the decision of installing a wind turbine, wind resource assessment for the specific site should be made based on a potential to produce electricity over an annual basis (Hayter and Kandt, 2011). Table 8.2 gives classification of wind resources. This table helps select an appropriate type of wind turbine. For example, for a class 3 wind resource a small wind turbine (100 kW or less), low wind speed turbine can be considered (Hayter and Kandt, 2011). If the site has a class 4 or greater wind resource, the site may be useful for larger, utility-scale wind turbines (Hayter and Kandt, 2011). Wind turbines installed at locations with lower wind resources are less likely to be cost-effective, but should be studied if the site is in a class 2 area and there are nearby pockets of class 3 resources (Hayter and Kandt, 2011). The global wind atlas website (https://globalwindatlas.info/) can be used to assess wind resource for the desired site. The website enables customized search for specific location, and it provides wind speed and wind power density map for the selected location.

8.5.3 Building-integrated/mounted wind turbine

Although it seems new approach, wind power has been an integral part of a building structure. There are plenty of examples, where windmills, wind-driven sawmills, and water pumps have been part of a home and the built environment. This makes the concept of building-integrated wind turbine (BIWT) historical. BIWT is a generic term used to include any wind turbine that can be incorporated within the built environment (i.e., close to or on buildings), does not need to be necessarily in an urban environment (Dutton et al., 2005).

Table 8.2 Classification of wind resources (Hayter and Kandt, 2011; NREL, 2009).

Wind power class	Resource potential	Wind speed at 50 m (m/s)	Wind power density at 50 m (W/m^2)
1	Poor	<5.6	
2	Marginal	5.6−6.4	
3	Fair	6.5−7.0	300−400
4	Good	7.0−7.5	400−500
5	Excellent	7.5−8.0	500−600
6	Outstanding	8.0−8.8	600−800
7	Superb	>8.8	800−1600

8.5.3.1 Building-integrated wind turbines

BIWTs are wind turbines capable of working in the close vicinity of buildings and exploiting, where possible, any augmentation that the building causes to the local wind flow. BIWTs can be incorporated within the building design (Dutton et al., 2005).

8.5.3.2 Building-mounted wind turbines

Building-mounted wind turbines (BMWTs) are physically connected to the building structure. The building is successfully used as a tower to support the wind turbine, facilitating a desirable wind flow (e.g., on top of a tower block). It is noted that if a BMWT is employed, the structure should be able to support the wind turbine loads and be able to address issues associated with noise and vibration (Dutton et al., 2005).

8.5.3.3 Building-augmented wind turbines

Building-augmented wind turbines (BAWTs) are integrated in such a way that the building is used to purposely change and augment the airflow into the wind turbine. Usually, in new building applications, the building is purposely designed to augment the airflow through the turbines. For retrofit applications, BAWTs are positioned in such a way that they are able to exploit any augmentation made possible by the existing building (Dutton et al., 2005).

There are several examples that demonstrate BIWT. Most of these buildings are high rise buildings. The Strata Tower, a 43-story complex residential building has three 5-blade, 9 m diameter, 19 kW custom wind turbines installed in its structure. The estimated annual energy production of 50MWh, which is enough to produce 8% of the towers annual energy consumption (Dymock and Dance, 2013). The Bahrain World Trade Center (Bahrain World Trade Center), which is fifty-story building, contains identical twin towers. The towers are connected by three bridges which hold three turbines. The turbines nominal generating power 225kW at 15-20 m/s wind speed, with a blade diameter 29.2 meters. Another prominent building that incorporates wind turbine to its structure is the Strata tower (Strata SE1). The 14-storey Kinetica building (Kinetica Apartment Building, 2010) features 4 VAWTs stacked vertically down the back of the building. These wind turbines provide elcttiricyt to the building, and any excess electricity produced goes back into the grid.

8.5.4 Optimizing building-integrated/mounted wind turbine devices

In rural areas, the landscape enables undisturbed airflow, and a wind speed of 4−6 m/s can be seen. This makes it relatively easy to select and install wind turbines in rural areas. In the built environment, however, average wind speeds are generally in the range of 2−4 m/s, which is low. This is because existing buildings tend to slow down wind. This implies that careful analysis of the built environment is required. However, in most cases, it is possible to find opportunities for exploiting wind power in the built environment. This needs skill and clever blending of prudently selected wind turbine technologies for specific building types. Care should also be taken in considering environmental conditions, such as noise. Each situation requires consideration of site-specific requirements and selection of wind turbine technologies accordingly. It should also be noted that the selected wind turbine technology should be optimized and guaranteed for reliable long-term energy yield at an acceptable cost (de Vries, 2008). Although site-specific considerations are crucial for the installation of wind turbines in the built environment, there are some unified approaches to certain generic situations (Dutton et al., 2005). One suggestion is that the use of a dedicated building-integrated/mounted wind turbine (BUWT) test facility would address the mentioned issues. According to Dutton et al. (2005), this dedicated BUWT testing facility should help in:

1. assessing different mounting configurations for BUWTs,
2. comparing the performance of different BUWTs,
3. comparing the performance of wind turbines optimized for use as BUWTs and turbines optimized for operation in the kind of enhanced flow fields experienced in wind concentrators or augmentation fields,
4. assessing velocity increase effects over different roof pitches and profiles (and validate computational fluid dynamics calculations of maximum location),
5. measuring and comparing noise and vibration characteristics of different combinations of turbines, mounting conditions, and effective isolation mounting systems
6. evaluating BAWT concepts,
7. supporting the development of appropriate standards for BUWT design, installation, operation, and performance estimation.

8.5.5 Small/micro wind turbines for building application

Small wind turbines are those that produce up to 100 kW, and are used to power a family home, farm, or small business, and are not connected to the grid [American Wind Energy Association (AWEA)]. Small/micro wind turbines are conventionally installed a bit away from buildings to avoid airflow disturbance due to the presence of buildings. General guidelines specified by the American Wind Energy Association (AWEA) suggest minimum land requirement for small wind turbine installation capable of powering 1 house to be 1 acre, which is 64×64 m (Dutton et al., 2005).

8.6 Heat pumps

Heat pumps use electrical energy to move heat from one place to another. Heat pumps are efficient because they do not generate heat (by burning oil or gas) but move it from one place to another. Heat pumps can be used both for heating and cooling because the cycle is reversible, that is, heat is moved from the outdoor to

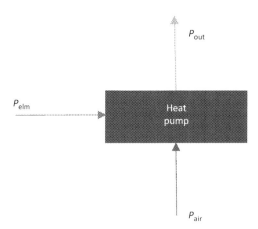

Figure 8.8 Schematic illustrating the performance of a heat pump.

indoor during heating season, and heat is moved from indoor to outdoor during cooling season. It is important to notice that the heat pumps are efficient because they move heat instead of generating heat. Fig. 8.8 illustrates why heat pumps are efficient. Let the heat pump use an electric motor that operates at power P_{elm} to extract heat at a rate P_{air} from the outdoor air and deliver heat at a rate of P_{out}.

The efficiency of such an ASHP is then given as:

$$\text{Efficiency} = \frac{\text{output}}{\text{input}} = \frac{P_{out}}{P_{elm}} = \frac{P_{elm} + P_{air}}{P_{elm}} = 1 + \frac{P_{air}}{P_{elm}} \qquad (8.1)$$

From the above expression, it is clear that for one input of energy, a heat pump always gives more than one output of energy. A heat pump can work efficiently even under cold temperatures. This is because even if the air is cold, heat can still be extracted. In fact, air at $-18°C$ contains about 85% of the heat it contained at $21°C$ (NRCAN, 2017). Performance of heat pump depends on the outside temperature. As the outside temperature drops, the performance of the heat pump drops too. In that case, the heat pump can supply only part of the heat required to keep the space comfortable. It is necessary to have supplementary heating mechanism. Heat pumps can be broadly classified into ASHP and GSHP.

8.6.1 Air-source heat pumps

An ASHP moves heat from the outdoor air and delivers it to indoor or vice versa, depending on the season. ASHPs can be single-packaged units containing both coils in one box. These types of ASHPs are installed on a roof of a building with the ductwork spreading through a wall. Large systems mainly for commercial buildings applications are installed in this way. Another type of ASHP is the split systems, which has an outdoor and an indoor component installed separately. Split heat pumps are usually small, hence known as mini split heat pumps (MSHPs). MSHPs are widely used in many parts of Europe and Asia. MSHP's ductless design allows for easy installation for both residential and commercial uses, in both new and retrofits. The constant increase in energy efficiency regulation allowed for steady growth of MSHPs, as people searched for cheaper means of cooling and/or heating. It is estimated that the ductless cooling system will account for more than 30% of the global energy-efficient HVAC system revenues by the year 2020 (Goetzler et al., 2011).

The MSHP's ductless design also allows for easy retrofit installation. The ductless design also means no duct losses. Duct losses can account for up to 30% of energy consumption in space heating, especially if the ducts are in an unconditioned space such as an attic (DOE). MSHP also allows for design flexibilities; its compact outdoor condensing unit and indoor fan unit can be placed in variety of ways, enabling selection of the most optimal positions. The indoor unit can be mounted on a wall or hung from a ceiling. Some models offer floor-standing option. In addition, multiple indoor coils can be installed for a single outdoor coil. This is called multisplit heat pump, and it allows for the conditioning of multiple zones in a building.

Generally, the only lines connecting the outdoor and the indoor unit are the refrigerant lines. Therefore, only a small hole in the wall, the size of 2–3 in., is needed to run the conduits that house the refrigerant lines. The refrigerant lines can be covered with an insulation material to minimize any unwanted heat transfer to and from the building.

ASHPs can be disadvantageous in an extremely cold environment during the winter months. If the temperature drops significantly below freezing, ASHPs simply cannot provide enough space heating as a stand-alone device. During these months, it would become necessary to use other means of space heating, such as electric baseboard, heating oil, and/or natural gas, if it is available.

Another type of heat pump is air-to-water heat pump (AWHP). AWHP can be used at homes with hydronic heating systems. In heating mode, the AWHP moves heat from the outside air and transfers it to the water in the hydronic heating system. In cooling mode, the process is reversed, that is, the AWHP extracts heat from the water in the hydronic system and transfers it to the outdoor.

8.6.2 Ground-source heat pumps

GSHPs utilize the nearly constant temperature of the ground as the heat exchange medium. Even though the ambient temperature varies significantly around the world, from baking heat in the summer to subzero temperatures in the winter, in most parts, a few meters below the earth's surface the ground temperature is nearly constant. Depending on the location, ground temperatures range from 7°C to 21°C (Kim et al., 2014). Ground temperature is cooler than the ambient air in the summer and warmer than the ambient air in the winter. GSHPs take advantage of this; they exchange heat with the ground. As any heat pump, GSHPs can be used for both heating and cooling. Compared to ASHPs, GSHPs are quieter, last longer, need little maintenance, and most importantly, they do not depend on the outdoor ambient air temperature.

One drawback of GSHPs may be their initial cost, which is estimated at about 50%–100% more than that of ASHP. This additional cost is associated with the costs of digging a ground loop into the land. Although there is greater initial cost of installing GSHPs, they are more efficient when it comes to heating, which results in higher fuel savings and lower energy bills.

8.6.3 Working principles of heat pumps

The working principle of heat pumps is similar to that of a refrigerator or an air conditioner. A heat pump circulates a refrigerant through evaporation and condensation cycles to transfer heat. Generally, a heat pump has two cycles: the heating and cooling cycles. If the heat pump is an air-source heat, it has one more additional cycle, that is, a defrost cycle. The following sections describe each cycle.

8.6.3.1 The heating cycle

A heat pump moves heat from the outdoor to the indoor during this cycle (Fig. 8.9). The steps of the cycle are as follows:

1. As the liquid refrigerant passes through the expansion device, it is converted into a low-pressure mixture of liquid and vapor. This low-pressure mixture enters the outdoor coil, which in a heating mode, serves as an evaporator. As the low-pressure mixture passes through the outdoor coil, it boils and becomes a low-temperature vapor.
2. The low-temperature vapor passes through a reversing valve to the accumulator, where any liquid remained is collected. The compressor then compresses the vapor and reduces its volume. The vapor heats up as it is compressed.
3. In the final step, the compressed hot gas is sent to the indoor coil, which acts as a condenser. The heat from the hot gas is transferred to the indoor space and the gas starts to cool and condense becoming a liquid. This liquid refrigerant then passes through the expansion device, and the cycle is repeated.

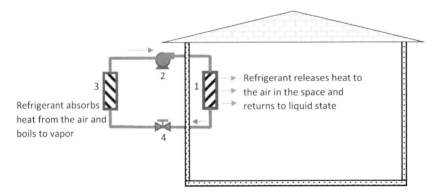

Figure 8.9 A heat pump working in heating cycle. 1: indoor coil (condenser), 2: compressor, 3: outdoor coil (evaporator), 4: expansion valve.

8.6.3.2 The cooling cycle

When a heat pump works in the cooling mode, the previously described cycle is reversed (Fig. 8.10). In this case, the indoor coil works as the evaporator, and the outdoor coil works as the condenser.

1. When the liquid refrigerant passes through the expansion device, it is converted into a low-pressure mixture of liquid and vapor. This mixture then passes through the indoor coil absorbing heat from the indoor space and consequently boiling.
2. Next, the vapor passes through the reversing valve to the accumulator, which collects the remaining liquid. The vapor is then directed to the compressor where it is compressed, its volume is reduced, and becomes hot gas.
3. Finally, the hot gas is directed to the outdoor coil, where the heat is rejected to the outdoor, consequently condensing the refrigerant into liquid.
4. The condensed liquid returns to the expansion device, and the cycle is repeated.

8.6.3.3 The defrost cycle

In case of an ASHP, that is, air-to-air heat pump, it may be necessary that the pump operates in an additional cycle known as the defrost cycle. When an ASHP is operating in a heating mode and if the outdoor temperature drops near or below freezing temperature, moisture over the outside coil may condense and freeze on the coil. This ice buildup may be substantial, depending on the outdoor temperature and the amount of moisture in the air. This is undesirable because it will affect the performance of the heat pump. It reduces the ability of the refrigerant to transfer heat. Hence, there is a need to remove the frost from the outside coil. This is accomplished through the defrost cycle. When the defrost cycle starts, the reversing valve switches to cooling mode in order to send hot gas from indoor to the outdoor. At the same time, the outdoor fan is also stopped from operating in order to reduce the amount of heat required to melt the frost. There are two methods to determine when the heat pump must go to defrost cycle. The first is the time–temperature

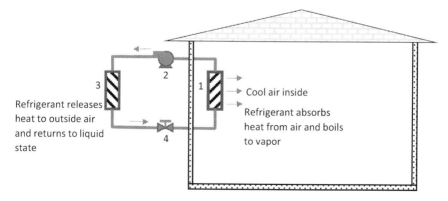

Figure 8.10 A heat pump working in cooling cycle. 1: indoor coil (evaporator), 2: compressor, 3: outdoor coil (condenser), 4: expansion valve.

defrost method. This method accomplishes the defrost cycle by employing a preset temperature (based on sensor reading located on outdoor coil) or preset predetermined time interval. When that temperature is sensed, or predetermined time interval is reached, the defrost cycle begins. The second method is the demand-frost method. In this method, airflow, refrigerant pressure, air temperature, coil temperature, and differential pressure across the outdoor coil are monitored to determine frost accumulation. The first method, which uses preset temperature or time interval, may result in unnecessary defrost cycles reducing the overall seasonal performance of the heat pump. The second method is, therefore, more efficient.

8.6.4 Performance measures

There are four heat pump performance measures: coefficient of performance (COP), energy efficiency ratio (EER), seasonal energy efficiency ratio (SEER), and heating seasonal performance factor (HSPF)

COP is the ratio of the output power to the input power, that is,

$$\text{COP} = \frac{\text{output power}}{\text{input power}} \tag{8.2}$$

COP, which is a dimensionless number, is therefore a measure of efficiency; the higher the number, the more efficient the heat pump is.

The EER is the ratio of output cooling energy (in BTUs) to electrical input energy (in Watt-hour).

$$\text{EER} = \frac{\text{output cooling energy in BTU}}{\text{input electrical energy in Wh}} \tag{8.3}$$

Therefore, it has dimensions of BTU/Wh. It is also noted that it is measured over time.

The SEER is the ratio of output cooling energy (in BTUs) to electrical input energy (in Watt-hour). SEER is a measurement of how the system behaves over a season where the outdoor temperature fluctuates.

$$\text{SEER} = \frac{\text{output cooling energy over a season in BTU}}{\text{input electrical energy over the same season in Wh}} \tag{8.4}$$

HSPF is a measure of the efficiency of a system and the units are the same to that of the SEER. HSPF, however, measures the efficiency of the system in heating mode, not in cooling mode.

8.7 Biomass

Although not necessarily carbon neutral, biomass is considered a renewable energy source. This is because biofuel sources such as crops and trees can be planted and harvested frequently, as short as annually to the timescale of a human lifetime. Use of wood biomass for space heating is not new. People have used wood for thousands of years to create heat by burning wood and other organic materials in their fireplaces, wood stoves, and campsites. Recent technological advances have created highly efficient wood stoves, pellet stoves, fireplace inserts. Today, biomass technology has evolved from burning just solid woods to highly processed wood pellets that burn at very high temperatures, making the heating system more efficient than traditional fireplaces or wood stoves. Biomass can be used to replace fossil fuels to heat single-family homes as well as large industrial facilities. Biomass heating systems can replace existing conventional heating infrastructure if there is a steady supply of biomass and on-site storage.

Biomass can also be used for generating electricity in a system called combined heat and power (CHP). Biomass heat and power (CHP) is approximately 75%−80% efficient, while generation of electricity is only 20%−25% efficient (Biomass Energy Resource Center, 2009). For larger commercial or industrial heating systems, wood chip and pellet boilers are used for space heating. Fort Drum, a US Army installation located in northern New York State is home to a 60 MW biomass power plant. Logging residues, noncommercial components such as tree tops and branches, are processed for delivery as part of harvesting operations and provide the primary feedstock to power the plant (DOE, 2017). One US State, Vermont, aims to meet its 90% energy demand from renewable energy resource by 2050. To achieve this goal, Vermont plans to get significant portion of renewable energy from biomass, mostly sourced from forests, with a small percentage of agricultural crops such as willow and grasses (Schlossberg, 2016). Biomass has also been found to be a better option of energy generation in remote rural areas. Stephen et al. (2016) used The Nuxalk First Nation Bella Coola community as a case to determine the viability of biomass for remote communities. Biomass energy has the potential to reduce heat costs, reduce the cost of electricity subsidization for electrical utilities, reduce GHG emissions, and increase energy independence of remote communities in Canada (Stephen et al., 2016).

8.8 Summary

This chapter presented the available renewable energy technologies that can be used in the building sector. Use of solar, wind, biomass, and heat pump technologies in the residential building has been discussed. Examples of such technologies deployed for building application have been provided. Advantages and disadvantages of each technology have been given. Selection guidelines have also been outlined.

References

Adamson, K.-A., O'Donnell, A., 2012. Thermal Storage for HVAC in Commercial Buildings, District Cooling and Heating, Utility and Grid Support Applications, and High-Temperature Storage at CSP Facilities. PikeResearch, New York.

Aksamija, A., 2015. High-performance building envelopes: design methods for energy efficient facades. In: BEST4 Conference, Kansas City, MO.

Alvarez, G., Arce, J., Lira, L., Heras, M.R., 2004. Thermal performance of an air solar collector with an absorber plate made of recyclable aluminum cans. Sol. Energy 77, 107−113.

Ammari, H.D., Nimir, Y.L., 2003. Experimental and theoretical evaluation of the performance of a tar solar water heater. Energy Convers. Manag. 44 (19), 3037−3055.

Athienitis, A.K., Getu, H., Yang, T., 2014. Computational fluid dynamics (CFD) analysis of air based building integrated photovoltaic/thermal (BIPV/T) systems for efficient performance. In: eSIM 2014, Ottawa, Canada.

Bahrain World Trade Center. [Online]. Bahrain World Trade Center Has Giant Wind Turbines! Available from: <https://inhabitat.com/bahrain-world-trade-center-has-wind-turbines/>.

Basics of wind energy. [Online]. <https://www.awea.org/wind-101/basics-of-wind-energy>.

Bertsch, S.S., Groll, E.A., 2008. Two-stage air-source heat pump for residential heating and cooling applications in northern U.S. climates. Int. J. Refrig. 31 (7), 1282−1292.

Biomass Energy Resource Center, 2009. Biomass Energy: Efficiency, Scale, and Sustainability. Montpelier, VT.

Braunstein, A., Kornfeld, A., 1986. On the development of the solar photovoltaic and thermal (PVT) collector. IEEE Trans. Energy Convers. EC-1 (4), 31−33.

Chen, Y., Athienitis, A.K., Galal, K.E., Poissant, Y., 2007. Design and simulation for a solar house with building integrated photovoltaic-thermal system and thermal storage. In: ISES Solar World Congress 2007. ISES 2007, Beijing, China.

Chen, Y., Athienitis, A.K., Galal, K., 2010a. Modeling, design and thermal performance of a BIPV/T system thermally coupled with a ventilated concrete slab in a low energy solar house: Part 1, BIPV/T system and house energy concept. Sol. Energy 84, 1892−1907.

Chen, Y., Galal, K., Athienitis, A.K., 2010b. Modeling, design and thermal performance of a BIPV/T system thermally coupled with a ventilated concrete slab in a low energy solar house: Part 2, ventilated concrete slab. Sol. Energy 84 (11), 1908−1919.

Chow, T.T., 2010. A review on photovoltaic/thermal hybrid solar technology. Appl. Energy 87 (2), 365−379.

de Vries, E., 2008. Small wind turbnies: Driving performance. Renew. Energy World 11 (3).

Dean, B., Dulac, J., Petrichenko, K., Graham, P., 2016. Towards a Zero-Emission, Efficient, and Resilient Buildings and Construction Sector. Global Status Report. Global Alliance for Buildings and Construction (GABC).

Dembo, A., 2010. The archetype sustainable house: investigating its potentials to achieving the net-zero energy status based on the results of a detailed energy audit. Building .

Devarakonda, L., Mil'shtein, S., 2012. Limiting solar cell heat-up by quantizing high energy carriers. ISRN Renew. Energy.

Dincer, I., 1999. Evaluation and selection of energy storage systems for solar thermal applications. Int. J. Energy Res. 23 (12), 1017−1028.

Dincer, I., 2002. On thermal energy storage systems and applications in buildings. Energy Build. 34 (4), 377−388.

Dincer, I., 2010. Thermal Energy Storage: Systems and Applications, second ed Wiley, Chichester, UK.

DOE, 2017. Biomass Powering a Military Base in Upstate New York, [Online]. Available from: <https://www.energy.gov/eere/bioenergy/articles/biomass-powering-military-base-upstate-new-york>.

DOE, Ductless, mini-split heat pumps.

Doiron, M., O'Brien, W., Athienitis, A., 2011. Energy performance, comfort, and lessons learned from a near net zero energy solar house. ASHRAE Trans. 117.
Duffie, J.A., Beckman, W.A., 2013. Solar Engineering of Thermal Processes, fourth ed Wiley, Hoboken, NJ.
Dutton, A.G., Halliday, J.A., Blanch, M.J., 2005. The Feasibility of Building-Mounted/Integrated Wind Turbines (BUWTs): Achieving Their Potential for Carbon Emission Reductions. Final Report. Energy Research Unit, CCLRC, p. 118.
Dymock, B., Dance, S., 2013. Building integrated wind turbines - a pilot study. In: Proceedings of Meetings on Acoustics, Montreal, Canada.
Evacuated Tube Solar Collectors, 2019. [Online]. <http://www.apricus.com/html/solar_collector.htm> (accessed 30.10.19.).
Florschuetz, L.W., 1975. On heat rejection from terrestrial solar cell arrays with sunlight concentration.
Florschuetz, L.W., 1979. Extension of the Hottel-Whillier model to the analysis of combined photovoltaic/thermal flat plate collectors. Sol. Energy 22 (4), 361–366.
Gao, L., Zhao, J., Tang, Z., 2015. A review on borehole seasonal solar thermal energy storage. Energy Procedia 70, 209–218.
Getu, H., Fung, A., Dash, P., 2014. Performance evaluation of an air source heat pump coupled with a building integrated photovoltaic/Thermal (BIPV/T) system. In: ASME Eighth International Conference on Energy Sustainability, Boston, MA.
Goetzler, W., Zogg, R., Lisle, H., Burgos, J., 2011. Ground-source heat pumps: overview of market status, barriers to adoption, and options for overcoming barriers. In: Geothermal Energy: The Resource Under Our Feet.
Hailu, G., 2019. Seasonal solar thermal energy storage. In: Thermal Energy Battery with Nano-enhanced PCM.
Hailu, G., Dash, P., Fung, A.S., 2015. Performance evaluation of an air source heat pump coupled with a building-integrated photovoltaic/thermal (BIPV/T) system under cold climatic conditions. Energy Procedia 78, 1913–1918.
Hailu, G., Hayes, P., Masteller, M., 2017. Seasonal solar thermal energy sand-bed storage in a region with extended freezing periods: part I, experimental investigation. Energies 10 (11), 1873.
Hayter, S.J., Kandt, A., 2011. Renewable energy applications for existing buildings. Preprint: Applicazioni dell'energia rinnovabile su edifici esistenti. In: 48th AiCARR International Conference, Baveno, Italy.
Hendrie, S.D., 1979. Evaluation of Combined Photovoltaic/Thermal Collectors. Electric Power Research Institute, EPRI EA, Palo Alto, CA.
International Energy Agency, 2014. Technology Roadmap: Energy Storage. Paris, France.
Jamar, A., Majid, Z.A.A., Azmi, W.H., Norhafana, M., Razak, A.A., 2016. A review of water heating system for solar energy applications. Int. Commun. Heat. Mass. Transf.
Joshi, A.S., Dincer, I., Reddy, B.V., 2009. Performance analysis of photovoltaic systems: a review. Renew. Sustain. Energy Rev.
Kamel, R., Ekrami, N., Dash, P., Fung, A., Hailu, G., 2015. BIPV/T + ASHP: technologies for NZEBs. Energy Procedia.
Kim, H.K., Ao, S.I., Amouzegar, M.A., 2014. Transactions on Engineering Technologies: Special Issue of the World Congress on Engineering and Computer Science 2013.
Kinetica Apartment Building, [Online]. <https://www.flickr.com/photos/41845311@N06/4528728533>.
Lalović, B., Kiss, Z., Weakliem, H., 1986. A hybrid amorphous silicon photovoltaic and thermal solar collector. Sol. Cell.

Martínez-Rodríguez, G., Fuentes-Silva, A.L., Picón-Núñez, M., 2018. Solar thermal networks operating with evacuated-tube collectors. Energy.

Mbewe, D.J., Card, H.C., Card, D.C., 1985. A model of silicon solar cells for concentrator photovoltaic and photovoltaic/thermal system design. Sol. Energy.

Michael, J.J., Iniyan, S., Goic, R., 2015. Flat plate solar photovoltaic-thermal (PV/T) systems: a reference guide. Renew. Sustain. Energy Rev.

Morrison, G.L., Budihardjo, I., Behnia, M., 2004. Water-in-glass evacuated tube solar water heaters. Sol. Energy.

NRCAN, 2017. What is a heat pump and how does it work?

NREL, 2009. United States - WInd Resource Map.

O'Leary, M.J., Clements, L.D., 1980. Thermal-electric performance analysis for actively cooled, concentrating photovoltaic systems. Sol. Energy.

Oral, G.K., Yener, A.K., Bayazit, N.T., 2004. Building envelope design with the objective to ensure thermal, visual and acoustic comfort conditions. Build. Environ.

Papadopoulos, A.M., 2005. State of the art in thermal insulation materials and aims for future developments. Energy Build.

Raghuraman, P., 2010. Analytical predictions of liquid and air photovoltaic/thermal, flat-plate collector performance. J. Sol. Energy Eng.

Sadineni, S.B., Madala, S., Boehm, R.F., 2011. Passive building energy savings: a review of building envelope components. Renew. Sustain. Energy Rev.

Safa, A.A., Fung, A.S., Kumar, R., 2015. Performance of two-stage variable capacity air source heat pump: field performance results and TRNSYS simulation. Energy Build.

Schlossberg, J., 2016. Vermont Burning: The Future of Biomass Energy [Online]. Available from: <https://vermontindependent.net/vermont-burning-the-future-of-biomass-energy/>.

Shah, Y.T. (Ed.), 2018. Thermal Energy: Sources, Recovery, and Applications. CRC Press, Boca Raton, FL.

Simple Solar. [Online]. <https://www.simplesolar.ca/flat-plate-collectors-vs-evacuated-tube-collectors.html> (accessed 25.10.19.).

Skoplaki, E., Palyvos, J.A., 2009. On the temperature dependence of photovoltaic module electrical performance: a review of efficiency/power correlations. Sol. Energy.

Solar Panels Plus, 2014. All about solar: evacuated tubes or flat panels? [Online]. <http://www.solarpanelsplus.com/all-about-solar/evacuated-tubes-or-flat-plates/> (accessed 21.10.19.).

Stephen, J.D., et al., 2016. Biomass for residential and commercial heating in a remote Canadian aboriginal community. Renew. Energy.

Strata SE1. [Online]. Available from: <https://en.wikipedia.org/wiki/Strata_SE1>.

Tian, Y., Zhao, C.Y., 2013. A review of solar collectors and thermal energy storage in solar thermal applications. Appl. Energy 104, 538–553.

Tripanagnostopoulos, Y., 2007. Aspects and improvements of hybrid photovoltaic/thermal solar energy systems. Sol. Energy.

US DOE. Planning for home renewable energy systems. [Online]. <https://www.energy.gov/energysaver/buying-and-making-electricity/planning-home-renewable-energy-systems> (accessed 28.10.19.).

Vijayakumar, R.S.P.P., Kumar, S.S., Sakthivelu, S., 2017. Comparison of evacuated tube and flat plate solar collector – a review. World Wide J. Multidiscip. Res. Dev. 3 (2), 32–36.

Walker, A., 2016. Solar water heating. [Online]. <https://www.wbdg.org/resources/solar-water-heating> (accessed 28.10.19.).

Wolf, M., 1976. Performance analyses of combined heating and photovoltaic power systems for residences. Energy Convers.

Zhang, L., Jing, D., Zhao, L., Wei, J., Guo, L., 2012. Concentrating PV/T hybrid system for simultaneous electricity and usable heat generation: a review. Int. J. Photoenergy.

Zondag, H.A., 2008. Flat-plate PV-thermal collectors and systems: a review. Renew. Sustain. Energy Rev.

Zubriski, S.E., Dick, K.J., 2012. Measurement of the efficiency of evacuated tube solar collectors under various operating conditions. J. Green. Build.

Part 4

Energy efficiency in industrial sector

Energy efficiency and renewable energy sources for industrial sector

Kamil Kaygusuz
Department of Chemistry, Karadeniz Technical University, Trabzon, Turkey

Chapter Outline

9.1 Introduction 213
9.2 Global energy trends 215
9.3 Energy consumption and emissions in industry 216
    9.3.1 General trends 216
    9.3.2 Energy and carbon-intensive industrial sectors 220
9.4 Energy efficiency in industry for climate change mitigation 221
    9.4.1 The need for innovation 222
9.5 Energy efficiency and renewable sources in industry 224
    9.5.1 Bioenergy 226
    9.5.2 Solar heat 227
9.6 Case study in Turkey 229
    9.6.1 National Energy Efficiency Action Plan 229
    9.6.2 General overview 230
    9.6.3 Industry and technology 231
    9.6.4 Aim of the development plans 232
9.7 Policy options 233
    9.7.1 Lessons learned 233
    9.7.2 International agreements 234
    9.7.3 Procurement 235
9.8 Conclusions 236
Acknowledgment 237
References 237

9.1 Introduction

Energy is needed in industry for a number of technologies and processes, including crosscutting technologies such as steam, motors, compressed air, pumps, heating and cooling, as well as specific processes in energy-intensive sectors (chemicals, iron and steel, cement, pulp and paper, nonferrous metals, and food). Greenhouse gas (GHG) emission reductions in industry can be achieved in different ways. One option is to reduce the energy consumption of processes and technologies by implementing specific energy efficiency measures and state-of-the-art energy management systems, or by generating energy reusing industrial by-products. Besides investing in energy efficiency measures, carbon dioxide (CO_2) emissions can also be reduced substantially through increased material efficiency. This includes various options such as fuel substitution, as well as the substitution and reuse of production materials (Philibert, 2017; European Union, 2015; Kreith and Goswami, 2008; Henzler et al., 2017).

Industrial heat makes up two-thirds of industrial energy demand and almost one-fifth of global energy consumption. It also constitutes most of the direct industrial CO_2 emitted each year, as the vast majority of industrial heat originates from fossil fuel combustion. Yet despite these impressive figures, industrial heat is often missing from energy analyses. While industrial heat demand grows in the central global energy scenario, the underlying drivers are different depending on temperature requirements. Low- and medium-temperature heat (below 400°C) accounts for three-quarters of the total growth in heat demand in industry by 2040, driven by less energy-intensive industries (Asia-Pacific Economic Cooperation, 2013; Weiss and Spörk-Dür, 2019; IEA, 2018a).

This is a reversal of historical trends: in the last 25 years, high-temperature heat represented two-thirds of overall heat demand growth, driven by China's rapid development of heavy industries such as steel and cement. On the other hand, developing Asia continues to drive industrial heat demand growth. The growth in low- to medium-temperature needs in this region also alone represents about half of the global industrial heat demand increase (IEA, 2018b,c; Solrico, 2017; Horta, 2015).

Low-temperature heat use grows in most regions through 2040, except in the European Union and Japan. The outlook for high-temperature heat varies even more across regions, including among developing countries. It decreases in China with the country's shift to a less energy-intensive development pathway, whereas it increases in India as the country becomes, by large distance, the main global driver (Asia-Pacific Economic Cooperation, 2013; IEA, 2017a, 2018a).

As industrial heat demand continues to grow, so does its share in energy-related CO_2 emissions. First, industrial heat is often generated on-site, making it more difficult to regulate than a more centralized sector such as large thermal power generation. There is also limited policy focus in this area compared with other sectors (Philibert, 2017). Second, while heating needs for residential and commercial buildings are fairly standard, industrial heat encompasses a wide variety of temperature levels for diverse processes and end uses. For instance, cement kilns require high temperature, whereas drying or washing applications in the food industry operate at lower temperatures. Different technology and fuel options are available depending on the required temperature level. For example, low-temperature heat from a heat pump cannot be substituted for high-temperature heat from a gas boiler (IEA, 2017b).

Today's industrial heat demand relies mainly on fossil fuels, biomass and electricity, and only very small shares of renewables in certain sectors. Therefore decarbonization would require a dramatic shift in how industrial heat is generated. Yet this goal is instrumental to following a low-carbon development pathway as defined in the sustainable development scenario, a new global scenario providing an integrated way to achieve three critical policy goals simultaneously: climate stabilization, cleaner air, and universal access to modern energy. The best option for reducing energy use of industrial heat will depend on the specific use and required temperature (Philibert, 2017; IEA, 2017b).

Fuel switching can provide some benefits, for instance substituting gas for coal, but for more ambitious climate targets, more transformative solutions are needed. For example, under certain conditions, electrification can be a low-cost and sustainable option. Therefore heat pumps can be economical solutions for low- and medium-temperature needs. Electrification may also be possible for specific high-temperature industrial processes, such as electricity-based steel production (Philibert, 2017). On the other hand, direct renewable heat sources such as solar and geothermal energy can also be economical for applications below 400°C, but they are not easy to integrate in all industrial facilities. Bioenergy can be used for high-temperature heat demand, but is resource constrained and only economical and sustainable under certain operating conditions and in certain regions (Asia-Pacific Economic Cooperation, 2013; Weiss and Spörk-Dür, 2019; IEA, 2018a,b).

Industrial heat can be decarbonized through the deployment of carbon capture, utilization, and storage. This can include, for instance, technologies to remove CO_2 emissions from flue gas before recycling the CO_2 in industrial processes, such as for methanol production, or storing it permanently. Finally, end-use efficiency, through the use of modern equipment, improved insulation or heat recovery, can reduce final demand before the heat is even generated—often, limiting overall heat requirements is the first strategy adopted, before taking actions to decarbonize remaining heat use (IPCC, 2014; Hasanbeigi et al., 2014; International Renewable Energy Agency, 2015, 2016; IRENA and C2E2 Copenhagen Centre on Energy Efficiency, 2015; Energy Information Administration, 2019; Lovegrove et al., 2015; Ministry of Energy and Natural Resources, 2016a,b; Baron, 2016; Baron and Garret, 2018; Reinaud, 2008).

9.2 Global energy trends

Global energy demand rose by 1.9% in 2016 and the forces driving energy demand, led by strong economic growth, outpaced progress on energy efficiency. As a result, energy intensity fell by just 1.6% in 2016 (IEA, 2017a, 2018a,b,c; Solrico, 2017; Horta, 2015). Without energy efficiency progress, increased economic activity would have had a greater impact on the global energy system. Efficiency improvements made since 2000 prevented 12% additional energy use in 2017. Efficiency gains also prevented 12% more GHG emissions and 20% more fossil fuel imports, including over USD 30 billion in avoided oil imports in IEA countries (IEA, 2017a).

Implementation of energy efficiency policy has slowed, putting at risk the recent gains from energy efficiency. The International Energy Agency tracks three types of energy efficiency policy: mandatory codes and standards, market-based instruments, and incentives. In 2016, 34% of global energy use was covered by mandatory energy efficiency policies, but progress in implementing new policies was slow for a second year running. Utility obligation programs remained largely unchanged in 2016 (Philibert, 2017). Spending on energy efficiency incentives in 16 major economies was estimated to be around USD 25 billion (Asia-Pacific Economic Cooperation, 2013).

There is still huge potential for energy efficiency gains, and primary energy demand has grown by 39% and the global economy has grown by nearly 85% (IEA, 2017a). In the efficient world scenario assumes the adoption of all cost-effective energy efficiency opportunities between 2017 and 2040. On the other hand, the global economy doubles but there would be only a marginal increase in primary energy demand. On average, investments in the efficient world scenario pay back by a factor of 3 over the life of the measures (Asia-Pacific Economic Cooperation, 2013).

The efficient world scenario can deliver a peak in energy-related GHG emissions before 2020. Emissions would subsequently fall to levels 12% lower than today, providing over 40% of the abatement required to be in line with objectives in the Paris Agreement. Energy efficiency, combined with renewable energy (RE) and other measures, is therefore indispensable to achieving global climate targets. The efficient world scenario would also see reductions in air pollution, lower household spending on energy, enhanced energy security, and many other benefits (Asia-Pacific Economic Cooperation, 2013).

9.3 Energy consumption and emissions in industry

9.3.1 General trends

The total global final energy consumption in 2017 amounted to 13,972 million tons of oil equivalent (Mtoe) (see Fig. 9.1; Tables 9.1 and 9.2). With a share of

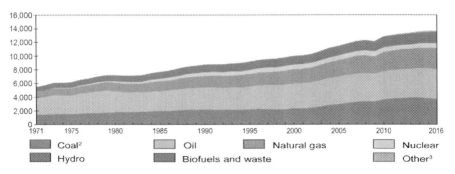

Figure 9.1 World Total Primary Energy Supply (TPES) from by fuel (Mtoe) (IEA, 2018a).

Table 9.1 World primary energy demand by fuel and industry (Mtoe).

	2000	2017	2030
TPED	10,027	13,972	16,167
Coal	2308	3750	3783
Oil	3665	4435	4830
Gas	2071	3107	3820
Nuclear	675	688	848
Hydro	225	353	458
Bioenergy	1022	1385	1691
Other renewables	60	254	736
Industry	1863	2855	3460
Coal	400	803	876
Oil	326	321	335
Gas	412	618	851
Electricity	462	768	987
Heat	101	140	148
Bioenergy	162	204	258
Other renewables	0	1	5

TPED, Total primary energy demand; Mtoe, million tons of oil equivalent.

Table 9.2 World renewable energy consumption in 2017 and 2030.

	2017	2030
Primary demand (Mtoe)	1334	2056
Share of global TPED	10%	15%
Tradition use of solid biomass (Mtoe)	658	596
Share of total bioenergy	48%	38%
Electricity Generation (TWh)	6351	10,917
Bioenergy	623	1039
Hydropower	4109	5012
Wind	1085	2707
Geothermal	87	162
Solar PV	435	1940
Concentrated solar power	11	54
Marine	1	4
Share of total generation	25%	38%
Heat consumption (Mtoe)	478	653
Industry	236	302
Buildings and other	242	351
Share of total heat demand	10%	13%
Biofuels (mboe/d)	1.8	4.4
Road transport	1.8	3.9
Share of total transport demand	3%	7%

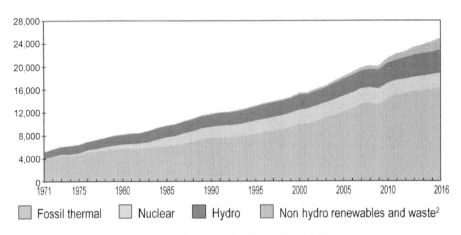

Figure 9.2 World electricity generation by fuel (TWh) (IEA, 2018a).

approximately 37%, the industrial sector consumes more energy than any other end-use sector (IEA, 2017a). In developing and emerging economies, the share can even be up to 50% (European Union, 2015). Most of the worldwide industrial energy consumption increase in 2017 was caused by China and India (64%). In those two countries, energy consumption in industry grew by 3.6% in 2017 and in Africa by 3.3%, whereas industrial energy consumption declined in the Middle East by 1.9% and in Latin America by 0.1% (IEA, 2017a).

Fig. 9.2 shows world electricity generation by fuel (TWh). As shown in Fig. 9.2, most of the electricity generation comes from nonrenewables and waste sources. The second fuel is hydropower. Electricity is increasingly the "fuel" of choice for society, but a dramatic transformation of the power sector is underway. Innovative technologies are disrupting traditional ways of producing, transporting, and storing electricity, creating opportunities for new actors and business models. Ensuring the reliable and secure provision of affordable electricity, while meeting environmental goals, is at the heart of the 21st century economy and is increasingly a central pillar of energy policy making.

Energy is needed in industry for a number of processes such as steam, cogeneration, process heating and cooling, lighting, and so on. Depending on economic and technological development, the composition of industries and other factors, the intensity and mix of fuels therefore differs across countries. According to the IEA's International Energy Outlook 2017 reference case, worldwide industrial sector energy consumption is assumed to increase on average by 1.3% per year between 2017 and 2040, where the average annual change in OECD countries is projected to be lower with around 0.8% and higher in non-OECD countries with around 1.5%. Nevertheless, in non-OECD countries the industry share of energy consumption is projected to decrease from 64% in 2017 to 59% in 2040, as many emerging countries are likely to shift their economic activities away from the energy-intensive manufacturing industry to other end-use sectors where energy use is more rapidly increasing (IEA, 2017a, 2018a; Energy Information Administration, 2019).

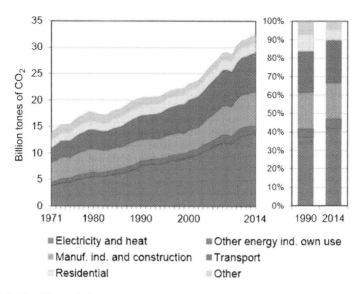

Figure 9.3 The CO₂ emissions by sector (IEA, 2017a).

Table 9.3 Global energy-related CO₂ emissions (million tons).

	2000	2017	2030	2040
By sector				
Power	9305	13,587	7839	3292
Industry	3922	6154	5936	5081
Transport	5757	7986	7326	5563
Buildings	2714	2997	2593	2202
Other	1424	1856	1788	1510
By fuel				
Coal	8951	14,448	8335	3855
Oil	9620	11,339	9501	6886
Gas	4551	6795	7645	6906
Total	23,123	32,581	37,748	42,475

The worldwide industrial CO_2 emissions were 14.39 $GtCO_2$ in 2014 (see Fig. 9.3; Table 9.3), comprised of direct energy-related emissions, indirect emissions from electricity and heat production, process emissions, and emissions from wastes, accounting for 44.4% of total global CO_2 emissions (IEA, 2018a). In industry, energy efficiency potentials exist in crosscutting technology systems, sector-specific processes, energy generation, and control systems for performance optimization (IEA, 2017a).

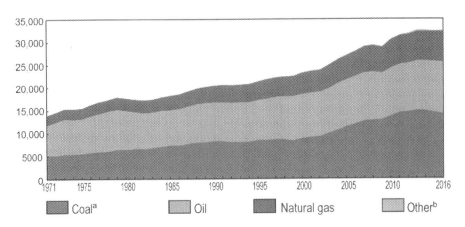

Figure 9.4 World CO_2 emissions from fuel combustion by fuel (Mt of CO_2) (IEA, 2018a). [a]Peat and oil shale are aggregate with coal. [b]Includes industrial waste and non-renewable municipal waste.

Fig. 9.4 shows the world CO_2 emissions from fuel combustion by fuel (Mt of CO_2) (IEA, 2018a). As shown in Fig. 9.4, the most of CO_2 emissions comes from coal burning and the second from oil. In 2016, GHG emissions from energy and industrial processes amounted to about 34 Gt of CO_2 equivalent (Gt CO_2-eq). Three-quarters of this is accounted for only eight source categories. By far the largest category is coal-fired power generation, with 2053 GW of capacity accounting for 27% of emissions. Buildings made up nearly 9% in 2017, followed by about 8% each for gas-fired power generation and petroleum-fueled cars (more than 1 billion cars). Emissions from cement production and oil and gas operations accounted for 7% each, with trucks (202 million vehicles) making up 6% and steel around 5% of the total.

9.3.2 Energy and carbon-intensive industrial sectors

According to IPCC's Fifth Assessment Report (IPCC, 2014), the major emitting industrial sectors are chemicals (plastics, fertilizer, and others), iron and steel, cement, pulp and paper, nonferrous metals (aluminum and others), and food processing (Philibert, 2017; IPCC, 2014):

- *Chemicals*: The largest energy-consuming industry sector is the basic chemical industry with a share of 15%–20% of the total delivered energy consumption in industry for non-OECD and OECD countries, respectively. Petrochemical feedstocks, such as ethylene, ammonia, adipic acid, and caprolactam used in producing plastics, fertilizer, and synthetic fibers, account for around 60% of the energy consumption (Energy Information Administration, 2019).
- *Iron and steel*: Iron and steel is the second largest energy consumer in industry. With nearly half of the world's steel production, China is the world's biggest steel-producing country (49.6% in 2015), followed by EU-28 with 10.2%, Japan with 6.5%, India with

5.5%, and the United States with 4.9% (IEA, 2018c). The energy intensity of the steel production varies across regions depending on the technology used. Due to the use of coal and coke, the conventional ironmaking process is the most emission-intensive part of steel production, accounting for 70%−80% of the emissions (IEA, 2017a)
- *Cement*: As the demand for cement is correlated with the construction industry, an increase of building and infrastructure construction induces a rise in the cement industry's energy use (Energy Information Administration, 2019). On the other hand, the CO_2 emissions in the cement sector are composed of emissions from fuel combustion for limestone, clay, and sand heating (around 40%), of process emissions from the calcination reaction (around 50%), and of emissions arising from grinding and transport (around 10%) (IPCC, 2014).
- *Pulp and paper*: The global demand for paper and variety of paper products is mainly driven by developing countries, leading to a steadily growing worldwide paper production (IPCC, 2014). Although the paper-producing process is very energy-intensive, nearly half of the needed electricity is provided through cogeneration in paper mills with wood waste products (Energy Information Administration, 2019).
- *Nonferrous metals*: The production of nonferrous metals, primarily aluminum, is very energy-intensive because of a high electricity demand. More than 80% of total GHG emissions in aluminum production are indirect electricity-related emissions (IPCC, 2014). Aluminum can be produced from raw materials or recycling, where the latter one can save up to 95% of energy needed for primary aluminum production. China currently produces more aluminum than any other country in the world, however, with recycling rates much lower (21% in 2012) than in the United States (57%) (Energy Information Administration, 2019).
- *Food*: In food processing, most energy is used for drying, cooling, storage, and food and beverage processing purposes (Horta, 2015). Due to a growing world population, the demand for food keeps rising, making the food industry one of the major GHG emitting industry sectors. However, the demand for food and hence energy use and emissions could be drastically reduced by avoiding wasted food, which is estimated at around one-third of the food produced.

9.4 Energy efficiency in industry for climate change mitigation

As GHG emissions reach a new record high in 2018 according to World Meteorological Organization, our time to switch to low-carbon energy systems is running out. Our energy system has to change rapidly to adapt to the actions and new policies to be taken according to 1.5°C Paris Agreement on climate change. Common target of the countries around the world should be low-carbon energy system that will replace fossil fuels by RE sources (IEA, 2018a). Increasing energy demand by industrial sectors makes reaching this target even more variable. Another challenge is matching the variable renewable production to different load profiles in energy consumption. Energy storage is the key technology to bridge the gap between supply and demand of such intermittent energy sources and realize ambitious goals of the low-carbon future. However, the potential of energy storage sometimes is underestimated, because it can be a hidden technology in the whole energy system. The environmental and economic benefits of energy storage need to be emphasized and demonstrated more (IEA, 2018a; Solrico, 2017).

9.4.1 The need for innovation

Fossil fuel use in industry and its associated CO_2 emissions can be reduced by various means, from improved energy efficiency to carbon capture and reuse or storage, as well as by novel manufacturing processes and ultimately by using different materials in industry and other end-use sectors—for example, more wood in construction. However, increased uptake of energy from renewables appears to be an ideal means to reduce fossil fuel consumption and emissions in a variety of industrial sectors of growing relevance (Solrico, 2017).

Various renewable sources, such as bioenergy, solar radiation, and geothermal energy, can be used to produce heat for industrial purposes, but the availability of these resources is neither spatially nor temporally uniform (Philibert, 2017). For example, in both temperate and hot but humid areas, the opportunities to cost-effectively collect solar heat at temperatures beyond $\sim 150°C$ are scarce, but biomass resources can be important (Kreith and Goswami, 2008). Conversely, in hot and arid areas, biomass resources are usually scarce, but concentrating solar power systems are able to efficiently collect the sun's energy at high temperatures (IEA, 2017a).

The energy used in industry can also be sourced from renewables through electricity generation, either from dedicated facilities or from the grid, or any combination of both. For remote, off-grid industrial facilities, these sources can replace the fossil fuels used to generate power. Furthermore, heat can be generated from electricity using a variety of technologies. The pillars of industrial emission reductions in the energy technology scenarios are energy efficiency, innovative processes, and energy storage technologies. Fuel and feedstock switching, and material efficiency, combining manufacturing material efficiency, interindustry material synergies, decreased end-use material intensity, and postconsumer recycling, also make small contributions.

Innovative processes account for 19% of cumulative CO_2 reductions in the industrial sectors. But, these processes, not yet fully commercialized, include new steelmaking processes, inert anodes for aluminum smelting, oxy-fueling kilns for clinker production in cement manufacturing, enhanced catalytic and biomass-based processes for chemical production, and integration of carbon capture and storage (CCS) in energy-intensive industrial processes (IEA, 2017a). Fig. 9.5 shows the global industry direct CO_2 emissions.

Climate-friendly scenarios rely particularly on ambitious development of bioenergy resources, CCS, and carbon capture and use technologies, or on the combination of both technology families with bioenergy carbon capture and storage (BECCS) (Kreith and Goswami, 2008). Bioenergy is sourced from organic material that stores sunlight as chemical energy as it grows through photosynthesis, which removes CO_2 from the atmosphere; therefore capturing and storing the CO_2 that is emitted when the bioenergy is consumed could possibly make the life-cycle

Energy efficiency and renewable energy sources for industrial sector 223

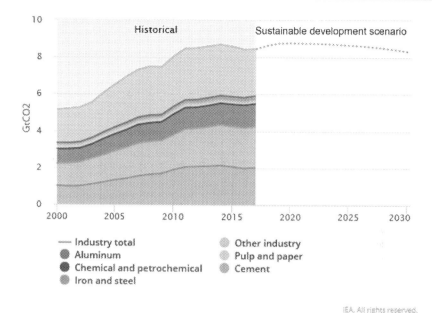

Figure 9.5 Global industry direct CO_2 emissions.

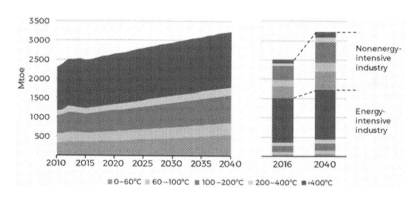

Figure 9.6 Global industrial demand by temperature level and sector.

emissions of BECCS negative. For this to happen, the combined amount of GHGs present in the entire supply chain and the GHGs that cannot be captured by CCS must total less than the amount that is captured and permanently stored (Philibert, 2017; Kreith and Goswami, 2008; Henzler et al., 2017; Weiss and Spörk-Dür, 2019). Fig. 9.6 shows the global industrial demand by temperature level and sector.

Achieving negative emissions in power and industry could compensate for temporary "overshoots" in emissions particularly hard-to-suppress emissions from other sectors (e.g., transport and industry), or both. However, CCS is proving slow and difficult to deploy, and it clearly does not reduce reliance on fossil fuels. Economic conditions for other low-carbon technologies may improve more rapidly than expected. Furthermore, BECCS is yet an unproven technology at scale, and there is a great degree of uncertainty surrounding its viability, although the uncertainties do not necessarily pertain to the CCS technology itself. Another question is what level of bioenergy resources might be sustainably available for use on a large scale by the energy sector (IEA, 2017a,b, 2018b,c), as all studies lead to the conclusion that the extensive application of BECCS needed to achieve Paris Agreement objectives would stretch the possibilities offered by bioenergy to their maximum. The supply of sustainable bioenergy will need to grow from today's 1512 Mtoe to around 3480 Mtoe under the sustainable development scenarios (Philibert, 2017; IEA, 2017a; Energy Information Administration, 2019).

The role of renewables in the power sector, and to some extent in the buildings sector, has been investigated extensively. This is less the case for industry, for which renewables have been the topic of only a limited number of studies. To date, most energy-related innovation in the industry has been to improve energy efficiency rather than to reduce GHG emissions, as energy has always had a cost whereas emissions have not (Philibert, 2017). Moreover, industries have considered electricity a costly source of heat, and rightly so, as electricity has been produced mostly in thermal plants from combusting fossil fuels at an efficiency rarely exceeding 50%. Electricity could thus compete with fossil fuels only if it was running more efficient devices such as heat pumps. However, recent and rapid cost reductions in some renewable electricity generation technologies have led to the emergence of new, affordable options that have not been considered in many studies (International Renewable Energy Agency, 2015, 2016; IRENA and C2E2 Copenhagen Centre on Energy Efficiency, 2015; Energy Information Administration, 2019; Lovegrove et al., 2015; Ministry of Energy and Natural Resources, 2016a,b; Baron, 2016).

9.5 Energy efficiency and renewable sources in industry

The combined application of RE and energy efficiency in industry is a natural marriage insofar as industry operators who have the foresight to convert their plants from using fossil fuels to renewable fuels are very likely to maximize the value of the renewable fuel by maximizing the efficiency of its use in their plants. Governments can create regulatory and business environments that promote development of RE and energy efficiency in industry, and industry will respond by developing business models configured to extract maximum value for the business from the opportunity available. There is, therefore, a need to broaden our thinking to include the goal of efficient use of RE in industry to achieve maximum value. This value flows through to the whole economy and, ultimately, the planet and its inhabitants (Philibert, 2017; IEA, 2017b,c).

Fig. 9.7 shows the share and breakdown of heat demand in industry. As shown in Fig. 9.7, three-quarters of the energy used in industry is process heat: the rest is for mechanical work and electricity (computers, lighting, and so on). About 30% of

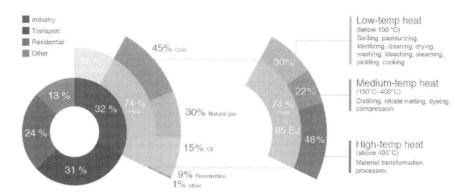

Figure 9.7 Share and breakdown of heat demand in industry (IEA, 2018a; IEA, 2017a).

process heat is "low-temperature heat" (below 150°C), 22% is "medium-temperature heat" (150°C–400°C), and 48% is "high-temperature heat" (above 400°C) as shown in Fig. 9.7. About 10% of process heat is estimated to be based on electricity.

While there are many examples of the combined application of renewables and efficiency in industry, their penetration to date has not been extensive. The applications considered most likely to achieve significant penetration in the middle term are the use of (Philibert, 2017; IEA, 2017b, 2018a,c):

- biomass energy for process heat
- biomass as a petrochemical feedstock
- solar thermal systems for process heat
- heat pumps for process heat

It has been suggested that RE, if used efficiently, has the potential to supply 23% of final energy use in the global manufacturing industry, and up to 14% of fossil feedstock can be replaced by biomass. Together, this equates to 21% of total final energy use.

The combined application of RE and energy efficiency in industry is a natural marriage insofar as industry operators who have the foresight to convert their plants from using fossil fuels to renewable fuels are very likely to maximize the value of the renewable fuel by maximizing the efficiency of its use in their plants. There are also a number of synergies and similarities that include (Philibert, 2017; IEA, 2017b, 2018c):

- Improvements in end-use energy efficiency reduce the cost of delivering end-use services by RE. The money saved through this can help finance additional efficiency improvements and deployment of RE technologies.
- Lower end-use energy requirements increase the opportunity for RE sources that have low energy density, such as solar, or of low energy content, such as low-temperature solar heat, to meet full energy-service requirements.
- Achievement of targets for increasing the share of renewables in an economy's energy portfolio by reducing total energy consumption as well as by increasing the percentage of RE.
- Reduction of the energy intensity in an industrial plant, both through the introduction of new technology and energy end-use efficiency gains, enhances the opportunity to introduce RE.

- Similarity of the obstacles to implementation that are often encountered and the regulatory frameworks and incentives required to ensure successful development of both RE and energy efficiency technologies in industry.

While RE and energy efficiency may be the twin pillars of a sustainable energy future, they are in fact fundamentally different concepts. Thus RE is energy that either renews itself. Energy efficiency also involves using the least amount of energy to do a job or achieve an objective. There is, therefore, a need to broaden our thinking to include the efficient use of RE in industry to achieve maximum value for the industrial end user, the community, the economy, and, ultimately, the planet and its inhabitants.

To this end, we believe that the combined use of RE and energy efficiency in industry needs to focus on how such combination can maximize the benefits that can be achieved. Such benefits include, but are not necessarily limited to (Philibert, 2017):

- minimizing the specific energy consumption required for production,
- maximizing revenues and economic value for an industrial company,
- minimizing the use of fossil fuels,
- reducing GHG emissions,
- managing waste disposal,
- minimizing environmental impacts,
- job creation,
- improvement of industrial working conditions and safety.

The ways in which RE and energy efficiency initiatives can be combined to achieve these objectives may be quite different depending on which are targeted by a particular industry or industrial plant.

9.5.1 Bioenergy

With an estimated consumption of 196 Mtoe in 2017, biomass is by far the largest RE source in industry today. Of this consumption, 58 Mtoe were used by the pulp and paper industry, 29 Mtoe by the food and tobacco branch, 7.7 Mtoe by the wood and wood products industry, 5.1 Mtoe by nonmetallic minerals such as the cement industry, 3.2 Mtoe by the iron and steel industry, and 2.71 by the chemical industry. Bioenergy consumption is most evident in industry sectors that produce biomass residues on-site suitable for fuel use. In other industries where this is not the case, bioenergy is less used because biomass fuel supply chains need to be mobilized (IEA, 2017a).

Brazil, India, the United States, some developing countries, and the European Union are the largest industrial consumers of bioenergy. The power sector led the growth, with renewables-based electricity generation increasing by 7%, almost 450 TWh, equivalent to Brazil's entire electricity demand. This was faster than the 6% average annual growth since 2010. Shares shown in Fig. 9.8 do not, however, include biomass used to generate electricity (total 720 Mtoe) and commercial heat (total 120 Mtoe) consumed globally by industries. Neither do they include the share of bioenergy consumed on-site to transform biomass into biofuels (IEA, 2017a).

On the other hand, Turkish bioenergy potential is very high for electricity generation from wood and agricultural crop residues (see Fig. 9.9). Biomass can become a reliable and renewable local energy source to replace conventional fossil fuels in local industries and to reduce reliance on overloaded electricity grids. In this perspective, many medium-to-large agricultural, wood processing industries in developing

Energy efficiency and renewable energy sources for industrial sector 227

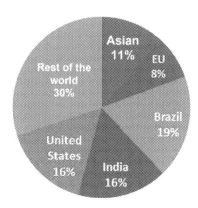

Figure 9.8 Country/regional shares of global biomass use in industry (IEA, 2017a).

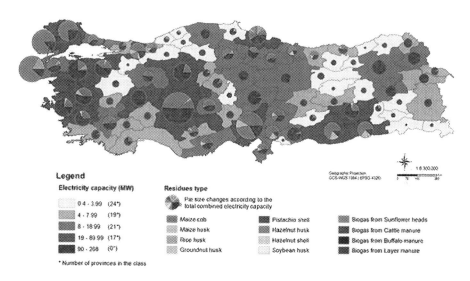

Figure 9.9 Electricity capacity generation (MW) from crop residues in Turkey.

countries and emerging economies are well placed to benefit from the successful development of biomass to energy. Turkey has rich biomass potential while limited sources of petroleum-based fuel made the subject of producing quality energy and productive usage. Among the RE sources, woody biomass seems to be the interesting because its share of the total energy production of Turkey is high, at about 10%, and the techniques for converting it to useful energy are not necessarily sophisticated.

9.5.2 Solar heat

Activity is accelerating with respect to solar heat for industries. While the deployment of small-scale solar water heating systems is slowing, large-scale solar-supported district heating systems and industrial applications are quickening. A recent study has identified over 130 companies in at least 22 countries worldwide that have realized more than 500 industrial plants with an overall combined installed collector area of 416,414 m^2 for process heat. This represents an installed capacity of only 280 MW$_{th}$, likely to produce 560 GWh$_{th}$ of heat per year at most, assuming a relatively high capacity factor of 2000 full load hours (Weiss and Spörk-Dür, 2019; Solrico, 2017).

The vast majority of projects use nonconcentrating technologies such as flat-plate collectors or evacuated tubes. These can be installed almost anywhere, as they use global solar irradiance, but they usually do not deliver usable heat above about 100°C. In recent years, however, new high-vacuum flat-plate collectors have been commercialized, which remain relatively efficient at temperatures up to 160°C (Horta, 2015). On the other hand, concentrating technologies, such as parabolic troughs, use direct irradiance only and are geographically limited to areas with good direct normal irradiance. But as these linear concentrating technologies can reach or even exceed 400°C, they could be a mean of supplying medium high-temperature process heat needs. Central receiver systems or "solar towers", which can achieve higher temperatures still, have so far developed in the power sector only (IEA, 2017a, 2018a).

Another important reason for this prevalence of nonconcentrating technologies rests in the relationship between solar heat cost and temperature level that makes competition with fossil fuels easier at low temperatures (Ministry of Energy and Natural Resources, 2016b) (Fig. 9.10). The food and beverage industry, the service

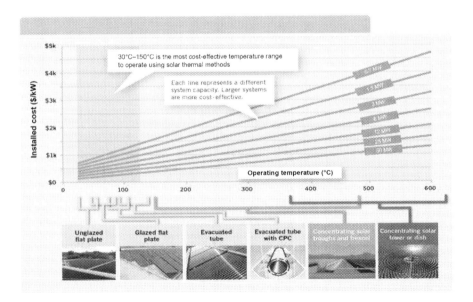

Figure 9.10 Cost of solar heat in relation to temperature and size of installation.

industry, and the textile industry, all of which mostly need low-temperature heat, are the main areas in which solar heat has been deployed.

Another example of low-temperature solar process heat is the 27.5 MW$_{th}$ system at mining company's Gaby copper mine in northern Chile, the largest such system in service so far: it has 39,300 m^2 of flat-plate collectors and 4000 m^3 of thermal energy storage, and supplies 85% of the process heat needed to refine copper. It was commissioned in 2013, in one of the world's best areas for concentrating or nonconcentrating solar power or heat (Weiss and Spörk-Dür, 2019).

There will likely be acceleration in deployment of solar process heat in upcoming years. In Oman, US-based start-up company Glass-point is currently building a much larger plant than all existing ones—not only individually, but taken altogether, as it will eventually reach a capacity of 1.0 GW$_{th}$. It will produce solar steam for Petroleum Development Oman's enhanced oil recovery operations, saving large natural gas consumption and associated CO_2 emissions. Given its sheer size, this unique innovation, based on a reinvention of parabolic trough technology, will also shift deployment of process heat technologies to solar concentrating ones (Weiss and Spörk-Dür, 2019; Solrico, 2017; Horta, 2015; IEA, 2017a).

9.6 Case study in Turkey

9.6.1 National Energy Efficiency Action Plan

Energy efficiency is an area that complements and crosscuts such national strategic goals as easing the burden of energy costs on the economy, ensuring energy supply security, alleviating risks arising from external dependency, transition to low-carbon economy, and protection of environment. The increased importance of sustainable development also increases the value of efforts in energy efficiency. This moves all countries toward improving energy efficiency and accelerates resolute action to that end.

The energy consumption in our country increases faster than in developed countries on account of such reasons as population growth, rising prosperity, strengthening service sector and industrialization. The primary energy consumption reached 120 Mtoe in 2015, representing an increase of 46% compared to 2005. With the import rate of 74% in 2015 in energy resources for primary energy supply, our country is in the category of high external-dependency countries. Under this scenario, according to the Republic of Turkey Ministry of Energy and Natural Resources, beginning in 2018 at the latest, there will be a supply shortage. This will be so even if the 13,762 MW private and 3475 MW public projects under construction after 2009 are completed. Fig. 9.11 shows the primary energy supply by resource type in Turkey (Ministry of Energy and Natural Resources, 2016a).

Premised on the mission to utilize energy and natural resources efficiently and environment-friendly to make the highest contribution to the national prosperity, and the vision to build a secure future in energy and natural resources, our country

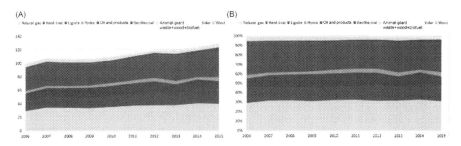

Figure 9.11 Primary energy supply by resource type, Turkey, 2006−15. (A) Mtoe and (B) Percentage (Ministry of Energy and Natural Resources, 2016a).

aims to increase efficiency in all processes from energy generation to end-use consumption.

In this context, the Energy Efficiency Law adopted in 2007 started a new transformation process. The Energy Efficiency Strategy issued in 2012 set energy efficiency goals for 2023, and the National Energy Efficiency Action Plan (NEEAP) was formulated for effective implementation and monitoring (Ministry of Energy and Natural Resources, 2016a,b).

Formulated in consideration of our country's current needs and good practices in the world, the NEEAP lays down the implementation steps, key performance indicators, how to implement, outputs, and potential impact. Since energy efficiency is a multidisciplinary area crosscutting many sectors and relevant to many stakeholders, it is necessary to build close cooperation among institutions and organizations in charge of implementing the actions defined in this plan and evaluating the outcomes. The General Directorate of Renewable Energy of the Ministry of Energy and Natural Resources will conduct the process of ensuring the said coordination and cooperation, monitoring, reporting, and validating the action plan.

Under the NEEAP that will be implemented in the period of 2017−23, it is aimed to reduce the primary energy consumption of Turkey by 14% by 2023 through 55 actions defined in six categories namely buildings and services, energy, transport, industry and technology, agriculture, and crosscutting (horizontal) areas. It is also projected to achieve savings of 23.9 Mtoe cumulatively by 2023, for which 10.9 billion USD of investment will be made (Table 9.1). The cumulative savings by 2033 will be USD 30.2 billion at 2017 prices, where the effect of certain savings will continue through 2040. The average payback period for actions is 7 years.

9.6.2 General overview

Turkey's gross domestic product (GDP) steadily increased in the period of 2005−15 except for 2009 and 2010. The cumulative growth in the period is by 65%, corresponding to an annual GDP growth 5.2%. The primary energy consumption in the same period grew by 46%, that is, lower than the GDP growth. This means that less energy is consumed to produce a unit of added value. On the other hand, the primary energy intensity index, a significant indicator of energy efficiency, decreased cumulatively by 23.1% in the period of 2000–15 on account of the measures taken, achieving an average annual improvement of 1.65%.The end-use energy intensity index in the same period (2000–15) went down by 21% corresponding to an average annual improvement of 1.5% (Ministry of Energy and Natural Resources, 2016b).

The energy intensity index is an important indicator of the contribution to the national economy through the energy savings and energy efficiency improvement in primary and end-use energy consumption in manufacturing, housing, and transport sectors. The breakdown of the indicator shows that, for the period of 2000–15, the manufacturing sector achieved an annual improvement rate of 1.8%, housing sector—1.9%, and transport sector—2.7% for an overall annual improvement of 2.1% in energy efficiency.

The energy efficiency measures implemented in the period of 2000–15 resulted cumulatively in 9.7 Mtoe of energy savings in the manufacturing sector, 7.1 mtoe in the housing sector, and 24.6 mtoe in the transport sector amounting a total of 41.5 mtoe. The fact that our energy intensity is high while our energy consumption per capita is lower than developed countries shows that Turkey has a significant potential for energy savings (see Fig. 9.6). In the period of 2005–14, when Turkey's GDP increased by 1 unit, the energy consumption increased only by 0.7 unit. In the same period, against the 1 unit of increase in GDP, France reduced its energy consumption by 1.1 unit, Germany by 0.7 unit, Japan by 3.3 units, and the United Kingdom by 2.0 units (Ministry of Energy and Natural Resources, 2016a,b).

The primary energy intensity in 2015 in Turkey is 0.12 toe per 1000 USD at 2010 prices, estimated on the basis of 2009-based new GDP series published by the Turkish Statistical Institute (TURKSTAT) on December 12, 2016. While this figure is lower than the world average of 0.18 toe, it is higher than the OECD average of 0.11 toe. The figure is 0.08 in Germany, 0.07 in Italy, and EU-28 average is 0.09 toe (Ministry of Energy and Natural Resources, 2016a).

9.6.3 Industry and technology

The industry with a share of 26% in GDP 2015 has continued in recent years to grow and drive the economic growth in our country as in many other countries. With the industry accounting for 32.4% of the end-use energy consumption and 47.6% of the net electricity in 2015, the Turkish economy is more "energy-intensive" compared to developed countries. Energy efficiency has become a priority area due to the fact that energy costs constitute one of the heaviest burdens on enterprises. Improvement in energy efficiency in industry offers significant opportunities to reduce energy consumption, as well as improve process efficiency, upgrade technological development levels, and reduce GHG emissions (Ministry of Energy and Natural Resources, 2016a).

The Energy Efficiency Law introduced requirements for industrial enterprises of a certain size to commission energy efficiency audits and establish energy management structures. In addition, various support mechanisms were introduced such as efficiency improvement projects and voluntary agreements. Investments for energy efficiency projects designed to save energy by at least 20% on the baseline, with a simple payback period of 5 years or shorter in manufacturing industry plants with minimum 500 toe of annual energy consumption are allowed to benefit incentives accorded to investments made in the fifth region (Ministry of Energy and Natural Resources, 2016a).

The energy efficiency strategy aims to reduce energy densities at rates not to be less than 10% to be set by sectoral cooperation in each subsector of the industry. To achieve such objective, various actions are defined such as promoting investments to improve energy efficiency and identifying savings potentials and measures that can be taken for energy efficiency in industrial subsectors.

Under this action plan, seven actions are defined to improve energy efficiency in the industry and technology sector in line with the strategic goals described previously. The said actions involve activities such as engaging in supporting activities, mapping the energy saving potential in industry, increasing the diversity of projects, defining new support mechanisms, scaling up cogeneration systems in large heat-using facilities, implementing environment-friendly design, and labeling system for appliances (Ministry of Energy and Natural Resources, 2016a,b).

9.6.4 Aim of the development plans

In Turkey, tenth development plan was formed for sustainable economic and environmental development. This economic road map for the period up to 2018 forecasts a 5.5% annual growth rate in the economy, increase in annual exports to $277 billion, an increase in Turkey's per capita income to $16,000, reducing unemployment to 7.2%, as well as lowering inflation to 4.5%. With the plan which named as "Energy Efficiency Target Strategy" is published in 2012, the energy intensity will be reduced at least 20% until the year 2023. To increase energy efficiency in industry, following actions were defined replacing low-efficient AC electric motors, which consume more than 70% of the electricity used in industry, with the high-efficient ones (Ministry of Energy and Natural Resources, 2016a).

Energy efficiency is highlighted as one of the measures to reduce production cost in industry. In addition to decrease energy intensity of the economy, current energy efficiency studies and activities will be continued. Natural resources to be used efficiently, wastes will be recovered to contribute to the economy. If environment-friendly approach has been used in practice applications, so the new business opportunities are used. Therefore some green growth opportunities are supported for the development of new products and technologies will be possible. According to the sustainable cities approach, environmental sensitivity and quality of life will be increased with applications such as waste and emission reduction in cities, increase energy, water and resource efficiency, recycling, noise and visual pollution prevention, using eco-friendly materials. In manufacturing and services, RE, eco-efficiency, and eco-friendly practices such as clean production technologies will be supported, and environment-friendly development of new products and branding will be encouraged

9.7 Policy options

In this section, lessons learned from the recent deployment of renewables in various industrial sectors are considered first to provide useful information to industrial decision/policy makers at national and possibly subnational levels. With respect to international trade in energy-intensive industries, such as chemicals, iron and steel, and cement, however, the main barrier for most products is likely to be competition from current GHG emitting technologies, as most substitutions with RE reveal a positive cost of carbon abatement (Philibert, 2017).

9.7.1 Lessons learned

Various barriers still hinder full RE deployment in the industry sector, but industrial decision/policy makers have a wide array of options available to overcome them. Eight issues were identified in the RE technology deployment industry study (International Energy Agency, 2018) that could influence industries in deciding whether to deploy RE production assets in their facilities. Diverse policy options can be implemented to surmount these barriers, which are summarized below (Philibert, 2017; Kreith and Goswami, 2008; International Energy Agency, 2018):

- *Energy supply regulatory regime*: In various jurisdictions, it is difficult to supply energy using independent producers and valorize energy through self-consumption and/or the right to sell excess energy produced. These are fundamental requirements for deploying renewables in industry.
- *Operability and integration*: Industrial installations and processes might not be adapted to integrate RE assets, especially for renewable heat, including biomass. Deep integration of renewables into processes can provide the best results, but also risks perturbing core processes. Switching from fossil fuels to renewables' production can add complexity and cost.
- *Investment*: RE projects require high upfront capital expenditures.
- *Return on investment*: Renewables' projects often show relatively low returns on investments; moreover, even if the profitability of assets with long technical life span is high, the payback time might be long compared with the company's core activities. This commonly used criterion may dissuade industrial decision makers.
- *Risk and insurance*: RE installations that use immature technologies or lack backup generation present a supply continuity risk, but also offer a hedge against fluctuations in market energy prices and often improve energy security. In some cases, faulty equipment can put the safety of the whole facility in jeopardy. More often, long-term continuity of the operation and solvency of the off-taker present a risk for RE investments.
- *Contractual scheme complexity*: The complexity of contracts between industrial customers, third-party power (or heat) producers, and utilities can deter investment.
- *Technology maturity*: While solar, wind, geothermal, and various biomass technologies are mature, some other options (power to gas, trigeneration) are less, so carry additional risks.
- *Awareness*: Industrial companies often lack awareness about technologies, possible incentives or guarantees, costs, and best practices.

RE portfolios and obligations to buy renewable power from independent power producers have been instrumental in ensuring rapid deployment and cost reductions in solar and wind power; similarly, obligations for shares of solar heat in domestic hot water supplies have proven effective in driving deployment of solar water heaters in a number of jurisdictions, where the obligations are now often integrated into strict building codes that either explicitly mandate the use of renewable heat or require it in practice. Such instruments should still be considered by governments and other public authorities at various levels (Philibert, 2017).

9.7.2 International agreements

Energy- and GHG-intensive industries, whose products are traded internationally, may not be in a position to support additional costs for process modification to reduce CO_2 emissions. Furthermore, imposing carbon prices or directly regulating emissions potentially puts carbon-constrained industries at a competitive disadvantage relative to their unconstrained competitors. Governments thus fear that uneven carbon constraints could enhance the competitiveness of noncarbon-constrained producers (Philibert, 2017).

This risk is usually characterized as "carbon leakage" as implementing uneven GHG emissions constraints could lead to an increase in emissions outside the given country or region. This could result from short-term deterioration in competitiveness, whereby carbon-constrained industrial products lose international market shares to the benefit of unconstrained competitors. In the longer term, differences in returns on capital associated with uneven emission mitigation actions provide incentives for firms to relocate their capital to countries with less stringent climate policies. Finally, reduced energy demand and lower prices in some countries may trigger higher energy demand and associated GHG emissions elsewhere (Philibert, 2017).

Carbon leakage is formally expressed as the ratio of emissions-increase outside the country that has implemented domestic mitigation policies, over the emissions-decrease within that country. This ratio could be above or below 100%: if below, it means that the global GHG emissions increase if emissions in the regions benefitting from a shift in production are greater than the emissions reduction in the regions implementing a constraint of any kind. This could happen, for example, if specific GHG emissions in the regions taking mitigation action were already lower than the specific emissions in the outside regions.

The Paris Agreement is significantly more encompassing than the Kyoto Protocol in including China, India, and other emerging economies. However, the heterogeneity of the current national pledges may still not provide enough security for industries in the most exposed sectors. On the other hand, cement is less at risk, as it is traded much less than ammonia and steel. With respect to chemicals, the situation varies according to the cost of shifting to renewables (International Energy Agency, 2018). If it is low, as for ammonia used as an industry feedstock, the risk of carbon leakage is low, while it cannot be ignored for methanol and high-value chemicals; the current shift in production. Finally, international competition and trading are intense in the iron and steel subsector (Philibert, 2017).

A global agreement to create a state of equality with respect to GHG emissions, for example, with a globally coordinated "single" carbon pricing system, would in principle solve this issue. While some economists consider it to be the "first-best" option, others point out that its adoption is very unlikely from a political standpoint, for it does not take inequalities in economic development into account: a single world price would be too high for some countries and too low for others. Global agreements by sector might be more realistic. Compared with countrywide, quantified targets, sector-wide approaches may (Asia-Pacific Economic Cooperation, 2013):

- require lighter monitoring and enforcement,
- more effectively link economic agents in these sectors and international investors,
- settle part of the abatement cost uncertainty inherent to uncertain economic growth,
- create emissions leakage from sectors covered to those unconstrained,
- complicate international negotiations with sector-specific technicalities.

9.7.3 Procurement

To ensure prompt deployment of innovative clean technologies based on renewables, public and private procurement of clean, carbon-free materials may be the most realistic short-term option. For example, the cost of steel represents only a small fraction of the overall cost of a vehicle. Manufacturers of brand products may want to bolster their green performance image they project to their customers and the general public, including their own stakeholders, whether out of personal conviction. Electric cars and plug-in hybrids could lead the transition toward "green steel", and RE developers could pay attention to the life-cycle emissions of wind turbines (Philibert, 2017).

Many developers have done this already, procuring green power and now turning their attention to the "gray energy" embodied in their products and to procuring preferably cleaner materials. It identified six areas in which suppliers can focus their clean energy efforts: agriculture, waste, packaging, deforestation, and product use and design (Philibert, 2017). Sustainable procurement of wood and paper-based products is not only a project of the World Resources Institute, but is also becoming a mandatory reference in industry and commerce, and for sustainable fisheries (Baron, 2016; Baron and Garret, 2018; Reinaud, 2008; International Energy Agency, 2019).

Public procurement may play a similar role. Public procurement accounted for 13% of the GDP of OECD countries in 2013, and even more in some emerging and developing economies. All jurisdictions, public services, and companies have buildings constructed for their own operations or for the public—schools, hospitals, social housing, and so on. They procure vehicles of all sorts, railways, bridges, roads, and other infrastructure, and therefore manage concrete, cement, and steel in massive quantities (Baron, 2016).

The primary objective of procurement is obviously to find and buy products and services that offer good value for taxpayers' money. However, as a government-operated instrument, public procurement should also be aligned with a country's broad policy objectives, balancing these objectives with its primary purpose of finding the best value for public money. Designing public procurement to promote low-carbon innovation can be justified on three grounds (Baron, 2016):

- Structural inefficiencies in government purchasing, for example, focusing on upfront acquisition costs when including operating costs could lead to a more environmentally conscious choice.
- Environmental market failure, for example, the absence of a price on CO_2 emissions due to political constraints, while an individual government may choose to include a CO_2 price to guide its own decisions.
- Insufficient support for innovation in light of positive externalities related to demonstrating and adopting new technologies, learning, and network externalities.

9.8 Conclusions

The combined application of RE and energy efficiency in industry is a natural marriage insofar as industry operators who have the foresight to convert their plants from using fossil fuels to renewable fuels are very likely to maximize the value of the renewable fuel by maximizing the efficiency of its use in their plants. Governments can create regulatory and business environments that promote development of RE and energy efficiency in industry, and industry will respond by developing business models configured to extract maximum value for the business from the opportunity available. There is, therefore, a need to broaden our thinking to include the goal of efficient use of RE in industry to achieve maximum value. This value flows through to the whole economy and, ultimately, the planet and its inhabitants.

While there are many examples of the combined application of RE and energy efficiency in industry, their penetration to date has not been extensive. The applications considered most likely to achieve significant penetration in the middle term are the use of: biomass energy for process heat, biomass as a petrochemical feedstock, solar thermal systems for process heat, heat pumps for process heat, and it has been suggested that RE, if used efficiently, has the potential to supply 23% of final energy use in the global manufacturing industry and up to 14% of fossil feedstock can be replaced by biomass. Together, this equates to 21% of total final energy use.

Reducing long-term GHG emissions of the industry sector is one of the toughest challenges of the energy transition. Combustion and process emissions from cement manufacturing, iron- and steelmaking, and chemical production are particularly problematic. This chapter considers a variety of current and forthcoming options to increase the uptake of renewables as one possible way to reduce industry sector energy and process CO_2 emissions. The main finding of this chapter is that the recent rapid cost reductions in solar PV and wind power may enable new options for greening the industry from electricity or the production of hydrogen rich chemicals and fuels. Simultaneously, electrification offers new flexibility options to better integrate large shares of variable renewables into power grids.

Turkey's main energy sources are natural gas and petrol; the country's import dependency on these sources are 98% and 92%, respectively. In 2013, Turkey spent almost USD 56 billion on energy imports (total imports approximately 252 billion). Due to the increasing demand, this figure is expected to rise to USD 106 billion by 2023. Industry's share in total final energy consumption is 39%, which make it the biggest energy consumer in Turkey and buildings consumption about 30% of total final energy. These two sectors also have the highest increase in energy demand. Industry's annual savings potential is USD 3.0 billion. This potential corresponds to 35% of industry's energy consumption in 2017. The share of energy costs in energy-intensive industrial sectors amounts to 20%–50% of total production costs. Turkey's strategic purposes for sustainable energy future are:

1. to reduce energy intensity and energy losses in industry and services sectors;
2. to decrease energy demand and carbon emissions of the buildings, to promote sustainable environment-friendly buildings using RE sources;
3. to provide market transformation of energy-efficient products;
4. to increase efficiency in production, transmission, and distribution of electricity, to decrease energy losses and harmful environment emissions;
5. to reduce unit fossil fuel consumption of motorized vehicles, to increase share of public transportation in highway, sea road, and railroad, and to prevent unnecessary fuel consumption in urban transportation;
6. to use energy effectively and efficiently in public sector

Acknowledgment

The author acknowledged to the Turkish Academy of Science for financial support during the preparation of this chapter.

References

Asia-Pacific Economic Cooperation (APEC), 2013. Best Practices in Energy Efficiency and Renewable Energy Technologies in the Industrial Sector. Hunterville, New Zealand.

Baron, R., 2016. 12–13 April The Role of Public Procurement in Low-Carbon Innovation. Background Paper for the 33rd Round Table on Sustainable Development. OECD, Paris.

Baron, R., Garret, J., 2018. 28–29 June Trade and Environment Interactions: Governance Issues. Background Paper for the 35th Round Table on Sustainable Development. OECD, Paris.

Energy Information Administration (EIA), 2019. International Energy Outlook 2019 with Projections to 2050. U.S. Department of Energy, Washington, DC.

European Union (EU), 2015. Energy Efficiency Trends and Policies in Industry. EU, Brussel.

Hasanbeigi, A., Arens, M., Price, L., 2014. Alternative emerging iron-making technologies for energy-efficiency and carbon dioxide emissions reduction: a technical review. Renew. Sustain. Energy Rev. 33, 645–658.
Henzler, M., Hercegfi, A., Barckhausen, A., 2017. Industrial Energy Efficiency and Material Substitution in Carbon-Intensive Sectors. Technology Committee, Bonn, Germany.
Horta, P., 2015. Available Solar Collector Technologies: Overview and Characteristics. IEA/SHC SHIP, Fraunhofer ISE, Freiburg, Germany.
IEA, 2017a. World Energy Outlook 2017. OECD/IEA, Paris.
IEA, 2017b. Energy Efficiency 2017. OECD/IEA, Paris.
IEA, 2018a. World Energy Balances 2018. OECD/IEA, Paris, <www.iea.org/statistics>.
IEA, 2018b. Energy Efficiency Indicators 2018 (Database). OECD/IEA, Paris.
IEA, 2018c. CO2 Emissions from Fuel Combustion (Database). OECD/IEA, Paris.
International Energy Agency (IEA), 2018. Energy Efficiency 2018: Analysis and Outlooks to 2040. OECD/IEA, Paris.
International Energy Agency (IEA), 2019. Global Energy & CO_2 Status Report: The Latest Trends in Energy and Emissions. OECD/IEA, Paris.
International Renewable Energy Agency (IRENA), 2016. Renewable Energy Benefits: Measuring the Economics. IRENA, Abu Dhabi, <www.irena.org/Documents>.
International Renewable Energy Agency (IRENA), 2015. Renewable Energy Options for the Industry Sector: Global and Regional Potential until 2030. IRENA, Abu Dhabi.
IPCC, 2014. Climate change 2014 – mitigation of climate change. Working Group III Contribution to the Fifth Assessment Report of the Intergovernmental Panel on Climate Change. <http://www.ipcc.ch/pdf/assessment-report/> (accessed 24.09.19.).
IRENA and C2E2 (Copenhagen Centre on Energy Efficiency), 2015. Synergies Between Renewable Energy and Energy Efficiency. IRENA, Abu Dhabi; C2E2, Copenhagen.
Kreith, F., Goswami, D.Y. (Eds.), 2008. Handbook of Energy Efficiency and Renewable Energy. CRC Press, Boca Raton, FL.
Lovegrove, K., Edwards, S., Jacobson, N., Jordan, J., Peterseim, J., Rutowitz, J., et al., 2015. Renewable Energy Options for Australian Industrial Gas Users. IT Power, Turner, Australia.
Ministry of Energy and Natural Resources (MENR), 2016a. Energy Statistics in Turkey. MENR, Ankara, Turkey.
Ministry of Energy and Natural Resources (MENR), 2016b. National Energy Efficiency Action Plan (NEEAP) 2017–2023. MENR, Ankara, Turkey.
Philibert, C., 2017. Renewable Energy for Industry: From Green Energy to Green Materials and Fuels. International Energy Agency, OECD/IEA, Paris.
Reinaud, J., 2008. Climate Policy and Carbon Leakage, IEA Information Paper. OECD Publishing, Paris.
Solrico, 2017. Solar thermal process heat: Surprisingly popular. Sun & Wind Energy. <www.sunwindenergy.com/> (accessed 20.09.19.).
Weiss, W., Spörk-Dür, M., 2019. Solar Heat Worldwide: Global Market Development and Trends in 2018, 2019 ed. IEA-SHC, Paris.

Energy efficiency in tourism sector: eco-innovation measures and energy

Sánchez-Ollero José-Luis[1], Sánchez-Cubo Francisco[1], Sánchez-Rivas García Javier[2] and Pablo-Romero-Gil-Delgado María[2]
[1]Department of Applied Economics, Faculty of Tourism, University of Malaga, Malaga, Spain, [2]Department of Economic Analysis and Political Economy, Faculty of Economics and Business, University of Sevilla, Sevilla, Spain

Chapter Outline

10.1 Introduction 239
10.2 State of the arts 240
10.3 Methods and data 242
10.4 Results and discussion 244
10.5 Conclusions 245
References 246

10.1 Introduction

The importance of the tourism industry for the Spanish economy is well known. This sector represents, according to the Spanish Tourism Satellite Account,[1] 11.2% of GDP and 13% of total employment. Unlike other sectors of the Spanish economy, the tourism sector withstood the worst years of the economic crisis thanks, above all, to international tourism. Despite this, it has had to suffer significant cuts in terms of employment, wages, and business performance (García-Pozo et al., 2015).

At the same time, tourists considered as consumers have drastically changed their requirements in terms of demand, introducing the environmental variable as one more element when deciding where and with what service providers they are going to contract their leisure or business trips (Sánchez-Ollero et al., 2014, 2016).

Many authors, Gutiérrez and Jiménez-Arellano (2011) and García-Pozo et al. (2018), among others, point out that growth in the Spanish economy in the 21st century has been characterized by a very moderate increase in labor productivity and effective stagnation in total-factor productivity. The determining cause of this phenomenon is the low technological effort made by Spanish companies: that is, a

---

[1] http://www.ine.es (accessed 05.06.19.).

Energy Services Fundamentals and Financing. DOI: https://doi.org/10.1016/B978-0-12-820592-1.00010-5
© 2021 Elsevier Inc. All rights reserved.

small percentage of companies engage in R&D, and their ability to turn this effort into innovation is modest.

In this context, eco-innovation comprises all the measures implemented by the actors or politicians involved in this issue; for example, those which contribute to reducing environmental damage by the development of new ideas, processes, or products. The European Union's Executive Agency for Small and Medium-sized Enterprises (EASME) (2017) suggested that "Eco-innovation is about changing consumption and production patterns and market uptake of technologies, products, and services to reduce our impact on the environment. Business and innovation come together to create sustainable solutions that make better use of precious resources, reduce the negative side effects of our economy on the environment, and create economic benefits and competitive advantage."[2]

The aim of this work is to determine if innovations investments are motivated by the need of energy reduction either for an environmental motivation or simply as an element of cost reduction in the hospitality firm.

10.2 State of the arts

The concept of environmental innovation, also known as eco-innovation, is recent and therefore in a continual process of development and review. For instance, the Oslo Manual of the OECD (1992) and OECD, Eurostat, European Commission (1997) defined the conventional understanding of innovation, but one of the most widely accepted definitions of eco-innovation among researchers in this field was provided by the Europe INNOVA project (2006). In this, eco-innovation is defined as "the creation of novel and competitively priced goods, processes, systems, services, and procedures designed to satisfy human needs and provide a better quality of life for everyone with a whole-life-cycle minimal use of natural resources (materials including energy and surface area) per unit output, and a minimal release of toxic substances."

In fact, modern interest in the concept began in the 2000s with the promotion of environmental technologies that contributed to the "Lisbon objectives" for growth and innovation, and to the "Gothenburg priorities" for sustainable development (Cainelli et al., 2011). During the same period, a new method to understand and analyze the concept of eco-innovation emerged that was based on three dimensions: target, mechanism, and impact. Its use has proliferated among companies and some governments in order to describe their contributions to sustainable development and to improve their competitiveness (OECD, 2010).

Eco-innovation has been defined in a variety of ways (Eco innovation Observatory, 2012; Fussler and James, 1996; Kemp, 2009; OECD, 2010; Wuppertal Institute, 2009), but none is clearly better than any other. As Rennings (2000) pointed out that innovation is different from invention, which is an idea or a model for a new improved product or process. For this author, in an economic sense, an

---

[2] http://ec.europa.eu/eaci/eco_en.htm (accessed 04.06.18.).

invention becomes an innovation when the improved product or process is first introduced to the market. In addition, Rennings identified the double externality problem, the regulatory push/pull effect, and the increasing importance of social and institutional innovation as three peculiarities of eco-innovation.

Other authors, as Kanerva et al. (2009), among others, did not define eco-innovation but suggested that environmental innovation, in its broadest form, includes any innovation that reduces environmental harm. In a broad sense, any innovation that reduces environmental damage is considered as eco-innovation.

All these definitions are easy to apply to the management of industrial sector companies. Traditionally, the concept of eco-innovation has been closely tied to the industrial sector in international studies, where the lack of indicators and information by which to design strategic plans to introduce eco-innovation measures has not been considered a limitation. The studies by Doran and Ryan (2012), Sierzchulaa et al. (2012), Horbach (2008), Brunnermeier and Cohen (2003), and Jaffe et al. (1995) are the examples of the analysis of eco-innovation in the industrial sector. However, its analysis in the hotel industry is quite complicated due to the specific characteristics of this sector. Thus Ludevid showed that tourists have a direct influence on the hotel sector, since their increasing environmental awareness has led to a greater demand for goods and services that respect the environment. It should also be noted that, unlike the industrial sector, environmental legislation in this field is almost nonexistent, and thus eco-innovative measures have been barely implemented by the hotel industry. This lack of legislation may be due to the fact that hotel activity is considered to have little environmental impact. Furthermore, as several authors have pointed out (e.g., Álvarez et al., 2001; García-Pozo et al., 2011), establishments in which the natural environment forms part of the tourism product, such as those specializing in sun and beach tourism or nature tourism, would be expected to be more concerned with eco-innovative measures. However, urban hotels, where the main client is the business traveler, are more concerned with the management of human resources and occupational health than with environmental management (Rodríguez-Antón et al., 2012). In fact, the type of tourism determines the type of eco-innovative measures the hotel is willing to implement. Thus depending on the characteristics of their clients, firms should pay more or less attention to environmental protection.

On the other hand, empirical studies have investigated the economic benefits that derive from environmental management and that are characterized by the use of variables related to hotel profitability (percentage of occupancy per room and beds, operating profit, market share, earnings growth, and so on), objective measures (gross operating profit per available room per day), and perceptual measures (Carmona-Moreno et al., 2004; López Gamero et al., 2013). To the best of our knowledge, studies on the hotel industry have mainly used financial variables, but have not addressed the issue of the impact on labor productivity of certain good environmental practices that could be considered as eco-innovative measures according to the definition of eco-innovation mentioned earlier. Comparisons of the economic performance of different groups of hotels have shown that hotels with the lowest values for all environmental management variables are on average significantly less profitable than the others. These empirical results suggest that environmental differentiation may be an order qualifier for hotels (Blanco et al., 2009).

According to Rodríguez-Antón et al. (2012), we found three reasons that account for environmental awareness within tourism firms. The first is based on the idea that the consumer who "pays to be green" leads the search for the forces underlying the relationship between the environment and firm's economic performance (Molina-Azorín et al., 2009). The second is based on the need to control the costs of hotel operations, their optimization, and the ability to provide a service that meets consumer needs (Judge and Douglas, 1998; Pan, 2003). The third lies in the guilt associated with the tourism activities due to the processes involved and the immense amount of water and material resources used, some of which are environmentally unfriendly (nondisposable plastic containers, nonrecyclable packaging, and so on) (Álvarez et al., 2001).

In this work, we intend to analyze in-depth these ideas that we understand are intimately intertwined in the firm's environmental actions. Not in vain, a saving in energy consumption means at the same time a saving in the costs of the company and a greater environmental respect.

10.3 Methods and data

The data used in the estimations were obtained from the Technological Innovation Panel (PITEC) database. PITEC is a panel-type database jointly prepared by the Spanish National Institute of Statistics and the Spanish Foundation for Science and Technology, which monitored the technological innovation activities of more than 12,000 Spanish companies between 2004 and 2014. The main users of PITEC are researchers who are interested in the field of innovation and economics. They access the data from the PITEC time series through the ICONO website, the Spanish R + D + I Observatory and use them in the empirical analysis of their studies. The data of the panel are also a key element of information for the public decision makers in their decision-making work regarding R & D & I policy.

The main objective of the survey on innovation in companies is to offer direct information regarding the innovation process in companies, compiling indicators that enable ascertaining the different aspects of this process (economic impact, innovative activities, cost, and so on). This large-scale study, apart from providing rich and varied information on the technological innovation process, may serve as the base framework for diverse specific studies on other aspects related to science and technology. Its methodology follows the guidelines proposed by the OECD for the collection and interpretation of data on innovation, better known as the Oslo Manual (OECD, 1992, 1997, 2005). The fact of using a methodology that is widely accepted on an international level enables reaching the objective of the international comparability of the results obtained, and providing our national experience in the study of innovation.

The survey uses the National Classification of Economic Activities, CNAE-2009, for encoding the activities of the companies, processing, and disseminating their data. This statistical research extends to all agricultural, industrial, construction, and service companies with at least 10 paid employees. According to the aims of this work, we have included the observations of Spanish companies classified in PITEC activity group 25 (hospitality), which correspond to the activities listed in the National Classification of Economic Activities, I, groups 55 and 56 (CNAE-20091).

The observations were treated as cross-sectional data by introducing dummy variables that represent the year of each observation and control the temporal nature of the data. This approach makes it possible to apply the methodology as described later. In addition, we eliminated companies with less than four observations in this period as well as any clearly abnormal observations according to criteria similar to those established by Lööf and Heshmati (2006) and Raymond et al. (2010). Thus the database used comprised an unbalanced sample of 200 companies and 2047 observations from the period analyzed.

The PITEC also uses the methodological framework established for the development of the community innovation survey developed by Eurostat and the innovation classification criteria contained in the Oslo Manual (OECD & Eurostat, 2005). Thus the data provided by PITEC can be compared with those of other European Union countries.

For achieving the aims of this study, we applied a logit multinomial model. This model is an extension of the binomial logit model in which the discrete dependent variable (Y) takes more than two values. These values $(j = 1, \ldots, J)$ can represent the different independent alternatives subject to the choice of an individual or to different groups in which the individuals are classified. Thus if we assume that (N_i) is the number of individuals $(i = 1, \ldots, N)$, the probability that an individual (i) with characteristics (X_i) will be classified in group (j) is as follows:

$$P_{ij} = P\left(Y = \frac{j}{X_j}\right)$$
$$= \frac{1}{1 + e^{-X_i \beta}} + U_i$$
$$= \frac{e^{X_i \beta}}{1 + e^{X_i \beta}} + U_i$$

where the parameters to be estimated are the (K) coefficients (β) associated to each of the (X) individual characteristics. In this way, the model can be written as:

$$P_{ij} = \Lambda(X_i \beta) + U_i$$

where Λ represents the logistic distribution function, U_i is a random variable that is distributed normal $N(0, \delta^2)$, X_i represents the (K) variables or characteristics that are fixed in the sampling, the dependent variable (P_{ij}) can take the values zero or unity, and represents the probability that an individual with characteristics (X_i) will be classified in the group (j).

Likewise, to assure the identification of the system of (J) equations that contain the (K) parameters of the model, we estimate ($J-1$) equations or set of parameters, establishing, accordingly, the standardization criterion that ($\beta_j = 0$).

It is convenient to take into account that the estimated vectors of coefficients β_j ($j = 1, \ldots J-1$) do not directly quantify the effect on the probability of belonging to each group of the increase in one unit in each of the variables, which make up (X_i). In contrast, the sign of each of these coefficients does indicate the direction of the change in the corresponding probability. Given this problem, a very convenient way to interpret the results of model estimation is to give the relative probability (odds ratio) of an alternative or group ($Y = j$, with $j = 1, \ldots, J-1$) with respect to the reference alternative ($Y = J$).

$$\text{Odds ratio} = \frac{P(Y=j)}{P(Y=J)} = e^{\beta_j/X_i}$$

In this case, the coefficients of the base alternative take the value 1, and the relative probability provides an idea of the effect that the change of a unit in a variable has on the probability with respect to the base alternative.

In our specification of the model, the dependent variable is a dummy variable that takes the value 1 if it is relevant to reduce energy consumption as an objective of business innovation and 0 when it is not relevant. The model includes variables that take into account the size and location of the firm, if belongs to a chain, receives or not public funding, and a set of dummy variables referred to the year of the data. In addition, three variables have been introduced that measure innovation in products, processes, and organizational structure of the company.

10.4 Results and discussion

The descriptive statistics and the results of applying the logit model show some interesting items. Regarding the location of the firm, almost 29% of the firms are based in the region of Madrid, more than double that in the second region in this ranking, Catalonia, in which 13.78% of the firms are located. If we compare this data with Andalusia, the most populated region of Spain with approximately 8.4 million inhabitants (Catalonia has 7.5 million inhabitants and Madrid has 6.5 million), we can determine the huge regional difference in this area. The rest of firms are distributed by the other 15 Spanish regions.

If we take into account the size of the firm, large firms represent 58.72% of the sample, and 47% of the total belongs to a chain. These two characteristics are important in terms of environmental proactivity, application of quality models, and increases in the profitability of tourism companies as shown in various scientific studies (e.g., Sánchez-Ollero et al., 2016; García-Pozo et al., 2015).

According to our results, the public funding the firms receive is scarce not reaching 2% of the companies in the best of cases. The lack of European funds is particularly noteworthy, since less than 0.4% of companies receive such funding, despite the efforts of the European Union to promote innovative initiatives and the availability of funds. Maybe it is the cause that in terms of innovation, only 14.17% of firms decide to undertake innovative measures for energy reduction and invest in innovation range between 9.42% and 16.61% of the firms.

The relevance of reduction on energy consumption as goal of innovative measures implementation is significantly influenced by process innovation, production innovation, and organizational innovation. Process innovation is particularly relevant; it shows an odds ratio of 32:34, five times more than innovation in products and much higher than the one registered by organizational change. The marginal effects reflect the link between the goal of reducing energy consumption and the firm's actions: if a firm realizes innovations in process, product, or organizational, respectively, the probability of that goal is relevant for the firm will increase in a 0.4804, 0.1954, and 0.0479 in relation to other firms with different goals.

The analysis of the data highlights the differential fact of the region of Catalonia compared to the rest of the Spanish regions. The fact that a hotel company is based in this region increases the likelihood that the objective of reducing energy consumption will be relevant to innovation. For the rest of the regions, these results have not been statistically significant.

The results confirm that, as indicated previously, belonging to a hotel chain has positive results on innovation proactivity, as well as to receive public funding. The size of the company, contrary to expectations, has not been statistically significant, nor has the age of the company or the perception of European funds.

Taking into account the year of the data (by dummy variables), the analysis of the variables reflects two different scenarios before and during/after the economic crisis, more favorable for this type of innovation before 2008.

In the Spanish case, the economic crisis began to affect companies in 2007 although the impact was more severe as of 2008. The crisis period did not begin to remit until the end of 2013, being the year 2014 the first where we could talk about a clear recovery of the economy. According to our data, it seems clear that innovation is a process that only takes place, at least in the Spanish case, during periods of economic growth, not contemplating by companies to undertake innovations in times of economic crisis.

10.5 Conclusions

In this chapter, we have analyzed the influence that the objective of reducing energy consumption can have on innovation decisions by tourism companies. For this goal, we have worked with data contained in the PITEC database that provides detailed information of a large group of Spanish companies over the years. After a process of filtering of the results, we have worked with a sample of 2047 observations. By means of the descriptive analysis of the sample and the formulation of an econometric analysis through a multinomial logit model in which the dependent variable was if firm decides to undertake innovative measures for energy reduction, we should highlight the following conclusions:

- The importance of the economic cycle in innovative decision-making by the tourism company. In this sense, it seems clear that only in moments of economic growth do these companies consider investing in innovation.
- The comparative advantage that seems to be detected in the region of Catalonia in relation to other regions. It is known the strong industrial fabric in this region and, in particular, in the province of Barcelona with a strong presence of R + D + I companies in the region. This favorable environment for innovation seems to be decisive in the case of tourism companies.
- Although the goal of reducing energy seems to be present in all types of innovations, it is in the innovation of processes where the influence of that objective seems more relevant in the decision-making process. This data is differentiating with respect to other sectors of the economy, especially industrial sectors, where innovation in products has a greater presence. However, it is consistent with the very nature of the tourist product that is composed mostly of intangible services.
- Belonging to a hotel chain and being able to count on public financing are two elements that positively help to make innovation decisions with the aim of reducing energy.

Related to the first of these conclusions, one of the limitations of our work has been not being able to count on data 2014 and over. Through the next wave of data from PITEC, we will be able to better refine our analysis in the immediate future. On the other hand, the limitations of the data source in terms of the provision of sufficient observations for some regions prevent us from a deeper regional analysis. Going in-depth in this observation, it would be interesting to be able to also carry out a somewhat more local analysis, perhaps at the provincial level, especially for the cases of the main Spanish cities.

References

Álvarez, M., Burgos, J., Céspedes, J., 2001. An analysis of environmental management, organizational context and performance of Spanish hotels. Omega 29, 457–471.
Blanco, E., Rey-Maqueira, J., Lozano, J., 2009. Economic incentives for tourism firms to undertake voluntary environmental management. Tourism Manage. 30, 112–122.
Brunnermeier, S.B., Cohen, M.A., 2003. Determinants of environmental innovation in US manufacturing industries. J. Environ. Econ. Manage. 45, 278–293.
Cainelli, G., Mazzanti, M., Zoboli, R., 2011. Environmentally oriented innovative strategies and firm performance in services. Micro-evidence from Italy. Int. Rev. Appl. Econ. 25 (1), 61–85.
Carmona-Moreno, E., Céspedes-Lorente, J., Burgos-Jiménez, J., 2004. Environmental strategies in Spanish hotels: contextual factors and performance. Serv. Ind. J 24 (3), 101–103.
Doran, J., Ryan, G., 2012. Regulation and firm perception, eco innovation and firm performance. Europ. J. Innov. Manage. 15 (4), 421–441.
Eco innovation Observatory, 2012. Europe in Transition. Paving the Way to a Green Economy Through Eco Innovation. Annual Report 2012.

Europa INNOVA, 2006. Thematic Workshop, Lead Markets and Innovation, 29–30 June 2006, Munich, Germany.
Executive Agency for Small and Medium-Sized Enterprises (EASME), 2017. <http://ec.europa.eu/eaci/eco_en.htm> (accessed 02.06.19.).
Fussler, C., James, P., 1996. Driving eco-innovation; a breakthrough discipline for innovation and sustainability by Claude Fussler with Peter James, 1996. Pitman Publishing, 364 pp, ISBN 0 273 62207 2
García-Pozo, A., Sánchez-Ollero, J.L., Marchante-Lara, M., 2011. Applying a hedonic model to the analysis of campsite pricing in Spain. Int. J. Environ. Res. 5 (1), 11–22.
García-Pozo, A., Sánchez-Ollero, J.L., Marchante-Lara, M., 2015. Eco innovation and management: an empirical analysis of environmental good practices and labour productivity in the Spanish hotel industry. Innov. Manag. Policy Pract. 17 (1), 58–68. Available from: https://doi.org/10.1080/14479338.2015.1011057.
García-Pozo, A., Sánchez-Ollero, J.L., Ons-Cappa, M., 2018. Impact of introducing eco innovation measures on productivity in transport sector companies. Int. J. Sustain. Transport. Available from: https://doi.org/10.1080/15568318.2017.1414340.
Gutiérrez, T., Jiménez-Arellano, C., 2011. Un análisis sobre la evolución reciente de la productividad en España. Boletín Económico de Información Comercial Española (ICE), Madrid, pp. 33–45 (No. 3009).
Horbach, J., 2008. Determinants of environmental innovation – new evidence from German panel data sources. Res. Policy 37, 163–173.
Jaffe, A.B., Peterson, S.R., Stavins, R.N., 1995. Environmental regulation and the competitiveness of U.S. manufacturing: what does the evidence tell us? J. Econ. Lit. 33 (1), 132–163.
Judge, W., Douglas, T., 1998. Performance implications of incorporating natural environmental issues into the strategic planning process: an empirical assessment. J. Manage Stud 35 (2), 241–262.
Kanerva, M., Arundel, A., Kemp, R., 2009. Environmental Innovation: Using Qualitative Models to Identify Indicators for Policy. Working Paper Series No 2009-047. United Nations University, Maastricht, The Netherlands.
Kemp, R., 2009. From end-of-pipe to system innovation. In: DRUID Summer Conference, June2009, Copenhagen, Denmark.
Lööf, H., Heshmati, A., 2006. On the relationship between innovation and performance: a sensitivity analysis. Econ. Innov. New Technol. 15 (4–5), 317–344. Available from: http://dx.doi.org/10.1080/10438590500512810.
López Gamero, M.D., Molina-Azorín, J.F., Pereira-Moliner, J., Pertusa Ortega, E.M., Tarí, J. J., 2013. Gestión medioambiental y rentabilidad: una revisión de la literatura en el sector hotelero. Cuadernos Económicos del ICE, 86.
Molina-Azorín, J.F., Claver-Cortés, E., Pereira-Moliner, J., Tarí, J.J., 2009. Environmental practices and firm performance: an empirical analysis in the Spanish hotel industry. J. Clean. Prod. 17, 516–524.
OECD, 1992. Proposed Guidelines for Collecting and Interpreting Technological Innovation Data – Oslo Manual, Paris, France, first ed.Retrieved from: <http://www.oecd.org/officialdocuments/?hf = 10&b = 0&r = %2Bf%2Flastmodifieddate%2F1992&sl = official_documents&q = oslo + manual&s = desc(document_lastmodifieddate)> (accessed 16.06.19.).
OECD, 2010. Eco Innovation in Industry: Enabling Green Growth. Executive Summary. OECD, París.

OECD & Eurostat, 2005. Guidelines for Collecting and Interpreting Innovation Data, third ed. OECD Publishing, Paris, France. Available from: http://dx.doi.org/10.1787/9789264013100-en.

OECD, Eurostat, & European Commission, 1997. Proposed Guidelines for Collecting and Interpreting Technological Innovation Data, second ed. OECD Publishing, Paris, France. Available from: http://dx.doi.org/10.1787/9789264192263-en.

Pan, J., 2003. A comparative study on motivation for and experience with ISO 9000 and ISO 14000 certification among Far Eastern countries. Ind. Manage. Data. Syst. 103 (89), 564−578.

Raymond, W., Mohnen, P., Palm, F., Schim van der Loeff, S., 2010. Persistence of innovation in Dutch manufacturing: is it spurious? Rev. Econ. Stat. 92 (3), 495−504.

Rennings, K., 2000. Redefining innovation-eco innovation research and the contribution from ecological economics. Ecol. Econ. 32, 319−332. Available from: http://dx.doi.org/10.1016/S0921-8009(99)00112-3.

Rodríguez-Antón, J.M., Alonso-Almeida, M.M., Celemín, M.S., Rubio, L., 2012. Use of different sustainability management systems in the hospitality industry. The case of Spanish hotels. J. Clean. Prod 22, 76−84.

Sánchez-Ollero, J.L., García-Pozo, A., Marchante-Mera, A., 2014. How does respect for the environment affect final prices in the hospitality sector? A hedonic pricing approach. Cornell Hospital. Q. 55 (1), 31−39. Available from: https://doi.org/10.1177/1938965513500709.

Sánchez-Ollero, J.L., Ons-Cappa, M., Febrero-Paño, E., 2016. An analysis of environmental proactivity and its determinants in the hotel industry. Environ. Eng. Manage. J. 15 (7), 1437−1445.

Sierzchulaa, W., Bakkerb, S., Maatb, K., Weea, B., 2012. Technological diversity of emerging eco innovations: a case study of the automobile industry. J. Clean. Prod. 37, 211−220.

Wuppertal Institute, 2009. Eco Innovation: Putting the EU on the Path to a Resource and Energy Efficient Economy, No. 38. Wuppertal Institute, Wuppertal, Germany.

Part 5

Energy services markets: development and status quo

Energy service markets: status quo and development

Marc Ringel
Faculty of Economics and Law, Energy Economics, Nuertingen Geislingen University, Geislingen, Germany

Chapter Outline

11.1 Introduction 251
11.2 The European framework for energy services 253
    11.2.1 Legal framework 253
    11.2.2 The European Union energy service markets: market volume, offers, and barriers 255
11.3 The German energy service market 259
    11.3.1 Legal framework and information sources 259
    11.3.2 Market overview 262
11.4 Developments of segments of the service market 264
    11.4.1 Advice services 264
    11.4.2 Energy management 265
    11.4.3 Contracting 267
11.5 Market development 268
11.6 Conclusions: lessons learned from the German case 271
References 271

11.1 Introduction

The European Union (EU) has committed to a nearly full decarbonization of its economies by 2050 as contribution to the Paris Agreement and underlying basis of the "European Green Deal." This aim requires a comprehensive transformation of the energy and economic systems. Seventy-five percent of EU emissions are energy related (European Commission, 2018a). This implies that the decarbonization aim will have strong repercussions on the energy sector. To this end, the European Commission has fostered the use of renewable energies and energy savings on both demand and supply side. Energy savings exist across all sections of the value chain of energy provision and uses. On the demand side, the building sector alone represents 36% of total EU GHG emissions and 40% of total final energy consumption. The European Commission estimates that some 75% of existing buildings are energy inefficient (European Commission, 2016a,b,c).

This picture is mirrored broadly in Germany as one of the EU's biggest energy markets. Implementing energy efficiency solutions is often technically complex and asks for high up-front investments (European Commission, 2016b). Energy services such as contracting can help to overcome these barriers, but still require technical advice. Since the early 1990s, the German government has actively encouraged energy services such as advice, audits, contracting, and energy management systems to support industrial and private users in their strive to save energy. Today, energy service companies (ESCOs) offer a wide range of energy services, starting from low-level standardized telephone advice to complex energy management systems. With this, Germany is among the most developed and dynamic energy service markets in Europe.

Before we set out to analyze the regulatory framework and the market dynamics of this market in the European context, it is necessary to first define the term "energy services." Various definitions have been advanced recently. Fell (2017) identifies 27 different definitions for "energy services" that stretch over 173 individual examples of such services. His study uses the definition advanced by ISO EN15900 and the European Commission's Joint Research Center (JRC). The JRC defines energy services as "an agreed task or set of tasks designed to lead to an energy efficiency improvement and other agreed performance criteria" (DIN/ISO, 2010; JRC—Joint Research Centre, 2017; Fell, 2017). Typically, these services encompass energy contracting, metering and monitoring, project implementation, energy management systems, or energy advice/consultations such as energy audits (JRC—Joint Research Centre, 2017). Whereas this definition might be preferable from an analytical point of view, we will stick to the definition used in the European Energy Efficiency Directive (EED). Article 2(7) EED defines energy services as: "[the] physical benefit, utility or good derived from a combination of energy with energy-efficient technology or with action, which may include the operations, maintenance and control necessary to deliver the service, which is delivered on the basis of a contract and in normal circumstances has proven to result in verifiable and measurable or estimable energy efficiency improvement or primary energy savings" (European Commission, 2012). The reason for using this definition is that the German market situation closely operates in the EU context and has largely taken over the EED definition into national law.

It is noteworthy that both energy service definitions and their realm highly vary with the national contexts. Dedicated literature on energy advice remains limited. In a broader international context, the International Energy Agency (IEA—International Energy Agency, 2017, 2007) and the JRC (Bertoldi and Boza-Kiss, 2017; JRC—Joint Research Centre, 2017) regularly publish reviews of the energy efficiency market. Largely, these reviews focus on the more advanced services such as contracting (Marino et al., 2011), ESCOs (Bertoldi and Boza-Kiss, 2017), energy performance contracting (EPC) (Laurenz and Warneke, 2014), or sectorial reviews (Labanca et al., 2015). National reviews are available for the United States (Larsen et al., 2012; Stuart et al., 2014), Russia (Roshchanka and Evans, 2016), the United Kingdom (Nolden et al., 2016), and France (Duplessis et al., 2012). The north European energy service markets have been analyzed for Sweden (Kjeang et al., 2017), and in Finland prevailing barriers have been reviewed (Kangas et al., 2018). Following European provisions, all EU countries have to present regular market reviews in their National Energy Efficiency Action Plans (European Commission,

2013). These plans also deliver further evidence on the European markets for energy services, including markets in Germany. Finally, the German government regularly investigates the status of its energy efficiency initiatives.

Our field of analysis is the German market for energy services. In order to present and classify market developments, it is first necessary to detail the European framework conditions. The remainder of this contribution is organized as follows: in Section 11.2, we will present the European legal framework for energy services. This is followed by a short overview on the EU market situation, which helps to understand and frame the German case study. Section 11.3 presents a similar legal and stylized fact overview for the German market, before analyzing the submarkets in detail in Section 11.4. Section 11.5 discusses remaining barriers and opportunities for the market, before policy conclusions round off our presentation in Section 11.6.

11.2 The European framework for energy services

11.2.1 Legal framework

Despite the strong interest in having a comprehensive offer of energy services and providers available in all EU Member States, the European legislation is the field that is only slowly emerging. The 2006 Energy Service Directive (ESD) was the first legislative act to introduce energy services and put forward first definitions of key terms such as "energy services" or "ESCOs" (European Commission, 2006). These definitions have been taken advanced in the 2012 EED (European Commission, 2012). Despite its 2018 revision (European Commission, 2018b), following key provisions of the EED 2012 still continue to apply (European Parliament Research Service, 2019):

- overall EU objectives for energy efficiency and supporting national energy saving objectives (20% primary energy savings by 2020; 32% by 2030 in the case of the EU) that act as a trigger to boost taking up energy services;
- harmonized legal definitions for energy services, energy service providers, energy audit, EPC (Article 2 EED), the measurement for energy savings (Annex IV and V EED), and the minimum EPC specifications for tendering out by public authorities (Annex XIII EED);
- exemplary role of the public sector: 3% renovation of national central government building stocks and promotion of energy services in the public sector by the means of public procurement (Article 5 EED);
- energy efficiency obligation schemes for energy suppliers or alternative measures by the national governments (Article 7 EED). These obligations are designed to act as a trigger for increasing both offer and demand for energy services;
- mandatory energy audits for non-SMEs every 4 years and availability of information on audits, both again acting as stimuli for market development on both supply and demand side (Article 8 EED);
- obligation for EU Member States to evaluate barriers against energy services and take appropriate measures to remove regulatory and nonregulatory barriers (Article 19 EED);

- possibility to set up national energy efficiency funds (Article 20 EED), which can act as a demand pooling mechanism for soliciting energy services.

Based on market stakeholder feedback, Article 18 EED assembles a number of provisions that are supposed to remove barriers against energy services and advance the national markets. This comprises the following issues (Szomolányiová and Keegan, 2018):

- creating transparency on provisions and clauses in energy service contracts, notably on requirements to guarantee energy savings and the rights and obligations of involved parties;
- disseminating information on available public support for taking up energy services, such as subsidies, grants and loans, or other incentives such as tax deduction possibilities to support energy services;
- supporting the development of quality labels for energy service offers;
- setting up and disseminating a list of available qualified or certified energy service providers. This list has to detail the respective qualifications and certifications;
- establishing model contracts for EPCs in the public sector, based on the specifications of Annex XIII EED;
- publishing and disseminating information on EPC best practices;
- establishing and publicizing one or more contact point(s) that provide guidance and information on energy services to final customers.

Whereas the EED acts as crosscutting legal instrument to foster energy services, several sectorial Directives and regulations exist that directly or indirectly act as a catalyst for energy service markets. The Energy Performance of Buildings Directive (EPBD; Directive 2010/31, now updated 2018/844) (European Commission, 2018c) promotes energy efficiency in buildings. The EPBD sets out a series of obligations to improve energy efficiency in the built environment. These obligations comprise the issuing and public display of energy performance certificates, inspection rules for heating and air-conditioning systems, and overall energy performance requirements for both new buildings and existing buildings in case of a major renovation. As many of the obligations are best met by the means of engaging dedicated energy service providers, they again stimulate market demand for energy services. For renovation activities in the public sector, public procurement regulations apply through Directive 2014/24/EU, which again link to the EED provisions on the exemplary role of the public sector to deploy energy efficiency solutions.

Further to the primary legislation, secondary standards and accreditation procedures have been developed, notably by European standardization bodies such as CEN/CENELEC. These help to increase transparency and trust of final customers. The most important of these standards are (Horowitz and Bertoldi, 2015; Szomolányiová and Keegan, 2018):

- European Standard EN ISO 50001:2011 on energy management systems
- A family of standards in the framework of EN 16247-1 on energy audits (buildings, transport, processes, competences of energy auditors)
- Energy use in the framework of environmental management systems in the framework of EN ISO 14001:2015

- Definition of energy services in standard EN 15900:2010
- Definition on methods to determine energy savings through a series of bottom-up or top-down methodologies, following earlier works of the European Commission and the JRC (EN 16212)

It is important to point out that despite the strive for harmonization of the national markets, the European Commission did not push for a European standardization of the services offered. Rather, it supports integration of national markets by the means of research and market uptake projects in the framework of the EU's "Horizon 2020" and "Horizon Europe" framework programs for research. This underlines the view that energy service markets are still largely national or even regional markets. However, Member States have to provide a regular qualitative review regarding the present and future developments of their markets (see Section 11.3). That said, the combined market volume of national markets and the emergence of international energy service providers lead to a continuous growth of the market and act as a considerable driver for the economy. Against this background, we will present the present market volume, offers, and barriers in a stylized overview in the following section.

11.2.2 The European Union energy service markets: market volume, offers, and barriers

The JRC and several projects under Horizon 2020 monitor the progress, developments, and barriers of energy service markets in the EU (Boza-Kiss et al., 2017; Szomolányiová and Keegan, 2018). As the European markets are among the most dynamic and developed markets on a global scale, many scientific contributions analyze developments and barriers to draw conclusions for other regions.

As energy services are defined in a rather general manner, they comprise many individual services. These services are provided as stand-alone services or as packages, assembling two or more service aspects. Fig. 11.1 provides an overview of energy services offered in the EU. As many countries had already a developed service market before the legal definitions introduced in the ESD and the EED, this overview serves rather as a heuristic means to distinguish energy services rather than presenting a complete picture of existing offers.

Volume and turnover for EU energy service markets can only be given in an approximate manner. This is due to the differing national definitions of energy services. Bertoldi and Boza-Kiss (2017) report revenues of ESCOs in the size of €2.4 billion. This volume is expected to grow to €2.8 billion by 2024. These figures compare to a market size of €5.6 billion (2015) in the United States, which is expected to double by 2024 (Talon and Gartner, 2015). However, as Boza-Kiss et al. (2017) and Bertoldi and Boza-Kiss (2017) point out, these figures compare to a €41 billion advanced by the Economist Special Report Energy and Technology (The Economist, 2015), which in fact would be outstripping the market size of the United States (€5.8 billion) and China (€10.6 billion) by large. After significant growth during the last years, many national markets enter a phase of moderate growth or stabilization.

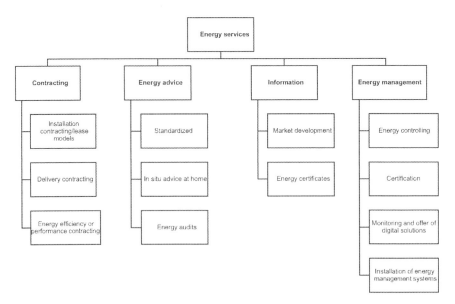

Figure 11.1 Analytic overview of energy services in the EU.
Source: Based on BfEE—Bundesstelle für Energieeffizienz, 2017. Untersuchung des Marktes für Energiedienstleistungen, Energieaudits und andere Energieeffizienzmaßnahmen: Endbericht BfEE 06/2015. Eschborn.

The European market for energy services can draw on a considerable number of offers. These offers are generally divided into offers for energy services (energy management, energy advice and auditing, provision of information) and energy contracting (energy supply contracting [ESC] and EPC). Still the lines between the services offered are often blurred, as each service provider or ESCO has a dedicated portfolio of services offered. Especially EPC offers are monitored closely as they can provide market support for delivering on the European energy and climate targets. Table 11.1 sums up the number of ESCOs, EPC providers, as well as status quo and outlook for the national ESCO markets of the EU28.

The offer is composed of international actors, national, and in many cases regional suppliers. It varies significantly between Member States, where a majority of markets can be categorized as small. This missing uptake of energy services is largely due to existing barriers that exist throughout all national markets in Europe. Table 11.2 presents an overview on these barriers, grouped into economic, organizational, policy, and regulatory, as well as technological and behavioral barriers. They act as impediment to taking up energy services. The overview outlines at which level (individual, communities, businesses, government, financial sector) these barriers exist.

In terms of a qualitative market review, Bertoldi et al. (2014) note that "Germany is [...] champion amongst the European ESCO markets in terms of maturity and market development," followed by very active markets in France, the United Kingdom, Austria, and the Czech Republic. On the other side of the

Table 11.1 Key figures on status quo and development of EU energy service markets.

Country	Number of ESCOs 2005	Number of ESCOs 2010	Number of EPC providers 2015	Number of EPC providers 2016	Level of development of the complete ESCO market	Expected development of EPC market
Austria	25	5–14	41	15–20	Excellent	Large public EPC projects are expected to further prevail, while the future of smaller public and private projects is less secure
Belgium	4	13–15	10–15	7(–20)	Moderate	Some growth both in ESC and EPC
Bulgaria	18	20	15	8–15	Preliminary	Unsure due to external barriers, but if removed, growth is expected
Croatia	1	2	10	5	Preliminary	Experts expect a boom in EPC, as the framework has improved, and more measures are pipelined
Cyprus	0	0	19	19	Initiation	Unsure due to barriers, but growth/kick-off is expected on the basis of recent efforts
Czech Republic	3	8–10	15	8–10	Excellent	Continued slow growth
Denmark	0	10	15–20	6–8	Well developed	ESC to develop, EPC is unsure (maybe starts in private sector)
Estonia	20	2	2–3	0	Nonexistent	Unsure, seems but some growth is expected
Finland	4	8	6–8	5–7	Moderate	Continued slow growth (mainly public sector)
France	n/a	10 + 100 small	300	10	Excellent	Continued growth
Germany	500–1000	250–500	c. 500	7–10	Excellent	Expectation for new (simplified EPCs), which may boost the market
Greece	0	2	47	1	Initiation	No development expected
Hungary	10–20	20–30	8–9	3–4	Preliminary	Unsure, dependent on external barriers
Ireland		15			n/a	Minor development of total ESCO market (no special focus on EPC)
Italy	20	50	200–300	4–5	Excellent	Continued slow growth, depending on the removal of barriers

(Continued)

Table 11.1 (Continued)

Country	Number of ESCOs 2005	Number of ESCOs 2010	Number of EPC providers 2015	Number of EPC providers 2016	Level of development of the complete ESCO market	Expected development of EPC market
Latvia	2	5	50–60	2–7	Preliminary	Expected to revive/grow and extend beyond multiapartment buildings
Lithuania	3	6	6	4–5	Preliminary	Due to the foreseen support, a slow growth and sectoral expansion are expected
Luxembourg	1–2	3–4	3–6	1	Moderate	ESC is expected to prevail, but some growth in EPC may be also seen
Malta	0	0	0	0	Nonexistent	With current circumstances, minor growth continued
Netherlands	0	50	100	15	Moderate	No change (no development) expected
Poland	8	2–10	3–4	10–15	Preliminary	Unsure
Portugal	n/a	10–12	15–20	10–15	Preliminary	Growth, as grants from EU dry out
Romania	2	14	20	<10	Preliminary	Depending on the removal of barriers, a growth is possible
Slovakia	10	5	8	10	Moderate	With the current conditions, no change
Slovenia	1	2–5	5–6	4–6	Preliminary	Stable or slow growth
Spain	10–15	>15	1000	20–30	Moderate	Significant growth based on the tenders already announced and based on the established frameworks + OP
Sweden	6–12	5–10	4–5	3	Preliminary	Overall decrease, with a possibility that ESC growth continues
United Kingdom	20	20	>50	25	Moderate	Further growth

Source: Based on Boza-Kiss, B., Bertoldi, P., Economidou, M., 2017. Energy Service Companies in the EU: JRC Status Review and Recommendations for Further Market Development with a Focus on Energy Performance Contracting. European Commission, Istra.

spectrum, market supply and dynamics still have to develop in Estonia, Malta, and Cyprus (Bertoldi et al., 2014). Given the frontrunner role of the German market, it can serve as point of analysis for successes and failures. This analysis can be used to draw lessons learned for market development in other countries.

11.3 The German energy service market

11.3.1 Legal framework and information sources

The German Federal government describes the German market as a "pluralistic, competitively organized and transparent market" (BMWi—Bundesministerium für Wirtschaft und Energie, 2011, 2014, 2017a). The market is competitively organized. Regulation is performed only in an indirect manner: whereas suppliers and offers are largely outside government legislation, quality standards are maintained by regulation of service outputs such as DIN/ISO norms for audits or energy management systems (BuildUp Skills, 2017). A push for high-quality offers is likewise safeguarded in an indirect manner: only "qualified" advice or service offers are eligible for government funding via the semipublic KfW bank or the Federal Office for Economic Affairs and Export Control (BAFA).

Market regulation is based on two pillars: (1) competition policy and (2) the German Energy Services Act (Energiedienstleistungsgesetz) of 2010, updated 2019. The Energy Services Act is largely deducted from European law, notably the EED. It transposes the EED definitions of energy services and mandatory energy audits for non-SMEs into national law. A revision of the law is presently ongoing to adapt it to the revised EED and introduce measures to stimulate further demand. The law installs the Federal Agency of Energy Efficiency (Bundesstelle Energieeffizienz, BfEE) as a unit within BAFA, which is tasked with the supervision and development of the German energy service markets. BfEE publishes an online list with energy service providers. This list provides information on contact data, services delivered, qualifications, experiences, and reference projects (BfEE—Bundesstelle für Energieeffizienz, 2019a). It is designed to increase market transparency and quality control (BAFA—Bundesamt für Wirtschaft und Ausfuhrkontrolle, 2017). Another dedicated list of experts that offer services in line with opening access to government funding, for example, for building refurbishment is maintained by the German energy agency (dena, 2019). Energy suppliers are obligated to refer to these lists on the annual electricity and gas bills.

Based on the ESD and EED, the German government reports on market develops in its National Energy Efficiency Action Plans (BMWi—Bundesministerium für Wirtschaft und Energie, 2011, 2014, 2017a). A more refined national analysis is undertaken by BfEE. BfEE has installed a regular external monitoring on annual basis to closely follow market developments (BfEE—Bundesstelle für Energieeffizienz, 2013, 2017, 2018, 2019b). In addition, evaluations of the market's individual segments are undertaken by the Federal Ministry of Economics and Energy on ad hoc basis (dena, 2017). Finally, KfW and the German energy agency

Table 11.2 Barriers against taking up energy services.

Field/barrier	Feature/description	Who is affected[a]
Economic		
Information barriers	Market fails to operate properly due to: imperfect information, incomplete markets (lack of knowledge, awareness, information)	I/B/C/G/F
Lack of appropriate market structure	Limited suppliers of energy efficiency solutions, such as ESCOs	I/B/C/G/F
Principal-agent problems	Imperfect competition and uncertainty; difficulty in proper pricing of energy efficiency services	I/B/C/F
Limited access to capital and high transaction costs	Lack of appropriate long-lasting financial and legal support; high costs for negotiating and enforcing energy efficiency solutions	I/B/C
Financial cost	High up-front investment costs; fear of additional service and maintenance costs	I/B/C
Perceived high risks	Energy efficiency seen as high-risk investment, thereby leading to high interest rates to cover risk factor	I/B/C/F
High uncertainty on payback	Up-front investments occur directly, whereas benefits only refinance these investments in future periods. This leads to high discount rates for future benefits, making energy efficiency investments less attractive	I/B/C/F
Split incentives/investor user dilemma	Landlord–tenant problem: investment costs, for example, building refurbishment would need to be shouldered by the property owner, whereas benefits (energy cost savings) would be fully on the tenant side	I/B/C
Organizational barriers		
Lack of agreement	For example, how dedicated provision of energy service should be measured and remunerated	I/B/C/G
Lack of supporting networks/structures	Missing fora/formats for gaining access to best practices in terms of technologies, policies or solutions	I/B/C/G
Missing qualifications or knowledge management	Unavailability of trained and qualified experts to implement energy efficiency solutions	B/C/G

Political and regulatory barriers

Missing or insufficient capacity to implement energy efficiency solutions	For example, insufficient staffing	B/C/G
Limited availability (e.g., program unavailability, inaccessibility)	Missing or insufficient support structures (e.g., energy agencies) to implement capacity building	I/B/C/G/F
Lack of supply chains, services and conventions	Missing standardization of applying, measuring or monitoring energy efficiency; missing standards on training and qualification schemes	I/B/C/G/F

Technological barriers

Limited supply of energy efficiency technologies	Limited availability of technological choice	I/B/C/
Technological "lock-in"	Path dependency on fossil fuels, for example, in coal territories or islands	I/B/C/G
Need of technological standardization	Missing technological solutions for metering and computing the large number of data to determine energy savings	I/B/C/
Communication and private data security	Delay in taking up energy efficiency solutions based on smart metering	I/B/C/G

Behavioral barriers

Cognitive biases in decision-making process	Potential factors are bounded rationality, resistance to change, confusion of choice (lack of professional advice)	I/B/C/G/F
Credibility and trust	Unwillingness to adopt unknown energy efficiency solution	I/B/C/G/F
Negative perceptions (negative values, not understanding)	Perceived idea that energy efficiency necessarily implies sufficiency or reduction of comfort	I/B/C
Negative word-of-mouth (i.e., negative information shared within a social network about the innovation)	For example, negative press reports on energy efficiency solutions (energy-efficient lighting, building refurbishment options)	I/B/C/G/F
Lacking information	See economic barrier, here however in the sense of nonawareness of saving options leading to suboptimal consumer choices	I/B/C

[a]*Legend:* I, individuals; B, business; C, communities; G, governments; F, financial sector and investors.

Source: Based on Kowalska-Pyzalska (2018), Sorrell (2015), Nygrén et al. (2015), Karakaya et al. (2014), Hu et al. (2015), Good et al. (2017), Gadenne et al. (2011), Negro et al. (2012), Pirlogea (2011).Kowalska-Pyzalska A., What makes consumers adopt to innovative energy services in the energy market? A review of incentives and barriers, Renew. Sustain. Energy Rev. 82, 2018, 3570–3581; Sorrell S. Reducing energy demand: a review of issues, challenges and approaches, Renew. Sustain. Energy Rev. 47, 2015, 74–82; Nygréen N.A. Kontio P., Lyytimäki J., Varho V. and Tapio P., Early adopters boosting the diffusion of sustainable small-scale energy solutions, Renew. Sustain. Energy Rev. 46, 2015, 79–87; Karakaya E., Hidalgo A. and Nuur C. Diffusion of eco-innovations: a review, Renew. Sustain. Energy Rev. 33, 2014, 392–399; Hu Z., Kim J-h, Wang J. and Byrne J. Review of dynamic pricing programs in the U.S. and Europe: status quo and policy recommendations, Renew. Sustain. Energy Rev. 42, 2015, 743–751; Good N., Ellis K.A. and Mancarella P., Review and classification of barriers and enablers of demand response in the smart grid, Renew. Sustain. Energy Rev. 72, 2017, 57–72; Gadenne D., Sharma B., Kerr D. and Smith T., The influence of consumers' environmental beliefs and attitudes on energy saving behaviours, Energy Policy 39 (12), 2011, 7684–7694; Geddes A., Schmidt T.S. and Steffen B., The multiple roles of state investment banks in low-carbon energy finance. An analysis of Australia, the UK and Germany, Energy Policy 115, 2018, 158–170; Negro S.O. Alkemade F. and Hekkert M.P. Why does renewable energy diffuse so slowly? A review of innovation system problems, Renew. Sustain. Energy Rev. 16 (6), 2012, 3836–3846; Pîrlogea, C., 2011. Barriers to investment in energy from renewable sources. Available at: <http://www.management.ase.ro/reveconomia/2011-1/12.pdf> (accessed 21.06.19.).

(dena) monitor and analyse the developments of the energy services markets (Bornemann, 2017). Their aim is to better align their funding programs to available energy services and to take into account developments of supply and demand.

11.3.2 Market overview

The German energy service market achieved a turnover of some €9 billion in 2018 (BfEE—Bundesstelle für Energieeffizienz, 2019b). The market is divided into three segments: energy management, energy contracting as well as energy audits, and counseling. Contracting accounts for some 85% of this turnover, energy management for 5%, and energy advice for 10% or €900 million. The border between these segments remains fluid, as many suppliers are active in different segments. Overall, some 7,400 companies are active in the market, employing staff of about 57,870. Table 11.3 presents the repartition of turnover and suppliers into the different market segments.

Due to different surveying concepts, market developments cannot be compared over a longer time period. Offermann et al. (2013) report a market turnover of some €186−333m in 2007. Whereas we caution against a direct comparison to the recent figures due to methodological differences, this helps to exemplify the rapid growth of the market over the last decade. Whereas the market continues to be robust, its dynamics can be characterized as stagnating over the last 3 years (Fig. 11.2).

Market supply is characterized by a large number of small or medium suppliers. These suppliers act mostly within a given region or radius. Most suppliers provide energy services as add-on to existing core products and services. This is notably the case in the energy advice markets, where largely architects, craftsmen, or building engineers offer advice on top of their core business (Ringel, 2018). Utilities and international companies, often with a background in facility and installation management, largely focus on the more complex or investment intensive segments such as contracting.

Concerning the availability of offers, BfEE maintains that energy services are available area-wide in Germany. Largely due to local and national offers, there is

Table 11.3 Market indicators for the German energy service market 2018.

Market segment	Turnover (mEUR)	Number of companies	People employed	Full-time equivalent
Energy advice	390	6000	12,000	4500
Contracting	7900	500	38,000	25,000
Energy management	470	900	6970	4600
Total	8760	7400	57,870	34,100

Source: Based on data from BfEE—Bundesstelle für Energieeffizienz, 2018. Empirische Untersuchung des Marktes für Energiedienstleistungen, Energieaudits und andere Energieeffizienzmaßnahmen: Endbericht BfEE 04/2017. Eschborn.

Energy service markets: status quo and development 263

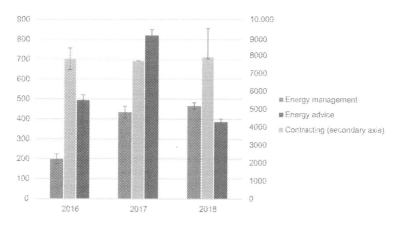

Figure 11.2 Development of energy service turnover 2015−18.
Source: Based on data from Kantar Emnid, ifeu and Prognos, 2018. Einfluss staatlicher Instrumente auf den Markt für Energiedienstleistungen: Kurzanalyse im Rahmen von Studie BfEE 04/2017. Eschborn; BfEE—Bundesstelle für Energieeffizienz, 2019b. Empirische Untersuchung des Marktes für Energiedienstleistungen, Energieaudits und andere Energieeffizienzmaßnahmen: Endbericht BfEE 2018. Eschborn.

no region in the country where the number of offers falls below 160 energy auditors, 20 contracting suppliers, and 100 energy management providers. However, a clear concentration of offers exists in the Western state of North Rhine-Westphalia and notably the Southern states of Bavaria and Baden-Wuerttemberg. Here, the supply average mentioned earlier exceeds by minimum 40%.

Turning to the customer side, demand for energy services can be clustered into three main segments: (1) industry, focusing on audits, energy management systems, and contracting solutions, (2) private entities, notably looking for energy advice on buildings and building refurbishment, and (3) the public sector with a strong demand for more developed energy services such as contracting for public buildings or street lighting. The analysis in the German National Energy Efficiency Action Plan recognizes a certain reticence on the demand side to engage in energy services (BMWi—Bundesministerium für Wirtschaft und Energie, 2017a). This analysis is confirmed by the 2019 BfEE market analysis (BfEE—Bundesstelle für Energieeffizienz, 2019b), which estimates that some 50% of potential demand is not realized. This weakness on the demand side is not attributed to a single cause, but rather to a combination of barriers as discussed previously in Section 11.2.2. In addition, consumer preferences for low investment costs and quick payback periods persist (Achtnicht and Madlener, 2014). Finally, some segments of the service markets still face regulatory disincentives to engage in energy services. The German government has started several initiatives and provisions to boost market demand and increase market dynamics. These initiatives are tailor made to each segment of the energy service markets. We will now turn to discuss these segments in further detail.

11.4 Developments of segments of the service market

11.4.1 Advice services

The market monitoring of the German government follows broader definition of energy services than that of the EED. This can be explained by the fact that some offers on the German market existed well before European legislation. This is notably the case with "energy advice" services, which had been introduced as a reaction to the 1970s oil crises and existed before more complex services entered the market. In consequence, there is a certain overlap with energy audits. The two cannot be separated clearly, as the advice component ("what could be done") merges into the audit component ("what should be done"), or implementation ("how can it be done") (Ringel, 2018).

The submarket for energy advice comprises consultancy services and energy audits. Some 12,500–13,500 companies are active in this market. The major suppliers of advice services are architects or civil engineers (31%), followed by craftsmen (27%) and consulting engineers (18%) (BfEE—Bundesstelle für Energieeffizienz, 2018, 2019b). These consultancies are mainly small-to-medium-sized enterprises (SMEs), with about 75% having five or fewer employees. To gain political leverage, they are organized in several industry associations such as DEN (Deutsche Energieberater Netzwerk) with about 700 members or GIH (Gebaeudeenergieberater Industrie Handwerk e.V) with about 2500 members.

In 2015 (latest figures available), between 335,000 and 375,000 advice sessions took place. These were mainly call-in or "stationary" advice sessions (25%), followed by energy checks for residential buildings (22%) and energy audits for industry (18%). For half of the suppliers, annual turnover reached €80,000; for another 15%, €150,000. On average, the advice component represents about 38% of the income generated. This figure varies considerably given the highly diverse main activities of suppliers. Fees vary from €1,500 for standard advice to nearly €11,000 for audits in the industrial sector (BfEE—Bundesstelle für Energieeffizienz, 2018). Energy advice is mainly a regional service, offered within a radius of some 100 km. Key demand groups for energy services are private home owners (56%), industry and service sectors (50%) and real estate (about 30%) (BfEE—Bundesstelle für Energieeffizienz, 2017).

The qualifications needed to offer energy advice are not standardized in Germany (Heinen et al., 2010). This implies that consultants or energy companies can offer energy advice based on various specializations and professional backgrounds. This in turn leads to wide differences in the quality of services provided. The government has so far been reticent to actively regulate qualification or training. However it stipulates that access to Germany's low-interest building efficiency loan schemes offered by the semipublic KfW bank is contingent upon receiving only "independent" and "qualified" advice (for a detailed discussion of the KfW schemes see Geddes et al., 2018). This creates a market push for suppliers who can prove that they have certain training and qualifications, that is suppliers which are enrolled in the aforementioned list of "qualified experts" as maintained by Germany's energy agency.

A key requirement for grant eligibility is that the advice received is "impartial," meaning independent of economic interests. That is, the advisor cannot act as a sales agent for the construction or refurbishment activity that might follow an advice session or for the energy package that might be recommended.

The latter implies that until recently, energy companies were not entitled to qualify as "listed" energy advisors. Likewise, to maintain "independence," advisors were not permitted to implement their recommendations. These regulations created subdivisions in the market, which are often not visible to consumers and hence to a certain extent trigger confusion and distrust in the available offers (Feser et al., 2015). In December 2017, the federal government mandated that the "impartiality" criterion could be satisfied by what amounts to an honor code, namely a self-declaration of "nonconflict of interest" (BMWi—Bundesministerium für Wirtschaft und Energie, 2017b). This regulatory change can be seen as an effort to boost market dynamics by increasing the existing offers. As a way of maintaining qualitative standards, the government has in parallel started a process to legally define the qualifications of energy advisors (BAFA—Bundesamt für Wirtschaft und Ausfuhrkontrolle, 2017).

This change in federal regulatory position has shaken the market and could potentially change its dynamics. Bigger energy companies are expected to enter what has been a small entrepreneurial market segment for "qualified" advice. The incumbent advisors will need to reposition themselves and adopt new business models. Here notably, the implementation of the EED acts as a market driver. Article 8 EED stipulates that non-SMEs have to undertake an energy audit every 4 years. At the time of implementation, this has caused a rush for energy audits by the affected industries. Many industrial clients have, however, opted for directly implementing energy management systems, so as to directly switch to a continuous energy monitoring rather than paying for 4-year-snapshots of their installations.

11.4.2 Energy management

About 900 companies offer energy management services (EMS) in Germany. These services notably comprise:

- certification and recertification of installations (ISO 50001, EMAS, ISO 14001);
- confirmation of energy efficiency measures to obtain federal tax credits;
- consultancy and planning of energy management systems (mostly ISO 500001);
- implementation of EMS, including software and maintenance;
- energy controlling.

As with the market for energy advice, the suppliers come from a highly heterogeneous background: some estimated that 47% are dedicated engineering offices, 18% are utilities, 8% are IT or software companies, further 8% are dedicated certification specialists and the remaining 19% come from largely different backgrounds (BfEE—Bundesstelle für Energieeffizienz, 2017). Given the largely heterogeneous backgrounds, the average turnover varied considerably between companies. Some 50% of (small and medium) suppliers had a turnover of up to €500,000. Utility and

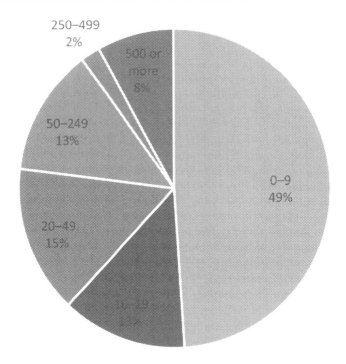

Figure 11.3 Employment structures in the EMS market segment.
Source: BfEE—Bundesstelle für Energieeffizienz, 2019b. Empirische Untersuchung des Marktes für Energiedienstleistungen, Energieaudits und andere Energieeffizienzmaßnahmen: Endbericht BfEE 2018. Eschborn.

contractors witnessed a turnover of €50m or more. The overall market is characterized by a dichotomy between a large majority of SME offers (90%) in contrast to large (international) suppliers (10%). This is mirrored in employment structures (Fig. 11.3).

The geographical spread of offers matches the other market segments, notably the advice and audit segment. Again, the Western and Southern federal states stand out in terms of availability of offers.

Given that this market segment is largely oriented to industry, price spans for the individual offers varied considerably. Whereas EMAS certification prices pivoted around €10,000, price spreads between €19,000 and €5000 were witnessed for energy controlling and €5000–€15,000 for certification with ISO 500001 (BfEE— Bundesstelle für Energieeffizienz, 2019b). This can be interpreted as reflecting the high span in terms of quality but also different scopes of services included in the EMS offers.

This strong discrepancy between offers is presently acknowledged as a sign of a "pluralistic market," as the federal government calls it. Still, it has to be noted, that uncertainty about prices and quality create confusion and uncertainty on the demand side. In order to lower these transaction costs, several associations call for a clearer standardization of products or more modular offers.

11.4.3 Contracting

So-called contracting services represent the largest segment of the German and indeed the European energy service markets. Three basic contracting models exist (see Fig. 11.1). These models originally developed from lease models and offer innovative financing solutions for (1) investing in new generation capacity or plants ("installation contracting"), (2) obtaining and bundling delivery of (renewable) energy ("delivery or supply contracting"), or (3) implementing energy efficiency solutions ("EPC,") through a dedicated third-party service provider. This provider is called ESCO. Apart from providing finance, ESCOs bundle technical, legal, and economic knowledge (CA EED—Concerted Action on the Energy Efficiency Directive, 2015). This allows their clients to shift the technical and financial risks to the ESCO. The ESCO will use the stream of income from the cost savings obtained (or the renewable energy generated and sold) to repay the costs of the project, including the costs of the investment. Part of the income stream will stay with the ESCO as remuneration for its services. The JRC of the European Commission identifies three characteristics of ESCOs (JRC—Joint Research Centre, 2019):

1. Guaranteeing energy savings and/or provision of the same level of energy service at lower cost. A performance guarantee can take several forms. It can revolve around the actual flow of monetary savings from a project; it can stipulate that the energy savings will be sufficient to repay monthly debt service costs; or it can guarantee that the same level of energy service is provided for less money.
2. Tying remuneration to actual cost/energy savings achieved or a certain form of energy delivered.
3. Providing or arranging financing for the operation of an energy system.

Most EPCs still focus on monetary savings, which allow for combining delivery and performance contracting. Increasingly though, EPCs providing a savings guarantee are concluded (JRC—Joint Research Centre, 2017).

The market for contracting in Germany is among the most developed ones in Europe. As such it can help to illustrate overall market dynamics. It also shows the pitfalls of the developed markets, where market development focuses only on the supply side.

Some 500 companies offer contracting services. They are organized in several associations, depending on the focus of their offers (e.g., heat delivery, EPC). The BfEE market report 2018 notes that still 60% of the offers relate to supply contracting, 22% to installation contracting, and only the remaining 18% to EPC (BfEE—Bundesstelle für Energieeffizienz, 2019b). The contracting market follows a continuous growth since 2005, albeit at a slower pace than in earlier years (Fig. 11.4). The clear plunge in growth rates starting in 2010 can be attributed to increased market uncertainty and distrust following a series of ESCO contracts, which were not redeemed due to bankruptcy of smaller suppliers and uncertainty about legal regulation of the market.

The present supply structure on the ESCO market is largely characterized by utilities (46%), large (international) ESCOs (26%), engineering offices (12%) as

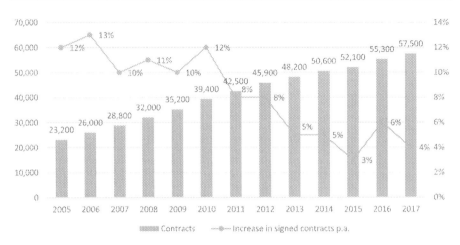

Figure 11.4 Development of ESCO contracts in Germany.
Source: Based on VfW—Verband für Wärmelieferung e.V, 2018. VfW - Tätigkeitsbericht 2018. <https://www.energiecontracting.de/6-verband/wir-ueber-uns/taetigkeitsberichte/VfW-Taetigkeitsbericht-2018-2.pdf> (accessed 25.07.19.).

well as facility managers, or diverse companies (each 8%). Overall market turnover represented some €7.9bn. Contract volumes pivot around €10m or in single cases well over this volume.

Key client on the demand side is the public sector. Notably at local level, ESCOs are used to implement projects regarding public building refurbishment (schools, hospitals, and administration buildings) as well as replacing outdated public lighting and traffic light infrastructure. Other focus groups are hotels and gastronomy, health installations (retirement homes), real estate companies, and the energy-intensive industry. It is interesting to note that regardless of the size of clients (SMEs vs non-SMEs), the motives for engaging in contracting are largely the same (Fig. 11.5).

Following the description of the status quo of Europe's biggest energy service market, we will now venture to discuss which further opportunities for market development seem promising to stimulate market dynamics.

11.5 Market development

The case study of the German market can serve to illustrate opportunities and pitfalls for energy service market development at large. Especially the continuous growth of the market combined with sinking growth rates merits further analysis as this is in line with overall projections for European and indeed global energy service markets (MWP Institute, 2015).

BfEE and KfW market analyses confirm the role of the German market as one of Europe's lead markets. Offers in all market segments are readily available area-wide across the territory. Consumers can chose between offers ranging from standardized

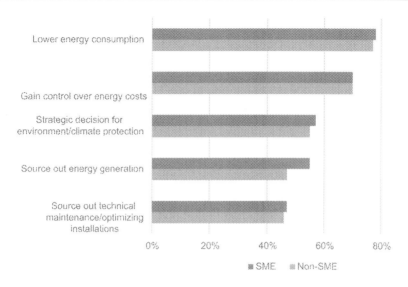

Figure 11.5 Motives for engaging in ESCO solutions.
Source: Based on BfEE—Bundesstelle für Energieeffizienz, 2017. Untersuchung des Marktes für Energiedienstleistungen, Energieaudits und andere Energieeffizienzmaßnahmen: Endbericht BfEE 06/2015. Eschborn; BfEE—Bundesstelle für Energieeffizienz, 2018. Empirische Untersuchung des Marktes für Energiedienstleistungen, Energieaudits und andere Energieeffizienzmaßnahmen: Endbericht BfEE 04/2017. Eschborn.

product such as basic energy advice to highly complex contracting solutions (BfEE—Bundesstelle für Energieeffizienz, 2019a; Bornemann, 2017). In this sense, the requirements of the EED regarding availability of offers are met in the German case. However, when looking at market structures, the market is facing a dichotomy between a multitude of small local suppliers on the one hand and few large international companies on the other. Energy services are often only part of a larger portfolio, especially in the case of SME offers. This allows for risk hedging and bundling business models. Yet, it also limits the capacity of supply and focuses too strongly on regional or local markets (Schüle and Döhrendahl, 2011). It remains to be seen whether this balance between a large number of SMEs on the one side and few large suppliers will remain the dominant market structure over the coming years or whether more comprehensive full-service offers by large companies will take the lead.

Turning to the demand side, market analysis suggests that only 50% of the potential demand for energy services is actually coming to market. This implies that contrary to the present focus on the supply side and availability, (missing) incentives on the demand side play a crucial role as well. The EED calls on EU Member States to systematically remove existing barriers that work against the deployment of energy services. However, some of these barriers remain inherent to the demand side, which so far is neglected.

This missing transparency and following uncertainly work against a stronger demand pull for energy services. The wide heterogeneity of offered services makes it hard for clients to choose a service that meets their needs and even harder to determine the appropriate price for it. In addition, the heterogeneity of qualifications reduces the transparency of available offers. Clients face uncertainty about their needs in terms of services, about what is fair to pay, and about who they can trust to provide it. These three barriers also exist in the wider energy service markets across Europe (Kowalska-Pyzalska, 2018). Article 18 EED asks Member States to set up lists of service providers. This is in principle an effective instrument to increase transparency. Still the German case with two parallel lists of providers (BfEE and dena) shows how quickly this transparency can be undermined. The case underlines the need for unambiguous guidance from the government to prevent market segmentation and gain trust with consumers.

Several instruments and market solutions are discussed to increase trust in the services and products offered. First, this concerns regulatory minimum quality criteria for the offers. Indirectly this is provided in the German market with only "independent" advice qualifying for government funding (Feser et al., 2015). However, no clear job and qualification descriptions exist for suppliers of energy services. Defining and legislating a harmonized set of minimum qualifications might increase trust and trigger additional demand for services. The German government has set out exploratory studies to define such criteria (BAFA—Bundesamt für Wirtschaft und Ausfuhrkontrolle, 2017). It will be interesting to monitor how this will affect the market in the coming years.

Second, a fixed and formalized pricing structure can help to overcome uncertainties. Experts suggested extending the current listing of experts to include a fixed price scheme for each of the services offered. These specifications already exist in Germany and several European countries for services provided by architects and engineers. The clear advantage of such packages lies in defining standardized energy services. Likely package configurations would include: a first contact advice package, a refurbishment advice package for homeowners, as well as packages for office buildings, or preaudit packages for industry. Each package would come with a clear task description, clear outcomes or goals for the advice and more or less fixed price components. Modularizing services in this way would mitigate the uncertainties that act as demand-side barriers (Ringel, 2018; Laurenz and Warneke, 2014).

Third, the case review shows the need for business associations and perhaps even governments to coordinate cooperation among market actors. A more active support or matching of actors might encourage small entrepreneurs to team up with other service providers to expand their offers. An interesting possibility for such cooperation emerges with the installation of local energy efficiency networks. These combine industry and service businesses, energy agencies, and energy advisors (Ringel et al., 2016). With increased digitalization of both buildings and the industrial sector, energy advisors will be challenged to offer a clear value added over the emerging smart meters and smart grids that analyze energy consumption data and deliver remote energy advice via artificial intelligence technology. It can be expected that these innovations will make their way into the public market and put further pressure on market actors to define and demonstrate their value added for the consumers.

11.6 Conclusions: lessons learned from the German case

This contribution sought to characterize the status quo and perspectives for energy service markets. Based on the review of one of the lead markets in Europe, we were able to present trends, barriers, and outlooks for the future. The findings can help identify policy actions to stimulate further market development. The market review confirms a growing offer and a high heterogeneity of supply. It also confirmed the persistence of multiple barriers, notably on the demand side. Apart from financial barriers, they can be summarized as a high transaction costs or "hassle factor" (Ringel, 2018) resulting from asymmetric information on the quality of services and a lack of market transparency. Regulatory measures are needed to reduce this "hassle factor" to free the full demand potential. These measures need to work on the following lines:

- Creating transparency in an only indirect manner may prove too weak. Rather than allowing only subsegments of "independent" offers access to government funding, this link should be more direct.
- This implies a clear (EU-harmonized) stance on qualification and certification of suppliers. Likewise, an EU-wide single list of suppliers should be established.
- The idea to define modular packages for services offered at fixed prices should be further investigated.
- Business organizations and governments should investigate in fostering further cooperation between market actors. A matching of supply and demand might be achieved by the means of energy efficiency networks that bring supply and demand together.

Summing up the status of European energy service markets, it can be concluded that they follow a robust trend of growth throughout all sectors. In order to trigger the full demand potential, further concertation and potentially legal framing will be needed.

References

Achtnicht, M., Madlener, R., 2014. Factors influencing German house owners' preferences on energy retrofits. Energy Policy 68, 254–263.
BAFA—Bundesamt für Wirtschaft und Ausfuhrkontrolle, 2017. Qualifikationsanforderungen in der Energieberatung. Available at: <https://www.evergabe-online.de/tenderdetails.html;jsessionid = F58A90474A4A24797EF352E3E7F7CE4F?0&id = 171094>.
Bertoldi, P., Boza-Kiss, B., 2017. Analysis of barriers and drivers for the development of the ESCO markets in Europe. Energy Policy 107, 345–355.
Bertoldi, P., Boza-Kiss, B., Panev, S., Labanca, N., 2014. The European ESCO Market Report 2013, vol. 23. European Commission, Ispra.
BfEE—Bundesstelle für Energieeffizienz, 2013. Marktanalyse und Marktbewertung sowie Erstellung eines Konzeptes zur Marktbeobachtung für ausgewählte Dienstleistungen im Bereich Energieeffizienz. Eschborn.
BfEE—Bundesstelle für Energieeffizienz, 2017. Untersuchung des Marktes für Energiedienstleistungen, Energieaudits und andere Energieeffizienzmaßnahmen: Endbericht BfEE 06/2015. Eschborn.

BfEE—Bundesstelle für Energieeffizienz, 2018. Empirische Untersuchung des Marktes für Energiedienstleistungen, Energieaudits und andere Energieeffizienzmaßnahmen: Endbericht BfEE 04/2017. Eschborn.

BfEE—Bundesstelle für Energieeffizienz, 2019a. Anbieterliste der Bundesstelle für Energieeffizienz. Available at: <https://www.bfee-online.de/BfEE/DE/Energiedienstlei stungen/Anbieterliste/anbieterliste_node.html>.

BfEE—Bundesstelle für Energieeffizienz, 2019b. Empirische Untersuchung des Marktes für Energiedienstleistungen, Energieaudits und andere Energieeffizienzmaßnahmen: Endbericht BfEE 2018. Eschborn.

BMWi—Bundesministerium für Wirtschaft und Energie, 2011. Second National Energy Efficiency Action Plan (NEEAP) of the Federal Republic of Germany. BMWi, Berlin.

BMWi—Bundesministerium für Wirtschaft und Energie, 2014. 3. Nationaler Energieeffizienz- Aktionsplan (NEEAP) 2014 der Bundesrepublik Deutschland. BMWi, Berlin.

BMWi—Bundesministerium für Wirtschaft und Energie, 2017a. Nationaler Energieeffizienz-Aktionsplan (NEEAP) 2017 der Bundesrepublik Deutschland, vol. 2017. BMWi, Berlin.

BMWi—Bundesministerium für Wirtschaft und Energie, 2017b. Richtlinie über die Förderung von Energieberatung im Mittelstand, BAnz AT 07.11.2017 B1. BMWi, Berlin.

Bornemann, A., 2017. Umfrage von dena und KfW unter Energieeffizienz-Experten. Available at: <https://www.energie-effizienz-experten.de/fileadmin/user_upload/Qualifizierte_Expertenliste_Landingpage/Umfrage_unter_EE-Experten.pdf> (accessed 21.02.18.).

Boza-Kiss, B., Bertoldi, P., Economidou, M., 2017. Energy Service Companies in the EU: JRC Status Review and Recommendations for Further Market Development with a Focus on Energy Performance Contracting. European Commission, Istra.

BuildUp Skills, 2017. Energy performance of buildings standards: past and future. Available at: <http://www.buildup.eu/en/news/overview-energy-performance-buildings-standards-past-and-future>.

CA EED—Concerted Action on the Energy Efficiency Directive, 2015. Energy Services and ESCOs, Energy Auditing, Solving Administrative Barriers. Brussels.

dena, 2017. Individual refurbishment roadmap for residential buildings. Available at: <https://www.dena.de/en/topics-projects/projects/buildings/individual-refurbishment-roadmap-for-residential-buildings/>.

dena, 2019. EnergieeffizienzExperten für Förderprogramme des Bundes. Available at: <https://www.energie-effizienz-experten.de/> (accessed 24.07.19.).

DIN/ISO, 2010. Energy efficiency services - definitions and requirements; German version EN 15900:2010. Available at: <https://www.beuth.de/de/norm/din-en-15900/124282925>.

Duplessis, B., Adnot, J., Dupont, M., Racapé, F., 2012. An empirical typology of energy services based on a well-developed market. France. Energy Policy 45, 268−276.

European Commission, 2006. Directive 2006/32/EC of the European Parliament and of the Council of 5 April 2006 on Energy End-Use Efficiency and Energy Services and Repealing Council Directive 93/76/EECamending Directives 2009/125/EC and 2010/30/EU and Repealing Directives 2004/8/EC and 2006/32/EC. European Commission. Brussels.

European Commission, 2012. Directive 2012/27/EU of the European Parliament and of the Council of 25 October 2012 on Energy Efficiency, Amending Directives 2009/125/EC

and 2010/30/EU and Repealing Directives 2004/8/EC and 2006/32/EC Text with EEA Relevance: EED. Brussels.
European Commission, 2013. Guidance for National Energy Efficiency Action Plans. European Commission, Brussels, *SWD(2013) 180 Final.*
European Commission, 2016a. Clean Energy For All Europeans. European Commission, Brussels, *COM(2016) 860 Final.*
European Commission, 2016b. Impact Assessment for the Amendment of the Energy Efficiency Directive. European Commission, Brussels, *SWD(2016)405.*
European Commission, 2016c. Impact Assessment for the amendment of the Energy Performance of Buildings Directive. European Commission, Brussels, SWD(2016)414.
European Commission, 2018a. A Clean Planet for All a European Strategic Long-Term Vision for a Prosperous, Modern, Competitive and Climate Neutral Economy. European Commission, Brussels, *COM(2018) 773 Final.*
European Commission, 2018b. Directive (EU) 2018/2002 of 11 December 2018 Amending Directive 2012/27/EU on Energy Efficiency. European Commission, Brussels.
European Commission, 2018c. Directive (EU) 2018/844 of 30 May 2018 Amending Directive 2010/31/EU on the Energy Performance of Buildings and Directive 2012/27/EU on Energy Efficiency. European Commission, Brussels.
European Parliament Research Service, 2019. Briefing. EU Legislation in Progress: Revised Energy Efficiency Directive. Brussels.
Fell, M.J., 2017. Energy services. A conceptual review. Energy Res. Soc. Sci. 27, 129–140.
Feser, D., Proeger, T., Bizer, K., 2015. Die Energieberatung als der zentrale Akteur bei der energetischen Gebäudesanierung? Z. für Energiewirtschaft 39 (2), 133–145.
Gadenne, D., Sharma, B., Kerr, D., Smith, T., 2011. The influence of consumers' environmental beliefs and attitudes on energy saving behaviours. Energy Policy 39 (12), 7684–7694.
Geddes, A., Schmidt, T.S., Steffen, B., 2018. The multiple roles of state investment banks in low-carbon energy finance. An analysis of Australia, the UK and Germany. Energy Policy 115, 158–170.
Good, N., Ellis, K.A., Mancarella, P., 2017. Review and classification of barriers and enablers of demand response in the smart grid. Renew. Sustain. Energy Rev. 72, 57–72.
Heinen, S., Frenz, M., Djaloeis, R., Schlick, C., 2010. Vocational training concepts and fields of activities of energy consulting in Germany. Available at: <https://www.researchgate.net/profile/Raymond_Djaloeis/publication/228521663_Vocational_Training_Concepts_and_Fields_of_Activities_of_Energy_Consulting_in_Germany/links/02e7e53394f21b4-ca0000000/Vocational-Training-Concepts-and-Fields-of-Activities-of-Energy-Consulting-in-Germany.pdf> (accessed 01.03.18.).
Horowitz, M.J., Bertoldi, P., 2015. A harmonized calculation model for transforming EU bottom-up energy efficiency indicators into empirical estimates of policy impacts. Energy Econ. 51, 135–148.
Hu, Z., Kim, J.-h, Wang, J., Byrne, J., 2015. Review of dynamic pricing programs in the U.S. and Europe: status quo and policy recommendations. Renew. Sustain. Energy Rev. 42, 743–751.
IEA—International Energy Agency, 2007. Mind the Gap - Quantifying Principal-Agent Problems in Energy Efficiency. OECD/IEA, Paris.
IEA—International Energy Agency, 2017. Energy Efficiency 2017. Market Report Series. OECD/IEA, Paris.

JRC—Joint Research Centre, 2017. Energy Service Companies in the EU. Status Review and Recommendations for Further Market Development With a Focus on Energy Performance Contracting. Ispra.
JRC—Joint Research Centre, 2019. Energy service companies. Available at: <https://ec.europa.eu/jrc/en/energy-efficiency/eed-support/energy-service-companies> (accessed 25.07.19.).
Kangas, H.-L., Lazarevic, D., Kivimaa, P., 2018. Technical skills, disinterest and non-functional regulation. Barriers to building energy efficiency in Finland viewed by energy service companies. Energy Policy 114, 63−76.
Kantar Emnid, ifeu and Prognos, 2018. Einfluss staatlicher Instrumente auf den Markt für Energiedienstleistungen: Kurzanalyse im Rahmen von Studie BfEE 04/2017. Eschborn.
Karakaya, E., Hidalgo, A., Nuur, C., 2014. Diffusion of eco-innovations: a review. Renew. Sustain. Energy Rev. 33, 392−399.
Kjeang, A., Palm, J., Venkatesh, G., 2017. Local energy advising in Sweden. Historical development and lessons for future policy-making. Sustainability 9 (12), 2275.
Kowalska-Pyzalska, A., 2018. What makes consumers adopt to innovative energy services in the energy market? A review of incentives and barriers. Renew. Sustain. Energy Rev. 82, 3570−3581.
Labanca, N., Suerkemper, F., Bertoldi, P., Irrek, W., Duplessis, B., 2015. Energy efficiency services for residential buildings. Market situation and existing potentials in the European Union. J. Clean. Prod. 109, 284−295.
Larsen, P.H., Goldman, C.A., Satchwell, A., 2012. Evolution of the U.S. energy service company industry. Market size and project performance from 1990−2008. Energy Policy 50, 802−820.
Laurenz, H., Warneke, J., 2014. Country Report on Uptake of the European Code of Conduct for the Energy Performance Contracting. Transparense Project. Available at: <http://www.transparense.eu/tmce/Germany/WP4_D4_08_Germany_BEA.pdf> (accessed 21.02.18.).
Marino, A., Bertoldi, P., Rezessy, S., Boza-Kiss, B., 2011. A snapshot of the European energy service market in 2010 and policy recommendations to foster a further market development. Energy Policy 39 (10), 6190−6198.
MWP Institute, 2015. The energy services market 2022. Available at: <http://www.mpw-net.de/fileadmin/media/mpw/Studien/MPW-Study_EDL-Market_2022.pdf> (accessed 21.02.18.).
Negro, S.O., Alkemade, F., Hekkert, M.P., 2012. Why does renewable energy diffuse so slowly? A review of innovation system problems. Renew. Sustain. Energy Rev. 16 (6), 3836−3846.
Nolden, C., Sorrell, S., Polzin, F., 2016. Catalysing the energy service market. The role of intermediaries. Energy Policy 98, 420−430.
Nygrén, N.A., Kontio, P., Lyytimäki, J., Varho, V., Tapio, P., 2015. Early adopters boosting the diffusion of sustainable small-scale energy solutions. Renew. Sustain. Energy Rev. 46, 79−87.
Offermann, R., Irrek, W., Duscha, M., Seefeldt, F., 2013. Monitoring the energy efficiency service market in Germany. In: ECEEE Summer Study Proceedings. Available at: <https://www.prognos.com/uploads/tx_atwpubdb/130000_Prognos_ECEEE_Monitoring_EnergyEfficiencyServices.pdf> (accessed 21.02.18.).
Pirlogea, C., 2011. Barriers to investment in energy from renewable sources. Available at: <http://www.management.ase.ro/reveconomia/2011-1/12.pdf> (accessed 21.06.19.).
Ringel, M., 2018. Energy advice in Germany: a market actors' perspective. Int. J. Energy Sect. Manag. 12 (4), 656−674.

Ringel, M., Schlomann, B., Krail, M., Rohde, C., 2016. Towards a green economy in Germany? The role of energy efficiency policies. Appl. Energy 179, 1293–1303.

Roshchanka, V., Evans, M., 2016. Scaling up the energy service company business. Market status and company feedback in the Russian Federation. J. Clean. Prod. 112, 3905–3914.

Schüle, R., Döhrendahl, E. (Eds.), 2011. Zukunft der Energieberatung in Deutschland. Wüstenrot-Stiftung, Ludwigsburg.

Sorrell, S., 2015. Reducing energy demand: a review of issues, challenges and approaches. Renew. Sustain. Energy Rev. 47, 74–82.

Stuart, E., Larsen, P.H., Goldman, C.A., Gilligan, D., 2014. A method to estimate the size and remaining market potential of the U.S. ESCO (energy service company) industry. Energy 77, 362–371.

Szomolányiová, J., Keegan, N., 2018. European Report on the Energy Efficiency Services Market and Quality. Available at: <https://qualitee.eu/wp-content/uploads/QualitEE_2-05_EuropeanReport-2018.pdf> (accessed 30.06.18.).

Talon, C., Gartner, J., 2015. Energy Service Company Market Overview Expanding ESCO Opportunities in the United States and Europe: Executive Summary. Navigant Consulting, Boulder, CO.

The Economist, 2015. Special Report Energy and Technology, pp. 1–10.

VfW—Verband für Wärmelieferung e.V, 2018. VfW - Tätigkeitsbericht 2018. Available at: <https://www.energiecontracting.de/6-verband/wir-ueber-uns/taetigkeitsberichte/VfW-Taetigkeitsbericht-2018-2.pdf> (accessed 25.07.19.).

Worldwide trends in energy market research

12

Esther Salmerón-Manzano[1], Alfredo Alcayde[2] and Francisco Manzano-Agugliaro[2]
[1]Faculty of Law, International University of La Rioja, Logroño, Spain, [2]Department of Engineering, ceiA3, University of Almeria, Almeria, Spain

Chapter Outline

12.1 Introduction 277
12.2 Data 278
12.3 Results 278
    12.3.1 Subjects from worldwide publications 278
    12.3.2 Journals metric analysis 279
    12.3.3 Countries, affiliations, and their main topics 281
    12.3.4 Keywords from worldwide publications 282
    12.3.5 Cluster analysis based on keywords 285
References 290

12.1 Introduction

Energy is a vital input for social and economic development (Banos et al., 2011). There are many kinds of energy and ways of generating it, just as there are many ways of consuming it and with different objectives (Manzano-Agugliaro et al., 2013), but all of them for the improvement of the quality of life understood in a broad way (Juaidi et al., 2019). The energy market is essentially the relationship between energy production and consumption (Fig. 12.1). As an example, the EU aims to fully integrate national energy markets in order to offer consumers and businesses more (and better) products and services, as well as more secure supplies. In this respect, progress has been made, such as consumers being able to change gas and electricity suppliers and suppliers being required to clearly explain the conditions (Asche et al., 2006). Rules on the management of national gas and electricity networks and markets have yet to be harmonized (Serrallés, 2006). It also remains to facilitate cross-border investments in energy infrastructure (Sencar et al., 2014).

    Scientific production on a given topic is an index of the amount of research carried out in that field (Chihib et al., 2019). In fact, there are those who define science as that which is published in scientific journals (Manzano-Agugliaro et al., 2019). Therefore, bibliometrics and its related sciences are a clear exponent not only of the quantity of scientific production but also of how it is distributed and allows us to analyze trends and progress in a particular research field (Salmerón-Manzano and

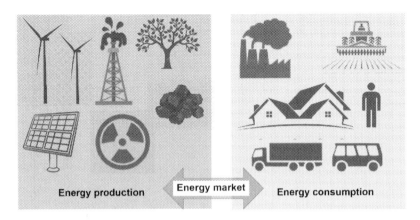

Figure 12.1 The energy market as a relationship between production and consumption.

Manzano-Agugliaro, 2019). Thus one of the research lines in this field is the analysis of the scientific communities, where it is possible to observe the publications clusters in certain fields related to the subject of study. This makes it possible to identify global trends in research in a particular field. This has been successfully applied in all spheres of science (Salmeron-Manzano and Manzano-Agugliaro, 2018).

12.2 Data

Scientific publications, if they are of high standard, are indexed in scientific databases. Although there are several scientific databases, and some of them specific to certain fields of knowledge, mainly two are usually used for bibliometric studies: Web of Knowledge and Scopus (Montoya et al., 2018). Studies indicate that Scopus has higher coverage in science areas (Mongeon and Paul-Hus, 2016), and for this reason it is the database used in this work.

In this work, all the scientific production data collected in the Scopus database with the search term "Energy Market" have been analyzed. So, the following search string was: TITLE-ABS-KEY ("Energy Market"), of which more than 15.353 results have been obtained in the period between 1958 and 2018. Fig. 12.2 shows the evolution, where from the 1990s onward, there has been an exponential growth, and it has become more noticeable since 2010, when more than 900 articles are published each year.

12.3 Results

12.3.1 Subjects from worldwide publications

Fig. 12.3 shows the Scopus classification by categories of the works published on the energy market. As expected, the energy category is the main category with

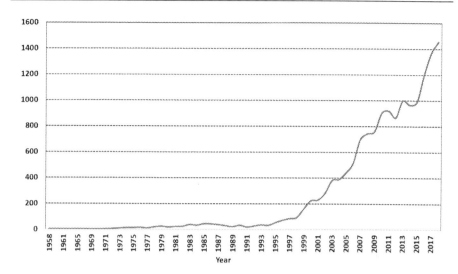

Figure 12.2 Development of world scientific production on "energy market."

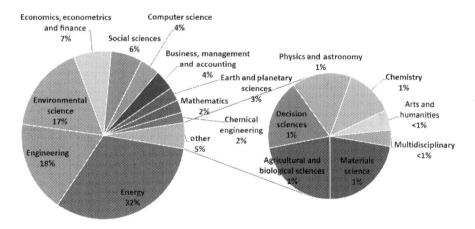

Figure 12.3 Distribution of scientific output by Scopus categories on the energy market.

32%, followed by engineering categories with 18%, and environmental sciences with 17%; adding these three categories results in 67% of world scientific output. The categories of economics, econometrics and finance and computer sciences are still very small, 7% and 4%, respectively. It should be noted that the social sciences category (6%) is mainly related to energy policy (Nilsson, 2005).

12.3.2 Journals metric analysis

This section includes the main journals where this topic is included. Table 12.1 shows the main bibliometric indexes of the most important journals on energy

Table 12.1 Bibliometric indexes of the main journals on energy market.

Source title	N	%	Impact factor (2018)	Citescore (2018)	TH index	Citations (C)	C/N
Energy Policy	1680	10.9	4.880	5.45	94	30,827	18.35
Energy Economics	940	6.1	4.151	5.04	83	17,880	19.02
Energy	614	4.0	5.537	6.20	60	13,875	22.60
Applied Energy	549	3.6	8.426	9.54	66	14,698	26.77
Oil and Gas Journal	449	2.9	0.072	0.04	7	253	0.56
Renewable Energy	268	1.7	5.439	6.19	44	7032	26.24
IEEE Transactions on Power Systems	146	1.0	6.807	8.94	51	7456	51.07
Utilities Policy	144	0.9	2.417	2.74	24	1666	11.57
Revue De L Energie	135	0.9	–	0.101	3	55	0.41
Renewable and Sustainable Energy Reviews	119	0.8	10.556	12.21	40	5935	49.87
International Journal of Global Energy Issues	104	0.7	–	0.54	10	317	3.05
Sustainability Switzerland	104	0.7	2.592	3.01	14	566	5.44
Energy and Environment	96	0.6	1.092	0.99	9	254	2.65
Geopolitics of Energy	89	0.6	–	0.02	3	29	0.33
Rynek Energii	85	0.6	–	0.15	5	125	1.47
Energies	76	0.5	2.707	3.18	11	526	6.92
Biomass and Bioenergy	74	0.5	3.537	3.96	27	2374	32.08
Energy Procedia	70	0.5	–	1.30	11	366	5.23

market. It can be seen that there is one journal that clearly stands out from the others, *Energy Policy* with 10% of total publications, followed by *Energy Economics* (6%), and *Energy* (4%). Regarding the impact factor and the citescore, three journals stand out: *Renewable and Sustainable Energy Reviews*, *Applied Energy*, and *IEEE Transactions on Power Systems*. But these indexes are of general nature for all works published in those journals. The thematic index (TH) is normally used to analyze the impact in a given field. Here the TH of the energy market is calculated for the journals, that is, only considering the articles on this specific topic. This index shows that the journals that produce the most publications in this field are the best ranked, for example, *Energy Policy* or *Energy Economics*. If the total number of citations (C) received by these works is analyzed and then the average citations per publication are calculated according to the journal, the last column

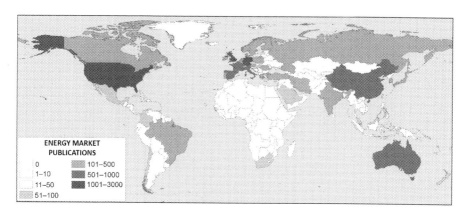

Figure 12.4 Worldwide geographical distribution of the scientific production of the energy market.

of the table (*C/N*) is obtained. With this last index, the work published in *IEEE Transactions on Power Systems* and *Renewable and Sustainable Energy Reviews* stand out, with 50 average citations per article, which is a lot even for journals with high impact index like these, which indicates that this topic is very significant. Note that *the Journal Biomass and Bioenergy*, shows a very high number of citations per article related to the energy market, more than 30, perhaps because of the fact that among all the mentioned journals it deals with a very specific topic within renewable energies such as biomass.

12.3.3 Countries, affiliations, and their main topics

If the scientific production of energy market by country is analyzed, it can be observed that this ranking is led by the United States with just over 19% of total publications, followed by Germany and the United Kingdom with just over 8% each. The third and fourth positions are held by China and Italy with 6.5% and 5%, respectively. Fig. 12.4 shows the geographical distribution of publications by country, where it can be seen that in almost the entire world there is scientific interest in this topic. Moreover, countries where there are no associated publications are almost an exception.

Fig. 12.5 shows the main institutions researching energy market. This ranking is led by University of Cambridge (United Kingdom), closely followed by Technical University of Denmark (Denmark), and University of California, Berkeley (United States). Of the top 20 institutions, 4 are from the United States, 3 from China, and 2 from the United Kingdom. Geographically, outside the European countries, and the United States and China an institution from Singapore stands out. Table 12.2 lists the three main keywords used in the works published by this institution, as a sample of their main interest. It is noted that almost all have the first three identical keywords: commerce, energy policy, and costs. If renewable energy resources is added to this list of words, only those referring to a geographical area would

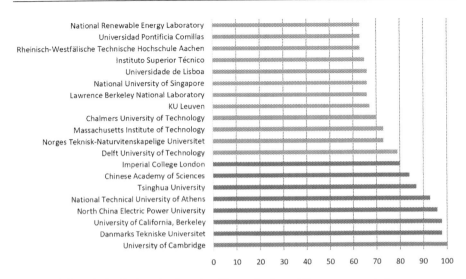

Figure 12.5 Main institutions in the scientific production of the energy market.

remain: China, United States, and Europe. After this, only a few of the local interests of the institutions are highlighted. For example, paper and pulp mills or carbon dioxide for the Chalmers University of Technology (Sweden).

12.3.4 Keywords from worldwide publications

The total set of keywords used in all articles related to the energy market results are shown in Table 12.3. It is to be expected that there are also those related to commerce, costs, and the market, but also those related to renewable energies such as renewable resource, renewable energy resources, wind power, or alternative energy. To get a visual idea of all the keywords, a cloud of keywords is made in Fig. 12.6, where the size indicates their relative importance. It is seen how the topics of commerce and energy policy actually stand out from the rest. It is also worth noting that those keywords are related to the type of energy as electricity, electricity supply, electric industry, or natural gas. Therefore, the energy market researches above all on issues related to electricity, to a large extent by renewable energies among which wind energy and then energy from natural gas stands out. With regard to keywords focused on geographical areas, the following has been observed: United States, Europe, Asia, and India, which also seems logical due to the interconnection of the electrical networks with neighboring countries. As an example, in Europe the connection is not only between countries of the European Union, since Spain sells electricity to Morocco in the South and the purchase to France in the North (Montoya et al., 2014). Finally, with regard to mathematical techniques, optimization appears among the main keywords; these techniques are a great aid to decision-making when there are conflicting objectives (Márquez et al., 2011).

Table 12.2 Main affiliations and their main keywords.

Affiliation	Country	1	2	3
University of Cambridge	United Kingdom	Commerce	Energy policy	Costs
Technical University of Denmark	Denmark	Commerce	Wind power	Energy policy
University of California, Berkeley	United States	Commerce	Costs	United States
North China Electric Power University	China	China	Commerce	Costs
National Technical University of Athens	Greece	Energy Policy	Commerce	Renewable energy resources
Tsinghua University	China	China	Commerce	Costs
Chinese Academy of Sciences	China	China	Energy policy	Costs
Imperial College London	United Kingdom	Commerce	Costs	Energy policy
Delft University of Technology	Netherlands	Commerce	Energy policy	Renewable energy resources
Norges Teknisk-Naturvitenskapelige Universitet	Norway	Commerce	Power markets	Costs
Massachusetts Institute of Technology	United States	Commerce	Costs	Energy policy
Chalmers University of Technology	Sweden	Carbon dioxide	Commerce	Energy efficiency/paper and pulp mills
KU Leuven	Netherlands	Commerce	Energy policy	Europe
Lawrence Berkeley National Laboratory	United States	United States	Energy policy	Energy efficiency
National University of Singapore	Singapore	Commerce	Costs	Energy policy
Universidade de Lisboa	Portugal	Commerce	Power markets	Costs
Instituto Superior Técnico	Portugal	Commerce	Power markets	Costs
Rheinisch-Westfälische Technische Hochschule Aachen	Germany	Commerce	Investments	Costs
Universidad Pontificia Comillas	Spain	Commerce	Costs	Electricity generation
National Renewable Energy Laboratory	United States	Commerce	United States	Costs

Table 12.3 Main keywords of the published works of energy market.

Keyword	N
Commerce	4288
Costs	2846
Energy policy	2711
Power markets	1875
Investments	1517
Energy efficiency	1352
Economics	1295
Renewable resource	1212
Price dynamics	1198
Renewable energy resources	1194
Electricity generation	1191
Electricity	1181
Optimization	1138
United States	1056
Wind power	1036
Marketing	966
Natural gas	937
Electric utilities	929
Alternative energy	878
Electricity supply	859
Electric industry	845
Europe	844

Figure 12.6 Cloud of keywords from the scientific production of the energy market.

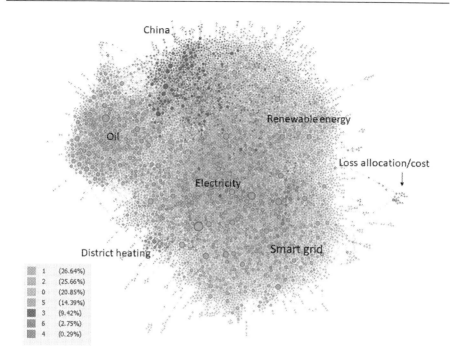

Figure 12.7 Community detection in the scientific production of the energy market.

12.3.5 Cluster analysis based on keywords

In the analysis of communities or clusters, the relations between scientific publications are established according to the citations between them and their keywords, allowing research to be grouped into different but at the same time related topics (Garrido-Cardenas et al., 2018). In our case, they have been grouped into seven communities, of which the first six account for 99.71% of world scientific production. Fig. 12.7 shows the relationship between the seven communities detected, which have been given names according to the most frequently mentioned keywords.

Community 1, smart grid, is the most numerous with 26.64% of the publications. The main keywords are grouped in Table 12.4, where it is observed that demand response, the electricity market, and energy storage occupy prominent places, along with that of renewable energy. So, the smart grid paradigm is based on the concepts of demand response, distributed generation, and distributed energy storage (Rahimi and Ipakchi, 2010).

Community 2 focuses on electricity market, as shown in Table 12.5, and accounts for 25.66% of publications. Wind energy plays an important role, because there are terms as: Wind power, Wind energy or Forecasting the latter being clearly related to electricity forecasts as a function of wind intensity (Hernández-Escobedo et al., 2018; Zapata-Sierra et al., 2019). However, it is remarkable that natural gas

Table 12.4 Main keywords: community 1, smart grid.

Keyword	N
Energy market	181
Demand response	169
Smart grid	137
Electricity market	96
Energy storage	88
Renewable energy	88
Electricity markets	87
Energy markets	78
Distributed generation	62
Optimization	55
Microgrid	52
Ancillary services	49
Wind power	49
Stochastic programming	44
Distributed energy resources	43
Bidding strategy	39
Electric vehicles	39
Game theory	38
Uncertainty	36
Virtual power plant	36

Table 12.5 Main keywords: community 2, electricity market.

Keywords	N
Electricity markets	108
Electricity market	93
Renewable energy	90
Energy markets	80
Electricity	76
Market power	52
Energy policy	50
Regulation	50
Electricity prices	48
Energy market	43
Wind power	39
Market design	38
Natural gas	31
Demand response	28
Forecasting	27
Wind energy	27
Energy	26
Real options	25
Deregulation	24
Renewables	24

Table 12.6 Main keywords: community 0, renewable energy.

Keywords	N
Renewable energy	133
Energy security	79
Energy policy	68
Energy	53
Energy efficiency	50
Natural gas	35
Russia	34
Energy markets	32
China	28
Sustainable development	28
Wind energy	28
Electricity	26
Energy transition	26
Climate change	25
Energy market	24
Electricity markets	23
Wind power	22
Sustainability	21
European Union	20
Electricity market	19

also appears within this community. Some authors think that shale gas resource development is not an option but a must for the continuance of our global energy market and economy (Melikoglu, 2014). In addition, for some countries such as Turkey, among primary energy sources, natural gas is the fastest growing one (Erdogdu, 2010).

Community 0, with 20.85% of scientific production, is focused on renewable energies and in particular on wind energy, energy security, and energy policy. Three geographical locations appear: Russia, China, and European Union (Table 12.6). Indeed, trade relations between Russia and the European Union are great on energy issues mainly because of the natural gas that is purchased by the countries of northern Europe from Russia (Konoplyanik, 2012). On the other hand, China will be central to the future of the global wind energy market (Changliang and Zhanfeng, 2009). It should be noted that the major development of renewable energy in China is due to the country's recent renewable energy legislation (Zhang et al., 2010).

Community 5 is focused on the oil market: oil prices, volatility, biofuels, or crude oil. It has a specific weight of 14.39% within all scientific production. Table 12.7 contains the main keywords of this cluster. The keyword of volatility is related to the volatility states of oil markets, and for this a mathematical technique called GARCH model (generalized autoregressive conditional heteroskedasticity) is

Table 12.7 Main keywords: community 5, oil prices.

Keyword	N
Oil prices	64
Energy markets	46
Natural gas	46
Oil price	46
Energy prices	38
Volatility	38
Biofuels	37
Cointegration	36
Crude oil	35
Energy	30
Energy market	30
Renewable energy	24
GARCH	21
Ethanol	20
China	19
Oil	19
Structural breaks	19
Biofuel	18
Forecasting	17
Asymmetry	16

used (Wang and Wu, 2012). In the literature, numerous works investigating the empirical properties of oil, natural gas, and electricity price volatilities are found (Efimova and Serletis, 2014).

Community 3 has China as its main target, and within this energy efficiency and energy consumption stands out. And although it accounts only for 9.42% of scientific production, it is a very important weight to be dedicated to a single country, of course the dimensions of both energy consumption and population amply justify it. Table 12.8 lists the main keywords. In 2003, China was the world's second largest energy consumer behind the United States (Crompton and Wu, 2005). Nowadays, China is the top CO2-emitting country (Ma et al., 2019). On the other hand, household energy consumption in rural areas of China is strongly related to the living standards and poverty alleviation (Wu et al., 2019).

The community 6 is mainly related to district heating, and it has a weight of 2.75 % among the energy market publications. The main keywords are summarized in Table 12.9. Note that biomass becomes relevant for example to replace existing systems based on district heating for heating or cooling (Perea-Moreno et al., 2017), or with cogeneration as profitable alternatives in a competitive market (Chicco and Mancarella, 2006) or combined heat and power, the latter term is also often referred to by its acronym (CHP). Broadly speaking, the decision to install cogeneration units depends mainly on the combination of personal economical and ecological concerns (De Paepe and Mertens, 2007). Thus this community also refers

Table 12.8 Main keywords: community 3, China.

Keywords	N
China	68
Energy efficiency	46
Energy consumption	38
Renewable energy	30
Energy intensity	28
Economic growth	27
Energy	21
Energy demand	21
Natural gas	19
Turkey	18
Energy policy	15
Energy prices	13
Cointegration	12
Electricity market	12
Regulation	12
Electricity	10
Climate change	9
Electricity demand	9
Energy market	9
Energy security	9

Table 12.9 Main keywords: community 6, district heating.

Keywords	N
District heating	36
Cogeneration	22
Combined heat and power	21
Process integration	15
CHP	14
Optimization	13
Energy market scenarios	12
Biorefinery	9
Combined heat and power (CHP)	9
Energy efficiency	9
Renewable energy	7
Biomass gasification	6
Industrial excess heat	6
Biomass	5
CO2 capture	5
Distributed generation	5
Heat pump	5
Kraft pulp mill	5
Optimization	5
Dynamic programming	4

Table 12.10 Main keywords: community 4, loss allocation/cost.

Keyword	N
Energy market	5
Loss allocation	5
Deregulation	3
Transmission loss allocation	3
Open access	2
Power system economics	2
Wheeling cost	2
Circuit analysis	1
Circulating current loss	1
Cost allocation	1
Deregulated energy market	1
Distributed generation	1
Electricity markets	1
Embedded wheeling cost methodologies	1
Energy pricing	1
Harmonic distortion	1
Hybrid energy market	1
Hybrid power market	1
Incremental wheeling cost	1
Industry consumer	1

to CO_2 capture when using biomass for heat or power generation, or even in CO_2 capture by oil refineries (Johansson et al., 2013).

Finally, community 4 is an incipient community, which is even far from the central nucleus of the energy market, see Fig. 12.7. Table 12.10 shows the keywords of this small cluster. In spite of not having a significant specific weight, it has been mentioned because from the electrical point of view it begins to have significance not only on the price of energy but also on the losses in transport or even the power quality (Sánchez et al., 2013). One of the main objectives of the energy market is to achieve a reduction of electricity prices to customers. Traditional energy systems are vertically integrated, that is, they have been organized into generation, transmission, and distribution companies. The transmission network is considered a monopoly, but a key issue for the transmission network is the satisfactory sharing of its costs between all participants in the energy market. As an example, power losses in transport are approximately 4% of the energy produced, which is a significant amount of the total system costs (Da Silva and de Carvalho Costa, 2003).

References

Asche, F., Osmundsen, P., Sandsmark, M., 2006. The UK market for natural gas, oil and electricity: are the prices decoupled? Energy J. 27–40.

Banos, R., Manzano-Agugliaro, F., Montoya, F.G., Gil, C., Alcayde, A., Gómez, J., 2011. Optimization methods applied to renewable and sustainable energy: a review. Renew. Sustain. Energy Rev. 15 (4), 1753−1766.

Changliang, X., Zhanfeng, S., 2009. Wind energy in China: current scenario and future perspectives. Renew. Sustain. Energy Rev. 13 (8), 1966−1974.

Chicco, G., Mancarella, P., 2006. From cogeneration to trigeneration: profitable alternatives in a competitive market. IEEE Trans. Energy Convers. 21 (1), 265−272.

Chihib, M., Salmerón-Manzano, E., Novas, N., Manzano-Agugliaro, F., 2019. Bibliometric maps of BIM and BIM in universities: a comparative analysis. Sustainability 11 (16), 4398.

Crompton, P., Wu, Y., 2005. Energy consumption in China: past trends and future directions. Energy Econ. 27 (1), 195−208.

Da Silva A.L., de Carvalho Costa, J.G., 2003. Transmission loss allocation: I. Single energy market. IEEE Transactions on Power Systems, 18 (4), 1389−1394. https://doi.org/10.1109/TPWRS.2003.818696

De Paepe, M., Mertens, D., 2007. Combined heat and power in a liberalised energy market. Energy Convers. Manag. 48 (9), 2542−2555.

Efimova, O., Serletis, A., 2014. Energy markets volatility modelling using GARCH. Energy Econ. 43, 264−273.

Erdogdu, E., 2010. Natural gas demand in Turkey. Appl. Energy 87 (1), 211−219.

Garrido-Cardenas, J.A., Manzano-Agugliaro, F., Acien-Fernandez, F.G., Molina-Grima, E., 2018. Microalgae research worldwide. Algal Res. 35, 50−60.

Hernández-Escobedo, Q., Perea-Moreno, A.J., Manzano-Agugliaro, F., 2018. Wind energy research in Mexico. Renew. Energy 123, 719−729.

Johansson, D., Franck, P.Å, Berntsson, T., 2013. CO2 capture in oil refineries: assessment of the capture avoidance costs associated with different heat supply options in a future energy market. Energy Convers. Manag. 66, 127−142.

Juaidi, A., AlFaris, F., Saeed, F., Salmeron-Manzano, E., Manzano-Agugliaro, F., 2019. Urban design to achieving the sustainable energy of residential neighbourhoods in arid climate. J. Clean. Prod. 228, 135−152.

Konoplyanik, A.A., 2012. Russian gas at European energy market: why adaptation is inevitable. Energy Strategy Rev. 1 (1), 42−56.

Ma, X., Wang, C., Dong, B., Gu, G., Chen, R., Li, Y., et al., 2019. Carbon emissions from energy consumption in China: its measurement and driving factors. Sci. Total. Environ. 648, 1411−1420.

Manzano-Agugliaro, F., Alcayde, A., Montoya, F.G., Zapata-Sierra, A., Gil, C., 2013. Scientific production of renewable energies worldwide: an overview. Renew. Sustain. Energy Rev. 18, 134−143.

Manzano-Agugliaro, F., Salmerón-Manzano, E., Alcayde, A., Garrido-Cardenas, J.A., 2019. Worldwide research trends in the recycling of materials. https://doi.org/10.1016/B978-0-12-803581-8.11518-8.

Márquez, A.L., Baños, R., Gil, C., Montoya, M.G., Manzano-Agugliaro, F., Montoya, F.G., 2011. Multi-objective crop planning using pareto-based evolutionary algorithms. Agric. Econ. 42 (6), 649−656.

Melikoglu, M., 2014. Shale gas: analysis of its role in the global energy market. Renew. Sustain. Energy Rev. 37, 460−468.

Mongeon, P., Paul-Hus, A., 2016. The journal coverage of Web of Science and Scopus: a comparative analysis. Scientometrics 106 (1), 213−228.

Montoya, F.G., Aguilera, M.J., Manzano-Agugliaro, F., 2014. Renewable energy production in Spain: a review. Renew. Sustain. Energy Rev. 33, 509—531.

Montoya, F.G., Alcayde, A., Baños, R., Manzano-Agugliaro, F., 2018. A fast method for identifying worldwide scientific collaborations using the Scopus database. Telemat. Inform. 35 (1), 168—185.

Nilsson, M., 2005. Learning, frames, and environmental policy integration: the case of Swedish energy policy. Environ. Plan. C Gov. Policy 23 (2), 207—226.

Perea-Moreno, A.J., Perea-Moreno, M.Á., Hernandez-Escobedo, Q., Manzano-Agugliaro, F., 2017. Towards forest sustainability in Mediterranean countries using biomass as fuel for heating. J. Clean. Prod. 156, 624—634.

Rahimi, F., Ipakchi, A., 2010. Demand response as a market resource under the smart grid paradigm. IEEE Trans. Smart Grid 1 (1), 82—88.

Salmeron-Manzano, E., Manzano-Agugliaro, F., 2018. The electric bicycle: Worldwide research trends. Energies 11 (7), 1894.

Salmerón-Manzano, E., Manzano-Agugliaro, F., 2019. The role of smart contracts in sustainability: worldwide research trends. Sustainability 11 (11), 3049.

Sánchez, P., Montoya, F.G., Manzano-Agugliaro, F., Gil, C., 2013. Genetic algorithm for S-transform optimisation in the analysis and classification of electrical signal perturbations. Expert. Syst. Appl. 40 (17), 6766—6777.

Sencar, M., Pozeb, V., Krope, T., 2014. Development of EU (European Union) energy market agenda and security of supply. Energy 77, 117—124.

Serrallés, R.J., 2006. Electric energy restructuring in the European Union: integration, subsidiarity and the challenge of harmonization. Energy Policy 34 (16), 2542—2551.

Wang, Y., Wu, C., 2012. Forecasting energy market volatility using GARCH models: can multivariate models beat univariate models? Energy Econ. 34 (6), 2167—2181.

Wu, S., Zheng, X., You, C., Wei, C., 2019. Household energy consumption in rural China: historical development, present pattern and policy implication. J. Clean. Prod. 211, 981—991.

Zapata-Sierra, A.J., Cama-Pinto, A., Montoya, F.G., Alcayde, A., Manzano-Agugliaro, F., 2019. Wind missing data arrangement using wavelet based techniques for getting maximum likelihood. Energy Convers. Manag. 185, 552—561.

Zhang, X., Ruoshui, W., Molin, H., Martinot, E., 2010. A study of the role played by renewable energies in China's sustainable energy supply. Energy 35 (11), 4392—4399.

Which aspects may prevent the development of energy service companies? The impact of barriers and country-specific conditions in different regions*

13

Marina Yesica Recalde
National Scientific and Technical Research Council (CONICET) / Environment and Development Program, Bariloche Foundation (BF), Argentina

Chapter Outline

13.1 Introduction 293
13.2 Which are the problems confronted by energy efficiency actions and policy instruments? 296
13.3 Which are the most relevant barriers confronted by energy service companies in different regions? 300
13.4 Removing barriers and promoting energy service companies 306
    13.4.1 Actions to remove economic and market barriers 306
    13.4.2 Actions to remove funding barriers 306
    13.4.3 Enabling frameworks for energy service companies and other energy efficiency actions 307
13.5 Lessons learned and conclusions 308
Acknowledgments 312
References 312
Further reading 315

13.1 Introduction

Along the last decades, energy efficiency has been remarked as an important contributor to energy security and energy transition, crucial for the achievement of the Paris Agreement goals, and having extremely important cobenefits for socioeconomic development (Zabaloy et al., 2019). However, despite its potential, there is still a huge gap in energy efficiency around the world. Energy conservation is far away from the levels it should be in order to achieve goals compatible with climate change objectives. As mentioned by Estache and Kaufmann (2011), this means that

---

* The views and opinions expressed in this chapter are those of the author and do not necessarily reflect the official policy or position of the institutions to which she belongs.

Energy Services Fundamentals and Financing. DOI: https://doi.org/10.1016/B978-0-12-820592-1.00013-0
© 2021 Elsevier Inc. All rights reserved.

the amount of energy used to fulfill different energy services (e.g., heating, cooling, lighting, transportation, producing goods) could (*and should*) be lower than it is today. As highlighted by the authors, the energy efficiency gap is defined as the difference between the efficiency levels that could be achieved by cost-effective energy efficiency investments and the (lower) efficiency levels currently observed.

This gap may be due to different reasons. Many of these reasons have been addressed in the economic literature as market fails. Indeed, as mentioned by Estache and Kaufmann (2011) between 1980s and 1990s, many studies estimated the relevance of different factors that could explain the nonadoption of desirable investment and consumption decisions. The authors argue that, as shown by many research papers, the energy efficiency gap may be reflecting an underestimation of private costs and misunderstood preferences of consumption (related to the price distortions), whereas other studies showed that the most important failure is the limited access to capital.

In a similar direction, Rosenow et al. (2017), argues that even though many energy efficiency actions are usually cost effective in the medium to long term, private stakeholders (firms or end users) face different problems (barriers) that prevent them to implement these actions. The identification of these problems is key for closing the energy efficiency gap, and for this purpose the implementation of public policies is crucial. It is straightforward that these public policies must include a range of instruments to overcome the different barriers.

In this context, energy services companies (ESCOs) may appear as effective instruments to overcome some energy efficiency barriers in different sectors (e.g., capacity, financing, economic feasibility). Energy performance contracts (EPCs) undertaken by ESCOs are demonstrated to be an efficient mechanism to promote investments in energy efficiency technologies in both public and private sectors (Fang et al., 2012). For instance, the European Commission in its Energy Efficiency Directive (2012/27/EU) proposes the development and implementation of ESCOs in buildings, as catalysts for the renovation of appliances.

As described by Fang et al. (2012), an ESCO can be defined as a *company that offers energy efficiency technologies for clients in the public, industrial, commercial, or residential sectors, including not only the development and design of energy efficiency measures, but also the installation and maintenance of energy efficient equipment, the monitoring and verification of the project's energy savings, and above all, a guarantee of the savings* (Vine, 2005; Ellis, 2010; Marino et al., 2010). In fact, one of the main differences between the traditional energy consultants or equipment suppliers and ESCOs lies in their capacity to finance, or arrange financing, for the operation and, especially the fact that their remuneration is directly tied to the energy savings achieved. In this sense, ESCOs can help to overcome economic, capacity, and financial barriers. The energy and climate conditions after 1970 energy crises, and particularly in the wake of climate change policies, negotiations and agreements, have increased the opportunities for ESCOs business.

According to the International Energy Agency, ESCOs market has been increasing along the last years. In 2018, the value of the global ESCO market grew 8%. The current three champions for ESCOs are China (representing 57% of total market in 2018), United States (27% out of total market) where ESCOs have been operating for well over 30 years, and Europe (11% of market share) which, nevertheless, is remarked to have an underdeveloped ESCO market compared to other major regions (IEA, 2019). There has been a rapid growth of the ESCOs market along last decades, fostered by favorable policies such as tax incentives and accounting systems that have accommodated ESCO projects, particularly in the Chinese case (IEA, 2018). However, the market growth has been recently reduced. Particularly, in China the market is expanding at a significantly slower pace compared with 2011−15, when the average annual growth rate was over 25%. This slowdown may be reflecting changes to the policy landscape and market dynamics, particularly reductions in subsidies and financial incentives (IEA, 2019). Nonetheless, the Chinese ESCOs market has succeeded in becoming mature and diversified enough, including a diversification of business models and actors in the market, evolving from EPC under the shared saving model to other contract models (guaranteed savings, outsourced energy services, and financial leasing, among others); and including other companies such as technology (hardware) providers, software developers, and building service management companies (IEA, 2019).

ESCOs have been usually equally implemented in two key end use sectors: Nonresidential buildings and industry, and there are interesting examples of ESCOs in residential building in Brazil and Korea. In the case of Japan, Brazil, Canadá, EU, and the United States, ESCOs have been more devoted to nonresidential buildings, whereas in China, Korea, India, Thailand, and México, they are more common in industrial sector.

Nevertheless, the implementation of ESCOs still faces different problems and barriers in many regions of the world. If ESCOs are to become an important policy instrument to contribute to closing the energy efficiency gap, it will be important to design a strategy to overcome their barriers. The literature on barriers and drivers for the development of ESCOs in different sectors and regions is vast. However, many of these papers do not differentiate barriers faced because of country-specific conditions, which cannot be modified by the policy maker, from specific ESCOs barriers that can be removed by additional energy policy. This is, from policy makers perspective, a crucial distinction to be made when designing the policy incentives.

This chapter, therefore, intends to contribute to the literature on ESCOs by identifying which are the border conditions that have affected the deployment of ESCOs around the world and which are the barriers for their implementation. The chapter is built on existing theoretical and empirical literature of energy policy design and ESCOs barriers. The chapter is divided into four additional sections. First, a theoretical discussion on the different barriers affecting the energy efficiency policy and their classification is presented. Second, a review of empirical studies of ESCOs barriers is presented. The next section presents some instruments to remove the barriers and the most common models for ESCOs. Finally, the chapter presents a review of the main aspects identified and some conclusions.

13.2 Which are the problems confronted by energy efficiency actions and policy instruments?

As recognized by different authors, closing the existing energy efficiency gap will require to enhance the efforts devoted to the promotion of energy efficiency actions. These aspects highlight the need for the implementation of policies, actions, and instruments to remove these problems and underpin the implementation of energy efficiency actions in the sectors of greatest energy, environmental, and economic relevance (e.g., buildings, industry, and transport).

In fact, as mentioned by the Office of Energy Efficiency & Renewable Energy from the US Department of Energy, "Energy efficiency policies and programs can help drive the implementation of projects that minimize or reduce energy use during the operation of a system or machine and/or production of good or service". Nonetheless, in some cases, even well-designed instruments may result in a bad performance of the energy efficiency actions and policies (Zabaloy et al., 2019). Therefore to increase the performance of the strategies and instruments implemented, it is crucial to identify and understand the key elements characteristics affecting the energy efficiency policy.

Different studies have focused in the evaluation of the reasons on why energy efficiency policies (even those apparently sound policies) have not been successful in closing the gap. Estache and Kaufmann (2011) resume these aspects (based on previous studies from IEA (2007), Fuller et al. (2009), among others) and conclude that both theory and evidence show that most of the fails are due to the underestimation of the relevance of basic agency issues and behavioral reactions.

First, governments usually underestimate the number of agency issues that arise in the implementation of energy efficiency policies, and do not consider that better information, communication, and education can reduce agency risks. Second, insufficient care in the design of the regulation can explain the stakeholder's lack of incentive to comply. Third, there is no obvious standard ranking of the instruments, as the cost-effectiveness of them can vary across designs, agents, sectors, countries, time, and the macro context (including global financial liquidity and the fiscal situation of the governments concerned with energy efficiency). Finally, the authors argue that there may be more than problems related to financing, prices, and enforcement behind the lack of success of energy efficiency policies; the problem may be that governments do not focus on the right problems.

The precedent arguments show that country-specific aspects have a crucial impact on the process for the elaboration of the energy efficiency policy. As stated by CEPAL/OLADE/GTZ (2003), the performance and success of any energy policy is the result of the combination of different factors: the international and national context; the instruments chosen and designed to overcome the existing barriers for the penetration of different technologies or practices; and the attitude of key stakeholders of the sector in reaction to the context conditions and to the implementation of the instruments. In this framework, the identification of barriers is a key step for the elaboration of the policy (Bouille et al., 2019a,b). Policy makers can remove these barriers with properly designed instruments and the selection of the instruments depends on the barrier that must be removed.

The literature on barriers to energy efficiency actions is vast, and there are studies for specific sectors and regions of the world (Langlois-Bertrand et al., 2015; Sorrell, 2004, 2011; Trianni et al., 2016; Yeatts et al., 2017; Vogel et al., 2015, among others). These barriers have different levels, and there is no consensus about their classification among different authors, ranging from barriers affecting the whole economy, to specific barriers affecting certain actions or instruments. Although there are gray areas between some of the categories of barriers shown in Fig. 13.1, especially in the case of border conditions, enabling frameworks and barriers, it may be very useful to identify them before designing the energy policy (Bouille et al., 2019a).

Border conditions can be highlighted as a first level and general barriers, even though they might not be catalogued as barriers. These elements set out the context of the policy and the energy efficiency actions and do not depend on the decision-making mechanism of the sector itself. The border conditions may be external (international oil market; negotiations in the framework of climate change, lack of access to certain technology; global economic conditions, international agreements) or national (institutional structure, general macroeconomic aspects—such as the foreign investment law, conditions with respect to the external sector, foreign trade, local financial market).

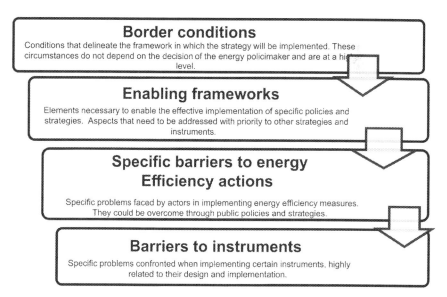

Figure 13.1 Definition and scope of different categories of barriers.
Source: Adapted from Bouille, D., Recalde, M., Di Sbroiavacca, N., Dubrovsky, H., Ruchansky, B., 2019a. GUIA METODOLOGICA PARA LA ELABORACIÓN DEL PLAN NACIONAL DE EFICIENCIA ENERGÉTICA ARGENTINA (PlanEEAr). Proyecto de Eficiencia Energética Argentina. GFA Consulting Group.

Border conditions are country and time specific, and their correct identification is crucial for the performance of the energy efficiency policy. As found by the IEA (2007) survey, incentive issues are context specific. Institutional and regulatory frameworks and approaches are relevant dimensions that need to be dealt with in choosing policies (Estache and Kaufmann, 2011). The use of correct and well-designed instruments could fail if the context of implementation is not properly considered (inadequately assessed barriers or edge conditions not considered) (Bouille et al., 2019a,b). Policy instruments emerge depending on the political dynamics of each country, and these dynamics clearly have an impact on the development of the policy instrument itself (Rosenow et al., 2016). As stressed by Estache and Kaufmann (2011), this means that for some policies, the concept of best practice can be quite misleading.

The second level of barriers is enabling frameworks or enabling environments, which have been broadly discussed in the literature. As defined by Bouille et al. (2019a,b) the enabling frameworks are those conditions that, if exist (*if absent*) they facilitate (*hinder*) the existence of an environment conducive to the implementation of energy efficiency actions, or for the implementation of certain instruments and specific policies. Necessary but not sufficient conditions for the implementation of energy efficiency actions. These aspects may be either out of the scope of the energy efficiency policy maker or not but, if absent they can (*and should*) be somehow modified by the national authorities.

The distinction between border conditions and enabling frameworks is not so clear, because some enabling frameworks can also be a boundary condition for the policy. Therefore in many opportunities in the literature, these concepts have been used in both directions.

Based on the definition from the Intergovernmental Panel on Climate Change, Nygaard et al. (2015a) define the enabling environment as "the entire range of institutional, regulatory and political framework conditions that are conducive to promoting and facilitating the transfer and diffusion of technologies, including the country-specific circumstances that encompass existing market and technological conditions, institutions, resources and practices, which can be subject to changes in response to government actions." The authors present a list and categorization of enabling environments, and stress that there are main areas that governments can influence directly or indirectly in order to modify the enabling conditions. Whenever these conditions are not present, the implementation of actions may be more complex. The relevance of enabling conditions for different energy and climate change policies has been deeply studied in the literature (Dyner et al., 2011; Duguma et al., 2014; Recalde, 2016; Recalde et al., 2015; Zabaloy et al., 2019). Similarly, its relevance for technology transfer (related to climate change policies and negotiations) has been demonstrated in different articles in a special Issue of the Climatic Change journal (Nygaard et al., 2015b; Ockwell et al., 2015; Trærup et al., 2015; Watson et al., 2015). As remarked by Ellis (2010), in many developing countries there are many overarching factors (*or enabling conditions*) that influence energy efficiency investments, such as energy prices, fluctuations, political and economic instability, and so on.

Specific barriers are the third level of barriers. They refer to the problems faced by actors to implement autonomous energy efficiency measures and raise the need for intervention through public policies and instruments for effective implementation (Bouille et al., 2019a,b). They are also defined by Ruchansky et al. (2011) as mechanisms that hinder investment in energy-efficient and economically profitable technologies. Policy makers can remove these specific barriers with properly selected and designed instruments, which depend on the barriers that must be removed.

The literature on barriers to energy efficiency actions is wide (Langlois-Bertrand et al., 2015; Sorrell, 2004; Sorrell et al., 2011; Trianni et al., 2016; Yeatts et al., 2017; Vogel et al., 2015, among others), and there are several empirical studies that evaluate specific barriers in particular sectors of the economy. It is straightforward that barriers are country and sectorial dependent, and that their intensity differs from one country to another because they depend on the market development, and policies implemented historically, among other aspects (Ruchansky et al., 2011). The latter aspect reinforces the relevance of the border conditions and context dependence of the actions mentioned before.

These specific barriers can be removed with properly designed instruments. For instance, financing barriers may require the implementation of soft loans; information barriers at industrial level usually require the implementation of audits; capacity barriers related to internal capacities can be removed by capacity buildings programs. Bearing in mind that a specific measure can be attained by more than one barrier, energy efficiency potentials require a well-targeted and comprehensive instrument mix (Jacobsson et al., 2017; Kern et al., 2017; Rosenow et al., 2016; Grubb et al., 2017; Purkus et al., 2017; Rosenow et al., 2017).

In this framework, ESCOs can be promoted by governments and used as instruments to remove different barriers to energy efficiency. ESCOs can help energy users, customers, companies, industries, and commercial sectors to improve efficiency of equipment by providing energy service (energy performance and/or credit risk) (Nurcahyanto, 2018). Depending on the specific model adopted, ESCOs can help to remove different barriers to energy efficiency actions. For instance, they can help to solve problems related to lack of internal technical capacity to identify and implement energy efficiency measures, reducing investment risks, or solving financial barriers when the ESCO model is based on financing. It may be frequent that consumers, even in the cases in which they have identified energy efficiency options, are adverse to making large investments in energy saving, due to lack of experience in the subject and/or uncertainty in the real achievement of savings that make their investment viable, as well as because they lack the financial capacity. The ESCOs can, therefore, overcome these barriers by carrying out energy studies, diagnose, and issue some recommendations, but more importantly they can develop, install, and finance projects specially designed to improve energy efficiency and reduce maintenance costs, throughout EPCs. In this context, the promotion of energy services market can be a specific policy action particularly relevant within an energy efficiency strategy or road map.

The last category of barriers, *barriers to instruments*, refers to the problems that confront certain instruments, depending on its characteristics and scope (Bouille et al., 2019a,b). These barriers are highly dependent on the context of implementation of the instrument and some design aspects. As it will be seen later, in the case of ESCOs different studies identify which are the most relevant barriers confronted by energy services hindering their development in some specific markets or regions.

13.3 Which are the most relevant barriers confronted by energy service companies in different regions?

Despite the considerable efforts devoted to the ESCOs penetration in the energy efficiency markets, many barriers have prevented their development, particularly in developing regions (Ellis, 2010). A long list of research articles and reports have been published regarding the analysis of barriers and drivers to ESCOs in different countries and regions of the world. Most of these articles focus in the analysis of the different factors that hinder the development of ESCOs in the whole economy, or in specific sectors.

Table 13.1 presents a review of some the most recent or relevant studies performed in this branch. It is important to note that, considering that the literature on barriers to ESCOs is very extensive, the table concentrates only on some articles. The selection of the case studies has been biased to developing countries as they are the cases in which ESCOs are, up-to-date, less developed; except for the case of China, which in 2018 represented 57% of Global ESCO market (IEA, 2019).

The objective of this review is twofold. On the one side, to identify if there are similar barriers in all countries, regardless their development stage or region. On the other side, to identify which of the barriers mentioned by the papers can be identified as barriers to the instrument, barriers to energy efficiency actions, enabling environments, or border conditions. Note that, in Table 13.1, aspects considered as border conditions and enabling frameworks have been indicated in italics. Barriers identified in each of the reviewed studies were categorized into economic and market barriers, funding barriers, institutional and regulatory barriers, capacity barriers, information barriers, and other barriers.

Some key aspects and conclusions can be driven from Table 13.1. As it may be expected, ESCOs confront more problems in developing countries than in developed ones which, as stressed by Ellis (2010), may be the key reason on why market penetration of ESCOs has been limited in those countries, in spite of the efforts made by governments and international institutions to promote ESCO development.

The most relevant categories of barriers found in the previous literature review are economic and market barriers, funding barriers, and institutional and regulatory ones. These barriers are resumed and discussed below:

1. *Economic and market barriers*: These barriers seem to be more relevant in the case of developing countries/regions than in developed ones. Within this category, the literature stresses the existence of high transaction costs and lack of governmental support, as well as other general aspects more related to enabling conditions rather than barriers (lack of

Table 13.1 Review of barriers to ESCOs mentioned in different research articles and reports.

Country	Barriers mentioned in the article	Authors
Developing regions	Economic and market barriers: High administrative and transaction costs; complex contract negotiations; lack of knowledge of ESCOs; lack of reliability in ESCOs; client preferences for in-house solutions, specific challenges of the EPC model (low profit margins, uncertainties, risks, cost of preparation of not successful bids); *Clients general preference to focus on increasing market share and production rather than energy efficiency; Lack of energy efficiency technologies available; low energy prices that disincentive energy efficiency investments.* Funding barriers: Difficulties to access financing; lack of ESCO awareness of how to access financing; lack of lender knowledge of ESCOs; commercial financing rules that are inconsistent with EPC; *scarcity of capital markets.* Institutional and regulatory barriers: Lack of a legal and institutional framework for ESCOs; accrediting mechanisms to certify ESCOs; *lack of political commitment with energy efficiency; lack or limited enforcement of energy policy; complicated government procurement rules; lack of governmental energy policy such as minimum energy performance standard* Capacity barriers: Lack of human resources within the ESCO; lack of technical capacity and experience to fully assess the benefits and risks of engaging in ESCO projects within clients; technical challenges associated with designing projects and procuring equipment. Information barriers: Lack of information, education, and training	Ellis (2010) and Nurcahyanto (2018)
Brazil, China, and India	Economic and market barriers: High transaction costs; lack of governmental support; preference for modernization investments; relevance of SMEs; *policies and practices that hinder energy efficiency options; lack of requirements for utilities to develop DSM programs; small size of the projects; lack of technical competence; lack of marketing for energy efficiency projects.* Funding barriers: ESCOs are treated as high credit risk; risk aversion and low risk-taking capacity of lenders; *scarcity of capital; most financial institutions focus*	Painuly et al. (2003)

(*Continued*)

Table 13.1 (Continued)

Country	Barriers mentioned in the article	Authors
	their appraisal mostly on revenue stream (cash-flow), rather than saving stream that energy efficiency projects offer; Currency risk. Institutional and regulatory barriers: Lack of familiarity with EPC concepts; lack of interest from utilities to participate in new institutional arrangements; weak legal contract enforcement. Other barriers: *Lack of awareness of the energy efficiency potential; cultural barriers.*	
Latin American Region	Economic and market barriers: Lack of (or insufficient) stakeholders in the supply side; energy efficiency investments aversion from stakeholders; insufficient guarantee provided by ESCOs; *low relevance to energy efficiency actions; impact of specific variables over the ESCOs market development: inflation, high interest rates, lack of consultants, and so on.* Funding barriers: High transaction costs (related to the market size); inexistence of financial market for EPC; high tax burdens; *financing problems: access and cost.* Institutional and regulatory barriers: *Difficulties to maintain long-term contracts.* Capacity barriers: Lack of market capacities; Lack of capacity in local ESCOs.	Carpio and Coviello (2013), Fundación Bariloche (2013)
Brazil, Colombia, Chile, Ecuador, and Uruguay	Economic and market barriers: Lack of certification schemes for ESCOs; lack of fiscal incentives to promote energy efficiency actions; *distortions in energy prices; inexistence of mandatory implementation rules for EMS; lack of long-term energy policies.* Funding barriers: Lack of financial options; lack of experience in project finance; *low developed capital market; lack of energy efficiency credit lines; high cost of financing options.* Institutional and regulatory barriers: Inexistence of an ESCOs association; *weakness of energy institutions; stability and permanence of public policies; lack of a well-designed market of guarantee.*	Blanco and Coviello (2015)

Mexico	Capacity barriers: Local capacities for energy efficiency projects. Other barriers: Cultural factors that reduce the openness to information from private corporations; *awareness of the relevance of energy efficiency and environmental aspects.* Economic and market barriers: High transactional costs; nonrecoverable expenses in energy audits when the client decide not to go on with the contract once the energy savings have been estimated; *higher cost of efficient technologies.* Funding barriers: Lack of credit history for ESCOs and the lack of experience in this market on the part of the banks and the high transaction costs involved in verifying a guaranteed minimum performance increase the cost of financing (vicious loop); risk that the customer/energy user will face financial difficulties; the risk that the savings generated will not be sufficient to repay the financing; absence of capitalization mechanisms—Equity of projects especially in SMEs; limited banking understanding of the ESCO model; lack of culture for project finance. Institutional and regulatory barriers: Lack of an institution that validates the initial parameters of the contract; absence of a legal referee for conflicts between the involved parts; low awareness of ESCOs models, which results in aversion; asymmetry of technical information between ESCOs, clients, and banks to confront the estimations; absence of ESCO certification, which is difficult for the selection of "good" ESCOs from clients and financial institutions.	CONUEE/GIZ (2012)
Chile	Economic and market barriers: Market size (low attractiveness for foreign investors). Other barriers: *Lack of interest in different sectors.*	Carpio and Altomonte (2015)
China	Economic and market barriers: The taxes and fees are high for ESCOs; low awareness about ESCOs; lack of confidence of ESCO services and solutions because of poor capabilities (ESCOs are small SMEs); a lack of skill and technical competence for energy efficiency in the ESCOs; absence of standardized procedures for energy audit and energy conservation measurement and verification which difficult the evaluation of the projects; *existing tax and fiscal system discourage energy efficiency.*	Da-li (2009)

(*Continued*)

Table 13.1 (Continued)

Country	Barriers mentioned in the article	Authors
	Funding barriers: Many of them are related to the fact that most ESCOs are SMEs; many new ESCOs in China could not demonstrate transparent financial and accounting systems required; shortage of bank finance available and the lack of loan category suitable for EPC projects; banks lack of appraisal ability to evaluate risks of EPC projects; lack of credit appraisal mechanism and the rating granted by the bank is low. Institutional and regulatory barriers: lack of governmental support; there are no laws to provide effective promotion and regulation for the implementation of EPC; administrative hurdles and high transaction costs for ESCOs. Other barriers: Lack of awareness of the energy efficiency potential due to information gap, managerial indifference, incompetent management, lack of interest and so on; *lack of aggressive marketing of energy efficiency projects; credit risk and improper societal creditworthiness environment.*	Bertoldi and Boza-Kiss (2017)
EU Member States and neighboring countries	Economic and market barriers: Lack of facilitators for ESCOs; lack of proper M&V of the practices to prove energy savings; lack of trust of clients in ESCOs related to lack of experience, lack of credible references. Funding barriers: Difficulties to account EPC projects; low awareness and motivation from banks; high transaction costs; lack of well-established partnerships between ESCOs and subcontractors; failed projects affected the markets. Institutional and regulatory barriers: *Unstable legislation about the sector* (Hungary, Slovenia, Italy, and Spain); lack of clear/official definition of ESCO (Netherlands, Croatia, Italy, and West Balkan countries); contradicting legislation; complexity of procurement laws and practices.	
Finland	Economic and market barriers: Lack of knowledge about the ESCO model. Funding barriers: High transaction costs.	Pätäri and Sinkkonen (2014) and Pätäri et al. (2016)
Italy (residential, tertiary, and industrial)	Economic and market barriers: *Energy price volatility (fuel purchased and heat sold).* Funding barriers: Difficulty of accessing loans. Institutional and regulatory barriers: *Uncertainty over future regulation sector limited long-term investments.*	Pätäri and Sinkkonen (2014)
Russia	Funding barriers: financing projects through a third party. Institutionalization and project tools: Risks and costs associated with the preparation of the tender documentation, quantifying energy savings.	Roshchanka and Evans (2016)

availability of energy efficiency technologies and higher cost of efficient technologies; local policies and practices that hinder energy efficiency options; high inflation rates; energy price volatility).

However, the most relevant aspects within this category seem to be related to the configuration of the ESCOs market itself, highly related (in most of the cases) to the lack of local experience on the topic and the scale of the market. Small market size, including both demand and supply side, have a negative influence in the attractiveness for foreign investors (which are usually the most experienced stakeholders) and increases the existence of SMEs in ESCOs. Lack of knowledge about the ESCO model in the market results in subutilization of their services and also impacts on financial barriers, whereas lack of reliability and confidence in ESCOs and insufficient guarantee provided by ESCOs reduce the potential demand for their services.

2. *Funding barriers*: The barriers in this category are directly linked to some of the barriers mentioned before and are linked to two problems: credit access and cost of loans. First, credit access may be due to a branch of reasons, such as: lack of ESCO awareness of how to access financing; limited banking understanding of the ESCO model; financing rules not compatible with EPC; inexistence of financial market. In the case of China, there is a special mention to the configuration of the ESCOs market, highly concentrated in SMEs that failed in demonstrating transparence in financial and accounting systems. Second, the cost of financing is usually increased in the case of ESCOs and EPC due to the lack of credit history for ESCOs and the lack of experience, as ESCOs are treated as high credit risk.

Funding barriers, particularly relevant in developing countries, are usually related to underdeveloped or not fully developed capital markets, which can be highlighted as an enabling (necessary) condition. This problem is reinforced by a risk aversion of banks to finance energy efficiency, which affects all the energy efficiency actions.

3. *Institutional and regulatory barriers*: In this case, problems relate to the lack of legal and institutional framework for ESCOs that validates the initial parameters of the contract, lack of interest from utilities to participate in new institutional arrangements; asymmetry of information between ESCOs, clients and banks; absence of ESCO certification to provide clear signals to clients; inexistence of laws to provide effective promotion; and regulation for the implementation of EPC.

There are highly relevant institutional and regulatory conditions that can be treated as both enabling or border conditions, as they affect the development of all energy efficiency actions and are especially relevant in developing regions: lack of political commitment; limited enforcement of energy policy; difficulties to maintain long-term contracts; uncertainty over future regulation sector limited long-term investments.

In addition to the aforementioned barriers, developing countries face other overarching factors, usually mentioned as barriers by the papers reviewed, but that in the framework of this chapter are known as border conditions or lack of enabling frameworks. These aspects (remarked in italics in Table 13.1) include conditions (or problems) that difficult (or even impede) almost all energy efficiency actions and investments in these developing countries such as political and economic instability, energy prices, lack of clear energy efficiency institutions, and governmental/political commitment with energy efficiency actions, among others. As stressed before, while border conditions cannot be easily removed by governmental programs and instruments but must be considered in detail, enabling frameworks are the aspects that need to be addressed in advance before removing any other barrier.

13.4 Removing barriers and promoting energy service companies

As previously mentioned, the only way to successfully incorporate ESCOs to energy efficiency scene, is throughout the removal of the existing barriers and the creation of enabling frameworks to this purpose. Ellis (2010) argue that "... developing countries governments have a key role to play in creating an environment favorable to ESCO development through financial support, energy-efficiency policy and removal of barriers, as well as the creation of an early market through demonstration programs. In the absence of this support, establishing a successful ESCO industry in any country will be challenging..." (Ellis, 2010, 47).

In this regard, countries have implemented different strategies to promote ESCOs. Following the framework introduced before, these policies and actions will be classified according to the barrier they will remove or the enabling condition that will create.

13.4.1 Actions to remove economic and market barriers

Problems confronted when promoting and implementing ESCOs market are usually the most relevant. Therefore countries should concentrate in developing a branch of instruments and actions to provide knowledge to stakeholders on the characteristics and opportunities of energy services and create market necessary conditions.

Probably, the first step may be related to capacity building programs, devoted both to the demand side and supply side. In the former, in order to increase the relevance of energy efficiency topic and help clients to understand how ESCOs can help. The latter will be especially important to remove the problem related to the lack of experience of local ESCOs in some critical topics such as EPC, contracts alternatives, funding opportunities, and so on. In this regard, it may be useful (at least at the very beginning) to promote the international partnerships and cooperation among ESCOs.

It may also be important to develop (or help developing) demonstration and ongoing projects. As argued by Ellis (2010), these demonstration programs were key in the development of energy services markets in developed countries, particularly in the public buildings/utilities. In a similar direction, Taylor et al. (2008) stress the role of public sector (schools, hospitals, governmental buildings, and so on) in North America.

13.4.2 Actions to remove funding barriers

In order to remove some of the barriers related to the difficulties to access credits and their high costs, countries implement different strategies to provide loans guarantees. Ellis (2010) argues that these guarantees are critical to foster credits from domestic financial institutions, and they can be backed either by public or international funds. These strategies can be very useful at the very beginning, in order to surpass the inexistence of credit history and the consequent reticence to lend money from banks. Some developing countries have implemented these strategies both with local and international support. For instance, Brazil has a loan guarantee fund for energy-efficient projects called PROESCO, in which the Brazilian National Development Bank (BNDES) shares up to 80% of the credit risk, whereas the remaining 20% is taken up by the intermediary bank, and the guarantee fee is paid by the borrower (Ellis, 2010). Another example of strategy to provide guarantees can be found in the case of the Loan Guarantee Special Fund in China, which may be one of the reasons explaining the success of the Chinese ESCOs market. In this latter case, the role of World Bank GEF has been critical.

Removing barriers associated to the limited knowledge of ESCOs market and bank's aversion to lend money to ESCOs requires the implementation of different actions, particularly in the first stage of the development of a national ESCOs market, when most of the stakeholders lack knowledge about the mechanism. For instance, establishing capacity building programs, both for banks and other credit stakeholders to educate about the ESCOs models; establishing partnerships between governmental and financial institutions to evaluate ESCOs projects; establishing networks with international institutions, national financing institutions can learn how to evaluate and deal with the risk of these loans.

13.4.3 Enabling frameworks for energy service companies and other energy efficiency actions

As mentioned before, enabling frameworks are those critical aspects that need to be addressed in advance as they generate a favorable environment for the development of ESCOs. In this sense, for instance, it may be advisable to have an energy efficiency law, with clear obligations to stakeholders. It is straightforward that the mere existence of the law may not be enough, and that enforcement of the regulation is needed (which is usually remarked as a problem in many developing countries). There are some relevant experiences of countries (such as European Union, China, Brazil, among others) that have established targets for reduction of global energy consumption, or special targets for some sectors of the economy, or even have included some requirements for utilities. These strategies are usually recommended for the promotion of all energy efficiency actions, not only to promote ESCOs. In addition, international experience shows that the existence of energy efficiency agencies can play a key role in the development of energy efficiency market (Bouille et al., 2019a,b), and therefore they may also play a crucial role for the promotion of ESCOs (Ellis, 2010).

One of the most remarked aspects in case of studies is the underdeveloped or not fully developed capital markets. Without a mature capital market, it will not be easy to promote ESCOs. Experience shows that without funding mechanisms, these industries will not success. As it has been remarked by research papers cited before, in many developing countries, national capital market cannot deal with this funding. Therefore international funding agencies play a key role, at least in the initial stage of the development of the ESCOs market. After an initial stage, it may be advisable that national governments go ahead with designing their own strategies for improving the funding of ESCOs locally.

It is important to remark that, based on their own characteristics, countries can decide which model of ESCOs to promote. The first model of ESCOs implemented globally was the *shared savings model* (Fig. 13.2A). This model consist of a contract between the client and ESCO, in which they arrange to share the energy savings based on some predetermined percentage for a certain number of years, after this period the project is over and the client retains any further savings afterward. Under this model, the ESCO usually finances the project, generally through third-party financing. Therefore while the client does not assume obligations, the ESCO assumes both the credit and the technological risk, and usually provides guarantees for the loan. The remuneration to the ESCO can be established as fixed remuneration (fixed percentage of savings), escalating remuneration (the percentage of ESCO decreases over the time of contract), and guided remuneration (ESCO charges a monthly fee and both stakeholders share savings over a certain threshold) (Ellis, 2010; Blanco and Coviello, 2015).

The *guaranteed savings model* (Fig. 13.2B), consists of a contract in which the client makes the upfront investment (or is responsible for acquiring the funding), and the ESCO guarantees a certain level of energy savings. The ESCO provides design, procurement, and construction services. Although the ESCO does not assume the responsibility for financing, it provides guarantees for the avoided energy performance: if the savings fall short of the ESCO forecast, the ESCO is obligated under the contract to compensate for this. In many cases, and depending on the availability of local fund, the guaranteed savings model is considered to be the lower interest rate option because the customer can find their own funding source. (Ellis, 2010; Blanco and Coviello, 2015).

It is important to note that although one of the advantages of the shared saving model is that it may help to deal with client's risk aversion and lack of funding capacity, it requires market confidence on ESCOs. In this regard, the previously mentioned market and financial barriers are common in this model, specially at the initial steps of ESCOs' market development. In order to provide funding for the projects, ESCOs are required to have funding capacity, and banks credibility is crucial. In these cases, some of the actions previously mentioned will be key: loan guarantees, credit lines, capacity building, providing information on ESCOs, and so on.

13.5 Lessons learned and conclusions

As stressed by Estache and Kaufmann (2011), the relevance of the context for the implementation of energy policies may reduce the weight of the concept of *best*

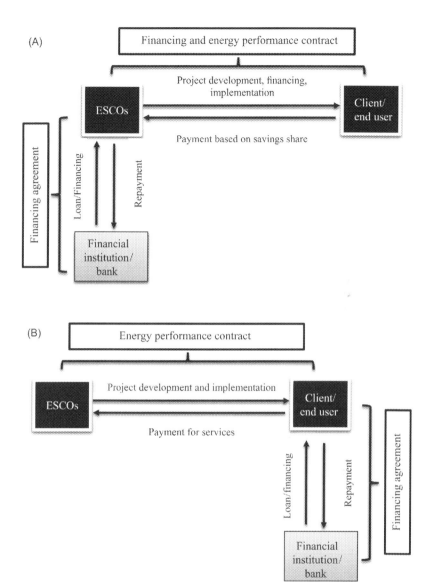

Figure 13.2 Basic ESCOs models. (A) ESCOs' shared savings model. (B) ESCOs' saving guarantee model.
Source: Adapted from Taylor, R.P., Govindarajalu, C., Levin, J., Meyer, A.S., Ward, W.A., 2008. Financing Energy Efficiency: Lessons From Brazil, China, India and Beyond. The World Bank, Washington, DC; Blanco, A., Coviello, M., 2015. Empresas de servicios energéticos en América Latina Un documento guía sobre su evolución y perspectivas. CEPAL, Naciones Unidas, octubre de 2015; Boza-kiss, b., Bertoldi, p., Economidou, M. 2017. Energy Service Companies in the EU. Status review and recommendations for further market development with a focus on Energy Performance Contracting. Science for Policy report by the Joint Research Centre (JRC), European Comission; Nurcahyanto T.U., Chapter 26: Development of energy service company (ESCO) market to promote energy efficiency programmes in developing countries, In: Sayigh A., (Ed), Transition Towards 100% Renewable Energy, Innovative Renewable Energy, 2018, 283−294. https://doi.org/10.1007/978-3-319-69844-1-26. <https://repositorio.cepal.org/bitstream/handle/11362/39008/1/S1500950_es.pdf>.

practices that has been so used in the literature of public policies. This aspect reinforces the value of the identification of border conditions and the promotion of enabling frameworks before promoting energy services markets. In fact, as shown by Fang et al. (2012), the impact of ESCOs promotion may differ significantly considering the development stage of the countries, and this may be highly related to the existence of informational, technical, financial, behavioral, and institutional barriers (*border and enabling conditions as classified in this chapter*) highly relevant in developing countries.

Nonetheless, even though solutions need to be customized based on the local institutional environment, contracting frameworks, and customs, they do not need to be entirely home grown, and lessons can be drawn from experiences in other countries (Ellis, 2010). Therefore it may be interesting to highlight some aspect that should be considered when promoting ESCOs in different countries and regions.

First, *border conditions* are highly relevant for the energy services markets. As in many other energy efficiency actions, the impact of macroeconomic, institutional, and legal frameworks will be crucial for ESCOs. Countries facing *international financing constraints*, or with a nonwell developed *internal financial market*, confront several problems for the deployment of energy services markets. This aspect is clear in different developing regions in the world, particularly in developing countries in which there are funding boundaries, and sometimes there are also international funding limitations (Blanco and Coviello, 2015; Nurcahyanto, 2018). *Institutional conditions* are also very relevant for the business cycle of the ESCOs and have a significant impact over the market. The relevance of this institutional border contexts has been evaluated and proven for many other efficiency (as well as renewables) investments (Recalde, 2018; Zabaloy et al., 2019). There is a clear negative impact of *weak institutional development*, which increases stakeholder's aversion to invest, in particular the impact of property rights and lack of clear enforcement of regulations.

Second, the experience shows that there are some *enabling environments/conditions* that can encourage the development of ESCOs. Most of these conditions do not differ from which is shown in the international experience for the implementation of any energy efficiency promotion instrument (Bouille et al., 2019a,b). Clearly, the presence of *global energy policy and long-term energy plan*, as well as the synergy between energy, industrial, and environmental policies, are key enabling conditions for any energy efficiency. The existence of an *effective institutional framework* for energy efficiency is also crucial and has been highlighted by different studies as a key barrier for ESCOs development.

The inexistence of *energy prices distortions* is also seen as an enabling framework. Blanco and Coviello (2015) found that in Latin American countries, energy services markets have developed more correctly in countries in which internal energy prices encourage energy efficiency. As stated by the authors, if the economic incentive is not present, the development of a market will not be possible, even with official rhetoric supporting or including energy efficiency in its long-term vision of the energy sector. It is important to note that, however, energy subsidies may be needed and important for other policies objectives, such as energy access, among others, that may affect socioeconomic development in these developing regions.

Information is, certainly, a very important condition for the performance of the energy services market. As stressed by Blanco and Coviello (2015) in order to overcome the barrier of lack of information about ESCOs and awareness about their potential, Latin American countries strategy was to adapt contracts to the characteristics and needs of each project and client, implying higher transaction cost for each individual operations. However, the authors argue that it could be a key element in introducing the business model into the market and creating trust. This is particularly remarked in the Colombian case.

Nurcahyanto (2018) stress the impact of the existence of *ESCOs associations* in developing countries in order to disseminate information and assist government for policy development. This aspect has been mentioned in many papers (see Table 13.1) as a barrier to ESCOs development, but as regards of its relevance to reduce some problems (e.g., stakeholders risk aversion and informational problems) it should be addressed in advance and therefore can be remarked as an enabling condition for energy services.

Third, regarding *specific barriers to ESCOs* there are many problems that ESCOs confront in different regions. *Market size* has been highlighted as one of the most important barriers to the market development, including not only the demand but also the supply side. In the demand side, the small quantity of stakeholders and companies requesting energy services reduces the attractiveness for foreign ESCOs investors to participate in the country (which are usually the most experienced companies). In the supply side, the lack of experience of ESCOs, the lack of knowledge about the ESCO model, and the lack of capacity in local ESCOs impact on the trust confidence of the process. This also impacts on other barrier mentioned by CONUEE/GIZ. (2012), *lack of credit history* for ESCOs, which increases the *cost of financing* (creating a vicious loop which may be difficult to overcome). The lack of ESCOs history on the country has negative impacts on their capacity to access to credits and increases the cost of credit. In this direction, policy recommendations stress the implementation of public policies oriented to identify the existing and potential market and instruments to create the market. As found by Nurcahyanto (2018), implementing a pilot project through government support can boost the ESCO market. In effect, Fang et al. (2012) stress that a key benefit from promoting ESCOs in the public sector buildings (e.g., hospitals, schools, government buildings) is that they may represent the largest single energy users of a country and exhibit a homogeneous consumption patterns, and therefore they offer huge potential for the implementation of energy efficiency projects, which will imply a posterior reduction in transaction costs and enlarge market activities.

The *lack of specific regulation for ESCOs* is also a main problem in many countries. This is highly related to the *lack of institutions that validates* the initial parameters of the contracts between the counterparts (this institutional problem was previously mentioned as an enabling framework). As suggested by Nurcahyanto (2018), enhancing standard and *regulation and accreditation of ESCO* is necessary for the reliable service in energy efficiency project, in order to reduce uncertainty and problems for distinguishing the energy performance contract in each sector (industrial, commercial, or public sector).

All the aspects discussed along this chapter highlight the complexity of the energy services markets deployment. This is particularly relevant for developing countries and stresses the need and potential of public policies, throughout other promotion instruments, to encourage the energy services market. Most of the successful cases have, to some extent, used public policy to promote the performance of energy services market. In an evaluation of the Latin American market for ESCOs Blanco and Coviello (2015) mention the existence of different public policies promoting the development of ESCOs, particularly throughout different energy efficiency lines and project-level initiatives that seek to employ performance-paid contracts, as well as developing special investment funds to overcome the financing barrier.

However, as mentioned in many opportunities along the chapter, there are no unique recipes, and energy policy must take into consideration country-specific characteristics. While ESCOs may be the most suitable solution for one country, it may not be the best answer for another one. Therefore countries should focus their efforts and funding for energy efficiency in the most convenient strategy. Furthermore, countries can choose to use either the conventional ESCOs model or any other model within country modifications (Fig. 13.2 and alternatives mentioned), according to their own characteristics (border conditions); as well as deciding in which sector (and which size of enterprises/firms) of the economy would be better to focus.

Acknowledgments

The author would like to thank Hilda Dubrovsky for her comments and suggestions, which were very helpful to improve the quality of this chapter.

References

Bertoldi, P., Boza-Kiss, B., 2017. Analysis of barriers and drivers for the development of the ESCO markets in Europe. Energy Policy 107 (2017), 345−355.

Blanco, A., Coviello, M., 2015. Empresas de servicios energéticos en América Latina Un documento guía sobre su evolución y perspectivas. CEPAL, Naciones Unidas, octubre de 2015. <https://repositorio.cepal.org/bitstream/handle/11362/39008/1/S1500950_es.pdf>.

Bouille, D., Recalde, M., Di Sbroiavacca, N., Dubrovsky, H., Ruchansky, B., 2019a. GUIA METODOLOGICA PARA LA ELABORACIÓN DEL PLAN NACIONAL DE EFICIENCIA ENERGÉTICA ARGENTINA (PlanEEAr). Proyecto de Eficiencia Energética Argentina. GFA Consulting Group.

Bouille, D., Recalde, M., Queiroz, T., 2019b. EXPERIENCIA INTERNACIONAL EN EL DESARROLLO DE PLANES Y ACCIONES DE EFICIENCIA ENERGÉTICA: Lecciones para el Plan Nacional de Eficiencia Energética Argentina (PlanEEAr). Proyecto de Eficiencia Energética Argentina. GFA Consulting Group.

Boza-kiss, b., Bertoldi, p., Economidou, M. 2017. Energy Service Companies in the EU. Status review and recommendations for further market development with a focus on Energy Performance Contracting. Science for Policy report by the Joint Research Centre (JRC), European Comission.

Carpio, C., Altomonte, H. 2015. Informe final del Grupo de Trabajo en Eficiencia Energética de la ECPA, Departamento de Desarrollo Sostenible de la Secretaría General de la Organización de los Estados Americanos.

Carpio, C., Coviello, M., 2013. Eficiencia energética en América Latina y el Caribe: avances y desafíos del último quinquenio. CEPAL. Naciones Unidas, noviembre de 2013. <https://repositorio.cepal.org/bitstream/handle/11362/4106/S2013957_es.pdf?sequence = 1&isAllowed = y>.

CEPAL/OLADE/GTZ, 2003. Energía y Desarrollo Sustentable en América Latina y el Caribe: Guía Para la Formulación de Políticas Energéticas. Santiago de Chile: CEPAL/OLADE/GTZ.

CONUEE/GIZ, 2012. Empresas de Servicios Energéticos (ESCO) − Generalidades, Perspectivas y Oportunidades en México, México, D.F., marzo de 2012. <https://energypedia.info/images/6/6e/ESCO_Perspectivas_y_oportunidades.pdf>.

Da-li, G., 2009. Energy service companies to improve energy efficiency in China: barriers and removal measures. Procedia Earth Planet. Sci. 1 (1), 1695−1704.

Duguma, L.A., Wambugu, S.W., Minang, P.A., Noordwijk, M., 2014. A systematic analysis of enabling condition for synergy between climate change mitigation and adaptation measures in developing countries. Environ. Sci. Policy 42, 138−148.

Dyner, I., Olaya, Y., Franco, C.J., 2011. An enabling framework for wind power in Colombia: what are the lessons from Latin America? Diffus. Renew. Energy Technol. Case Stud. Enabl. Framew. Dev. Ctries. 73−86.

Ellis, J., 2010. Energy Service Companies (ESCOs) in Developing Countries. International Institute for Sustainable Development, Manitoba, Canada.

Estache, A., Kaufmann, M., 2011. Theory and evidence on the economics of energy efficiency. Lessons for the Belgian building sector. Reflets et. Perspect. de. la. vie économique, tome l (3), 133−148. Available from: https://doi.org/10.3917/rpve.503.0133.

Fang, W.S., Miller, S., Yeh, C.-C., 2012. The effect of ESCOs on energy use. Energy Policy 51 (2012), 558−568.

Fuller, M.C., Portis, S., Kammen, D.M., 2009. Towards a low-carbon economy: municipal financing for energy efficiency and solar power. Environment 51 (1), 22−32.

Fundación Bariloche, 2013. ENERGÍA: UNA VISIÓN SOBRE LOS RETOS Y OPORTUNIDADES EN AMÉRICA LATINA Y EL CARIBE EFICIENCIA ENERGÉTICA.

Grubb, M., McDowall, W., Drummond, P., 2017. On order and complexity in innovations systems: conceptual frameworks for policy mixes in sustainability transitions. Energy Res. Soc. Sci. 33 (2017), 21−34. Available from: https://doi.org/10.1016/j.erss.2017.09.016.

IEA, 2007, Mind the Gap: Quantifying Principal-Agent Problems in Energy Efficiency. OECD Publishing, Paris, https://doi.org/10.1787/9789264038950-en.

IEA, 2018. Energy Service Companies (ESCOs). IEA, Paris. <https://www.iea.org/reports/energy-service-companies-escos-2>.

IEA, 2019. Energy Efficiency 2019. IEA Publications. International Energy Agency, Paris, October 2019. <https://webstore.iea.org/market-report-series-energy-efficiency-2019>.

Jacobsson, S., Bergek, A., Sandén, B., 2017. Improving the European Commission's analytical base for designing instrument mixes in the energy sector: market failures versus system weaknesses. Energy Res. Soc. Sci. 33 (2017), 11−20. Available from: https://doi.org/10.1016/j.erss.2017.09.009.

Kern, F., Kivimaa, P., Martiskainen, M., 2017. Policy packaging or policy patching? The development of complex energy efficiency policy mixes. Energy Res. Soc. Sci. 23, 11−25. Available from: https://doi.org/10.1016/j.erss.2016.11.002.

Langlois-Bertrand, S., Benhaddadi, M., Jegen, M., Pineau, P.-O., 2015. Political-institutional barriers to energy efficiency. Energy Strategy Rev. 8, 30−38.

Marino, A., Bertoldi, P., Rezessy, S., 2010. Energy Service Companies Market in Europe-Status Report 2010. Publications Office of the European Union, Luxembourg.

Nurcahyanto, T.U., 2018. Chapter 26: Development of energy service company (ESCO) market to promote energy efficiency programmes in developing countries. In: Sayigh, A. (Ed.), Transition Towards 100% Renewable Energy, Innovative Renewable Energy. pp. 283−294. Available from: https://doi.org/10.1007/978-3-319-69844-1-26.

Nygaard, I., Hansen, U.E., 2015b. The conceptual and practical challenges to technology categorization in the preparation of technology needs assessments. Clim. Change 131 (3), 371−385.

Ockwell, D., Sagar, A., de Coninck, H., 2015. Collaborative research and development (R&D) for climate technology transfer and uptake in developing countries: towards a needs driven approach. Clim. Change 131 (3), 401−415.

Painuly, J.P., Park, H., Lee, M.-K., Noh, J., 2003. Promoting energy efficiency financing and ESCOs in developing countries: mechanisms and barriers. J. Clean. Prod. 11 (2003), 659−665.

Pätäri, S., Sinkkonen, K., 2014. Energy service companies and energy performance contracting: is there a need to renew the business model? Insights from a Delphi study. J. Clean. Prod. 66, 264−271.

Pätäri, S., Annala, S., Jantunen, A., Viljainen, S., Sinkkonen, A., 2016. Enabling and hindering factors of diffusion of energy service companies in Finland—results of a Delphi study. Energy Effic. 9, 2016, 1447−1460. https://doi.org/10.1007/s12053-016-9433-z

Purkus, A., Gawel, E., Thrän, D., 2017. Addressing uncertainty in decarbonisation policy mixes − lessons learned from German and European bioenergy policy. Energy Res. Soc. Sci. 33 (2017), 82−94. Available from: https://doi.org/10.1016/j.erss.2017.09.020.

Recalde, M.Y., 2016. The different paths for renewable energies in Latin American Countries: the relevance of the enabling frameworks and the design of instruments, WIREs Energy Environ. 5 (3), 305−326.

Recalde, M. Y., Bouille, D., Girardin, L.O. octubre−diciembre 2015. Limitaciones para el desarrollo de energías renovables en Argentina: el rol de las condiciones de marco desde una perspectiva histórica, Revista Problemas del Desarrollo 183 (46), 89−115.

Rosenow, J., Fawcett, T., Eyre, N., Oikonomou, V., 2016. Energy efficiency and the policy mix. Build. Res. Inf. 44 (5-6), 562−574. Available from: https://doi.org/10.1080/09613218.2016.1138803.

Rosenow, J., Kern, F., Rogge, K., 2017. The need for comprehensive and well targeted instrument mixes to stimulate energy transitions: the case of energy efficiency policy. Energy Res. Soc. Sci. 33 (2017), 95−104. Available from: https://doi.org/10.1016/j.erss.2017.09.013.

Roshchanka, V., Evans, M., 2016. Scaling up the energy service company business: market status and company feedback in the Russian Federation. J. Clean. Prod. 112 (5), 3905−3914.

Ruchansky, B., Januzzi, G., Buen, O.D., Romero, A., 2011. Eficacia institucional de los programas nacionales de eficiencia energética: los casos del Brasil, Chile, México y el

Uruguay CEPAL, Santiago de Chile. <https://repositorio.cepal.org/bitstream/handle/11362/6355/2/S1100313_es.pdf>.

Sorrell, S., 2004. The Economics of Energy Efficiency: Barriers to Cost-Effective Investment. Edward Elgar, Cheltenham.

Sorrell, S., Mallett, A., Nye S., 2011. Barriers to Industrial Energy Efficiency: A Literature Review. Development Policy, Statistics and Research Branch Working Paper 10/2011. United Nations Industrial Development Organisation, Vienna.

Taylor, R.P., Govindarajalu, C., Levin, J., Meyer, A.S., Ward, W.A., 2008. Financing Energy Efficiency: Lessons From Brazil, China, India and Beyond. The World Bank, Washington, DC.

Trærup, S., Stephan, J., 2015. Technologies for adaptation to climate change. Examples from the agricultural and water sectors in Lebanon. Clim. Change 131 (3), 435−449.

Trianni, A., Cagno, E., Farné, S., 2016. Barriers, drivers and decision-making process for industrial energy efficiency: a broad study among manufacturing small and medium-sized enterprises. Appl. Energy 162, 1537−1551.

Vine, E., 2005. An international survey of the energy service company (ESCO) industry. Energy Policy 33, 691−704.

Vogel, J.A., Per Lundqvist, P., Arias, J., 2015. Categorizing barriers to energy efficiency in buildings. Energy Procedia 75, 2839−2845.

Watson, J., Byrne, R., Ockwell, D., Stua, M., 2015. Lessons from China: building technological capabilities for low carbon technology transfer and development. Clim. Change 131 (3), 387−399.

Yeatts, D., Auden, D., Cooksey, C., Chen, C.F., 2017. A systematic review of strategies for overcoming the barriers to energy-efficient technologies in buildings. Energy Res. Soc. Sci. 32, 76−85.

Zabaloy, F., Recalde, M., Guzowski, C., 2019. Are energy efficiency policies for household context dependent? A comparative study of Brazil, Chile, Colombia and Uruguay. Energy Res. Soc. Sci. 52 (2019), 41−54.

Further reading

Nygaard, I., Hansen, U.E., 2015. Overcoming Barriers to the Transfer and Diffusion of Climate Technologies, second ed. UNEP DTU Partnership. TNA Guidebook Series. <https://orbit.dtu.dk/ws/files/121688225/Overcoming_Barriers_2nd_ed.pdf>.

Index

Note: Page numbers followed by "*f*" and "*t*" refer to figures and tables, respectively.

A

Absorption chillers, 170
Active demand response (ADR) programs, 100–101
Active solar water heating system, 185
Adaptation, 81–83, 85–86, 91
Adaptive thermal comfort models, 85–87, 87*f*
Advice services, energy service markets, 264–265
Aggregator, 100–101
Agriculture, energy use in, 9–12
  agri-environment schemes, 11–12
  direct energy, 9–10
  effective field capacity, 11
  equivalent energy of inputs and outputs, 10*t*
  indirect energy, 9–10
  machine energy, 11
Air-based solar thermal systems, 184–185
Air-conditioning systems, 169, 194–195
Air handler, 170–171
Air-handling units (AHUs), 170–171
Air-source heat pump (ASHP), 169, 190, 200–201, 203–204
Air-to-water heat pump (AWHP), 201
American Wind Energy Association (AWEA), 199
Annual energy saving, 138–139
ASHRAE Global Thermal Comfort Database II, 86–87
ASHRAE psychrometric chart, 86*f*
ASHRAE 55 Standard on thermal comfort, 85–86
ASHRAE-thermal comfort zone, 9

B

Barriers to energy efficiency, 297
  barriers to instruments, 299–300
  border conditions, 297–298
  context dependence, 299
  definition and scope, 297*f*
  enabling frameworks, 298
  ESCOs, 299–306
    economic and market barriers, 300–306
    enabling frameworks for, 307–308
    funding barriers, 305–307
    institutional and regulatory barriers, 305
    other energy efficiency actions, 307–308
    removing barriers, 306–308
    review of, 301*t*
  specific barriers, 299, 311
Battery energy storage (BES) system, 171–173, 172*t*
Better energy communities, 48
Binomial logit model, 243–244
Bioenergy, 215, 222–224, 226–227, 227*f*
Bioenergy carbon capture and storage (BECCS), 222–224
Biogas technology, 17
Biomass, 6–7, 16, 19*t*, 101–103, 204–205, 226–227, 227*f*
  for development, 16
  direct burning of, 17–18
  for domestic needs, 16
  final energy projections, 20*t*
  for petroleum substitution, 16
  residues and current use, 18*t*
  resource utilization, 15*t*
Blow-through units, 170–171
Boiler systems, 168
Border conditions, 297–298
Briquette processes, 16–17
BSim, 135

Building-augmented wind turbines (BAWTs), 198
Building automation and controls (BAC), 99–100
Building energy, 81–84, 82f, 181–182
 biomass, 204–205
 energy-efficient building envelopes, 182–183
  climate-specific design, 183, 183t
  thermal resistance, increase of, 182–183
 envelope and materials, 93f, 94–96
 evolution of, 84f
 heat pumps, 199–204, 200f
  air-source heat pumps, 200–201
  cooling cycle, 203, 203f
  defrost cycle, 203–204
  ground-source heat pumps, 201–202
  heating cycle, 202, 202f
  performance measures, 204
  performance of, 200f
  working principles, 202–204, 202f
 passive to nearly zero-energy building design, 96–99, 97t
 renewable energy sources for, 183–191
  building-integrated photovoltaic thermal system, 188–191, 189f, 190f
  electrical/thermal loads analysis, 184
  local codes and requirements, consideration of, 184
  solar energy systems, 184–188
 smart buildings
  and home automation, 99–100
  to smart districts and cities, 100–103, 102f
 solar thermal energy storage, 191–196
  latent heat storage (LHS) systems, 195
  maturity of, 191f
  sensible cold storage system, 194–195
  sensible heat storage system, 192–194
  thermochemical storages, 195–196
  types of, 192–196
 thermal comfort, 83–91
  adaptive thermal comfort models, 87f
  ASHRAE Global Thermal Comfort Database II, 86–87
  ASHRAE psychrometric chart, limits in, 86f
  classical thermal comfort models, 87f
  percentage of people dissatisfied, 84–85
  standard effective temperature (SET), 84–85
  user behavior, 88–91, 90f
 weather conditions, climate change and growing urbanization, 91–93, 92f
 wind energy, 196–199
  building-augmented wind turbines, 198
  building-integrated wind turbine, 197–198
  building-mounted wind turbines, 198
  horizontal-axis wind turbines, 196–197
  optimizing building-integrated/mounted wind turbine devices, 198–199
  small/micro wind turbines, 199
  vertical-axis wind turbines, 196–197
  wind resource assessment, 197, 197t
  wind turbines, 196–197
Building energy management systems (BEMS), 99–100
Building energy model, 133–134
Building envelopes of energy-efficient, 182–183
 climate-specific design, 183, 183t
 thermal resistance, increase of, 182–183
Building-integrated/mounted wind turbine (BUWT), 197–198
 optimizing, 198–199
Building-integrated photovoltaic thermal (BIPV/T) system, 188–191, 189f, 190f
Building-integrated wind turbine (BIWT), 197–198
Building Loads Analysis, 135
Building-mounted wind turbines (BMWTs), 198
Bulgaria, financial schemes for energy efficiency, 62–65

C
Capacity building activities, 72
Carbon capture and storage (CCS), 222–224
Carbon dioxide (CO_2) emissions, 4, 125–126, 125t, 132, 139, 213–214, 219
 factor, 127–128, 128f
 from fuel combustion by fuel, 220, 220f
 global energy-related, 219t

Index 319

global industry direct, 223*f*
  by sector, 219*f*
Carbon-intensive industrial sectors, 220–221
  cement, 221
  chemical, 220
  food, 221
  iron, 220–221
  nonferrous metals, 221
  pulp and paper, 221
  steel, 220–221
Carbon leakage, 234
Case study in Turkey, industrial sector, 229–232
  development plans, 232
  gross domestic product (GDP), 230–231
  industry and technology, 231–232
  National Energy Efficiency Action Plan (NEEAP), 229–230, 230*f*
Cement industry, 221
Charcoal stoves, 16–17
Chemical industry, 220
Chilled-water systems, 169–170
Citizen energy communities, 41–43
Civic energy communities, 41
Classical thermal comfort models, 87*f*
Climate change, 4–5, 29
  mitigation, energy efficiency, 221–224
  weather conditions under, 91–93, 92*f*
Climate-sensitive design approach, 8–9
Climate-specific design, 183, 183*t*
Coefficient of performance (COP), 204
Combined heat and power (CHP) systems, 4–5, 19–20, 19*t*, 156–158, 205
  combustion turbines, 158
  compression-ignition (CI) engines, 158
  fuel cells, 158
  microturbines, 158
  network flow model, 157*f*
  power generation unit, 156–158
  prime movers, 156–158, 160*t*
  spark-ignition (SI) engines, 158
  steam turbines, 158
  US installed CHP facilities, 158, 159*f*
Commercial energy resources, 6–7
Compound semiconductor photovoltaic cells, 159–161
Computerized control system, 113
Concentrating solar collectors, 161

Contracting services, energy service markets, 267–268
Conventional geothermal installation, 117, 117*f*, 122
  initial investment for, 123*t*
Cook stoves improvement, 17
Cooling cycle, heat pumps, 203, 203*f*
Cooling systems, 169–170
Cost of financing, 305, 311
Credit access, 305
Crowdfunding
  advantages and barriers of, 63*t*
  definition of, 60
  modalities, 60
Cryogenic forms, 21
Cypriot net metering model, 45–46

D
Data mining, 88–89
Defrost cycle, heat pumps, 203–204
Delivery or supply contracting, 267
Demand-side management (DSM), 39–40, 100–101
Desiccant cooling cycles, 170
Designer's Simulation Toolkits, 135
Direct energy, 9–10
Direct-expansion (DX) systems, 169–170
Distributed generation, 39–41
DOE-2.1E, 135
Domestic hot water (DHW), 19–20, 89–91
Downward modulation events, 99–100
Draw-through units, 170–171
Dry cell batteries, 17–18

E
Earth energy designer (EED) software, 116–118, 116*t*, 122
Eco-innovation
  defined, 240–241
  discussion, 244–245
  methods and data, 242–244
  results, 244–245
  state of the arts, 240–242
Economic analysis
  ground-source heat pump system, 122–123
    conventional geothermal installation, initial investment for, 122, 123*t*

Economic analysis (*Continued*)
  suggested system, initial investment for, 122, 124*t*
  traditional geothermal system, annual expense of, 122−123, 124*t*
  of lighting fixtures' replacement with LED bulbs, 151*t*
  of solar panels installation, 152*t*
  on university campus buildings, 138−139
Economic and market barriers, 300−306
  ESCOs, actions to remove, 306
ECOTECT, 135
Effective field capacity, 11
Efficient bioenergy use, 14−19
  agricultural residues routes for development, 15*t*
  biogas technology, 17
  biomass, 16
    for development, 16
    for domestic needs, 16
    for petroleum substitution, 16
    resource utilization, 15*t*
  briquette processes, 16−17
  classifications of data requirements, 14*t*
  forest and tree management, 17−18, 18*t*
  improved cook stoves, 17
Electrical/thermal loads analysis, 184
Electricity Directive (EU) 2019/944, 39, 41−43
Electricity market, 286*t*
Electricity price, sensitivity analysis, 126, 127*f*
Electric rate, sensitivity analysis, 126−127, 127*f*
Electrification, 215
Electrolysis, 20
Embodied carbon (EC), 96
Embodied energy (EE), 96
Emitter-wrap-through PV cells, 159
Enair Energy, 120−121
Energy audit, university campus buildings, 137
  in cafeterias1 and 2, 139−141, 140*f*, 141*f*, 143*f*
  in health center, 147−148, 147*f*, 148*f*
  in Mechanical Engineering building, 141−144, 143*f*, 144*f*
  protocol, 133−134

in students' halls of residence, 148−150, 149*f*, 150*f*
  in university library, 145−146, 145*f*, 146*f*
Energy carrier in rural areas, 18*t*
Energy communities, 41, 101−103
Energy conservation, 133−134, 136, 144−145, 293−294
Energy consumption, 131−138, 141, 145−147, 218, 229
  energy and carbon-intensive industrial sectors, 220−221
  general trends, 216−220
  industrial sector, 216−221
  world electricity generation by fuel, 218*f*
  world primary energy demand, 217*t*
  world renewable energy consumption, 217*t*
  World Total Primary Energy Supply (TPES) from by fuel, 216*f*
Energy conversion systems, 168−171
  cooling systems, 169−170
  heating systems, 168−169
  ventilation systems, 170−171
Energy efficiency, 133−134, 229, 294
  actions and policy instruments, 296−300
  barriers to, 297, 299−306
    economic and market barriers, 300−306
    enabling frameworks for, 307−308
    funding barriers, 305−307
    institutional and regulatory barriers, 305
    other energy efficiency actions, 307−308
    removing barriers, 306−308
    review of, 301*t*
  eco-innovation, in tourism sector
    discussion, 244−245
    methods and data, 242−244
    results, 244−245
    state of the arts, 240−242
  financial schemes, in-country demonstrations
    advantages of, 63*t*
    barriers of, 63*t*
    capacity building activities, 72
    case study context, 68*f*
    case study countries, 62−67
    crowdfunding, 60
    energy performance contracting, 60

Index

green bonds, 60
guarantee funds, 61
key actors identification, 67–68
knowledge and best practices sharing, 68–72
  peer-to-peer learning, 69–71, 71t
  proposed methodology, 58–59, 59f
  revolving funds, 61
  soft loans, 61
  third-party financing, 62
gains, distributed generation, 39–41
increase of, 38–39
industrial sector, 224–229, 225f
  bioenergy, 226–227, 227f
  case study in Turkey, 229–232
  for climate change mitigation, 221–224
  solar heat, 227–229, 228f
Energy Efficiency Fund, 66–67
Energy Efficiency Law, 230–232
Energy Efficiency Obligation Scheme, 66
Energy efficiency ratio (EER), 204
Energy-efficient building envelopes, 182–183
  climate-specific design, 183, 183t
  thermal resistance, increase of, 182–183
Energy Express, 135
Energy flexibility of building, 99–100
Energy from Renewable Sources Act, 49
Energy generation systems, 156–167
  combined heat and power systems, 156–158
  geothermal energy, 163–166
  organic Rankine cycle systems, 162, 163f
  solar photovoltaic (PV) systems, 158–161
  solar thermal systems, 161–162
  wind turbine system, 167
Energy intensity index, 231
Energy management services (EMS), 265–266, 266f
Energy market research, worldwide trends in
  affiliations, 281–282, 283t
  cluster analysis, keywords, 285–290
    China, 289t
    community detection, 285f
    district heating, 289t
    electricity market, 286t
    loss allocation/cost, 290, 290t
    oil prices, 288t
    renewable energy, 287t
    smart grid, 286t
  countries, 281–282
  data, 278
  geographical distribution of scientific production, 281f
  institutions in scientific production, 281–282, 282f
  journals metric analysis, 279–281
  keywords from worldwide publications, 282–284, 284f, 284t
  production and consumption, 277, 278f
  results, 278–290
  subjects from worldwide publications, 278–279
Energy modeling software, 134–136
Energy performance building (EPB), 131–132
Energy performance contracting (EPC), 57, 60, 62, 294
  advantages and barriers of, 63t
Energy performance gap, 88–89
Energy Performance of Buildings Directive (EPBD), 83, 97–98, 254
Energy planning, 101–103
EnergyPlus, 135
Energy-related communities, 41
Energy saving in buildings, 7–9
Energy-saving strategies on university campus buildings
  building envelope of cafeteria, 140f
  centrographic approach, 132–133
  discussions, 139–152
  energy audit
    in cafeterias 1 and 2, 139–141, 140f, 141f, 143f
    in health center, 147–148, 147f, 148f
    in Mechanical Engineering building, 141–144, 143f, 144f
    protocol, 133–134
    in students' halls of residence, 148–150, 149f, 150f
    in university library, 145–146, 145f, 146f
  energy conservation, 136
  energy modeling software, 134–136
  materials and methods, 136–139
    data analysis, instrumentation and procedure for, 137–138
    data collection, procedure for, 137

Energy-saving strategies on university
    campus buildings (Continued)
  daylighting control for cafeteria 1, 141f,
    142t
  economic analysis, 138–139
  environmental impacts, assessment of,
    139
  study location, 136–137
  qualitative recommendation analysis,
    150–152
    light-emitting diode bulbs, lighting
      fixtures replacement with, 151–152,
      151t
    solar panels installation, 152, 152t
  results, 139–152
  sustainability, 133–134
Energy service companies (ESCOs), 66,
    251–252, 256, 267–268, 268f, 269f,
    294–295
  barriers to energy efficiency, 299–306
    economic and market barriers,
      300–306
    enabling frameworks for, 307–308
    funding barriers, 305–307
    institutional and regulatory barriers,
      305
    other energy efficiency actions,
      307–308
    removing barriers, 306–308
    review of, 301t
  border conditions, 310
  definition of, 294
  effective institutional framework, 310
  enabling environments/conditions, 310
  energy prices distortions, 310–311
  ESCOs associations, 311
  guaranteed savings model, 309f
  institutional conditions, 310
  lack of specific regulation for, 311–312
  regulation and accreditation of, 311–312
  shared savings model, 309f
2006 Energy Service Directive (ESD),
    253–254
Energy service markets, 251–253
  developments of, 264–268
    advice services, 264–265
    contracting services, 267–268
    energy management services (EMS),
      265–266, 266f

  European framework, 253–259, 257t,
    260t
    barriers against, energy services, 260t
    legal framework, 253–255
    market volume, offers, and barriers,
      255–259, 256f
    status quo and development, key
      figures on, 257t
  German energy service market, 259–263,
    262t, 263f
    development, 268–270
    ESCO contracts in, 268f
    information sources, 259–262
    legal framework, 259–262
    market overview, 262–263
Energy service provider, 38
Energy services, 3–5
  definition of, 38, 252
  energy and population growth, 5–7
  energy and sustainable development,
    25–28, 26t
    best practice tools, 26
    cleaner, leaner production processes, 26
    economic, social, and environment
      systems, 27t
    indicators for sustainable consumption
      and production, 28t
    innovation and ecodesign, 26
    issues, 25
    key variables defining facility
      sustainability, classification of, 27t
    management and measurement tools, 25
    openness and transparency, 26
    performance assessment tools, 25
    product stewardship, 26
    resources and productivity, 27f
    supply chain management, 26
  energy saving in buildings, 7–9
  energy use in agriculture, 9–12
    agri-environment schemes, 11–12
    direct energy, 9–10
    effective field capacity, 11
    equivalent energy of inputs and outputs,
      10t
    indirect energy, 9–10
    machine energy, 11
  global warming, 28–30, 29t, 30f
  recommendations, 30–31
  renewable energy technologies, 12–25

combined heat and power, 19–20
efficient bioenergy use, 14–19
hydrogen production, 20–21
hydropower generation, 21–22, 23t
solar energy, 12–14
wind energy, 22–25
in rural areas, 18t
Energy storage systems, 171–174, 221–222
battery, 171–173, 172t
thermal, 173–174
Energy supply regulatory regime, 233
Energy systems in buildings, 156–174
energy conversion systems, 168–171
cooling systems, 169–170
heating systems, 168–169
ventilation systems, 170–171
energy generation systems, 156–167
combined heat and power systems, 156–158
geothermal energy, 163–166
organic Rankine cycle systems, 162, 163f
solar photovoltaic (PV) systems, 158–161
solar thermal systems, 161–162
wind turbine system, 167
energy storage systems, 171–174
battery energy storage system, 171–173, 172t
thermal energy storage system, 173–174
Ener-Win, 135
Enthalpy, 114–115
Environmental analysis
of lighting fixtures' replacement with LED bulbs, 151t
of solar panels installation, 152t
Environmental evaluation, ground-source heat pump system, 123, 125t
Environmental impacts, assessment of, 139
Environmental innovation, 240–241
eQUEST, 134–139, 141, 145, 147–149
Equivalent energy of inputs and outputs, 10t
ESP-r, 135
EU member states greenhouse gas (GHG) emissions, 29t
European energy service market, 253–259, 257t, 260t
barriers against, energy services, 260t

legal framework, 253–255
market volume, offers, and barriers, 255–259, 256f
status quo and development, key figures on, 257t
European Green Deal, 251
European Union (EU), renewable energy communities in, 44–47
European Wind Energy Association, 22
Eurostat, 67
Evacuated tube solar collectors, 161, 194f
External cladding systems, 94
External revolving funds, 61

F
Fanger's method, 85–86
f-chart method, 119–120
Federal Agency of Energy Efficiency, 259
Feed-in premiums, 45
Feed-in tariffs, 44–45
Financial schemes for energy efficiency, in-country demonstrations
case study context, 68f
case study countries, 62–67
Bulgaria, 62–65
Greece, 65–66, 66f
Lithuania, 66–67
Spain, 67
innovative financing schemes, 59–62
advantages of, 63t
barriers of, 63t
crowdfunding, 60
energy performance contracting, 60
green bonds, 60
guarantee funds, 61
revolving funds, 61
soft loans, 61
third-party financing, 62
key actors identification, 67–68
descriptions and benefits, 70t
key stakeholders' interactions, 68, 69f
knowledge and best practices sharing, 68–72
capacity building activities, 72
peer-to-peer learning, 69–71, 71t
proposed methodology, 58–59, 59f
Financing bodies, 68, 70t
Fixed-plate heat exchangers, 171
Flat-plate solar collectors, 161, 188, 229

Food processing, 221
Forest management, efficient bioenergy use, 17–18, 18t
Fossil fuel, 222, 224
French draft NECPs, 49
Fuel switching, 215
Funding barriers, 305–307
 ESCOs, actions to remove, 306–307
Furnaces, 168–169

G
General trends, 216–220
Geothermal energy, 110–111, 116–118, 116t, 117f, 163–166
Geothermal heat pumps, 110–113, 126. *See also* Ground-source heat pump (GSHP) system
German energy service market, 259–263, 262t, 263f
 development, 268–270
 energy management services (EMS), 265–266
 ESCO contracts in, 268f
 information sources, 259–262
 legal framework, 259–262
 market overview, 262–263
German Energy Services Act, 259
German National Energy Efficiency Action Plan, 263
Global energy trends, 215–216
Global final energy consumption by sector, 156f
Global warming, 4, 28–30, 29t, 30f, 81–84, 91, 103
Gothenburg priorities, 240
Gray energy, 235
Greece, financial schemes for energy efficiency, 65–66, 66f
Green bonds, 60
 advantages and barriers of, 63t
Greenhouse gas (GHG) emissions, 4–5, 12, 29, 29t, 55, 110, 132–134, 181–182, 213–215, 234–235
Greenhouse soil, 11
Greenpeace, 22
Gross domestic product (GDP), 230–231
Ground-coupled heat pump (GCHP) systems, 164–165, 166f
Ground heat exchanger (GHE) loop, 164
Ground-source heat pump (GSHP) system, 164–165, 190–191, 201–202
 contribution of suggested installation, 121–122
 drawback of, 202
 earth connection, 164
 economic analysis, 122–123
  conventional geothermal installation, initial investment for, 122, 123t
  suggested system, initial investment for, 122, 124t
  traditional geothermal system, annual expense of, 122–123, 124t
 environmental evaluation, 123, 125t
 heat distribution, 164
 heat pump, 164
 initial investment and annual costs, economic comparison, 125f
 methodology, 111–116
 power module, 120–121
  photovoltaic solar energy, 120, 121t
  wind energy, 120–121, 121f, 122t
 principal characteristics of, 111t
 proposed solution, 111–115
  computerized system operation, 113–114, 114f
  decision-making in computerized system, 115f
  geothermal installation, 112f
  renewable module, 112–113, 113f
 sensitivity analysis, 126–128
  CO_2 emission factor, 127–128, 128f
  electricity price, 126, 127f
  electric rate, 126–127, 127f
 technical calculation, 116–122
 test procedure, 115–116
 thermal module, 116–120
  geothermal energy, 116–118, 116t, 117f
  thermal solar energy, 118–120, 118t, 119f, 120t
Groundwater heat pump (GWHP) systems, 164–165, 166f
Guaranteed savings model, 308
Guarantee funds, 61
 advantages and barriers of, 63t

H
Heating cycle, heat pumps, 202, 202f

Index

Heating seasonal performance factor (HSPF), 204
Heating systems, 168–169
Heating, ventilation, and air-conditioning (HVAC) systems, 88–89, 99–101, 132–134, 168
Heat pipes, 171
Heat pumps, 100–101, 102f, 199–204, 200f
  air-source heat pumps, 200–201
  cooling cycle, 203, 203f
  defrost cycle, 203–204
  ground-source heat pumps, 201–202
  heating cycle, 202, 202f
  performance measures, 204
  performance of, 200f
  working principles, 202–204, 202f
Heat recovery ventilation systems, 171
HEED, 135
Heterogeneity of energy efficiency, 56
Higher energy efficiency, 99–100
High-temperature heat, 224–225
High-vacuum flat-plate collectors, 228
Horizontal-axis wind turbine (HAWT), 167, 168t, 196–197
Horizontal closed loop systems, 164–165, 167f
Hourly analysis program, 135
Hybrid photovoltaic/thermal (PVT) collectors, 161–162, 188–189
Hydraulic energy, 13, 19t
Hydraulic system, 99–100
Hydroelectricity, 22
Hydrofluoroolefins (HFOs), 162
Hydrogen production, 20–21
  cost, large-scale production, 21
  electrolysis, 20
  storage, 21
Hydropower generation, 21–22, 23t
Hydropower plants, 22

I

Indirect energy, 9–10
Indoor environmental quality (IEQ), 88–89
Industrialization, 94–95
Industrial sector, 213–215
  case study in Turkey, 229–232
    development plans, 232
    gross domestic product (GDP), 230–231
    industry and technology, 231–232
    National Energy Efficiency Action Plan (NEEAP), 229–230, 230f
  energy consumption and emissions, 216–221
    energy and carbon-intensive industrial sectors, 220–221
    general trends, 216–220
    world electricity generation by fuel, 218f
    world primary energy demand, 217t
    world renewable energy consumption, 217t
    World Total Primary Energy Supply (TPES) from by fuel, 216f
  energy efficiency and renewable sources, 224–229, 225f
    bioenergy, 226–227, 227f
    solar heat, 227–229, 228f
  energy efficiency, for climate change mitigation, 221–224
    innovative processes, 222–224, 223f
  global energy trends, 215–216
  policy options, 233–236
    awareness, 233
    contractual scheme complexity, 233
    energy supply regulatory regime, 233
    international agreements, 234–235
    investment, 233
    operability and integration, 233
    procurement, 235–236
    return on investment, 233
    risk and insurance, 233
    technology maturity, 233
Infrastructure Fund of Funds ("InfraFoF"), 66
Innovative financing schemes for energy efficiency, 59–62
  advantages of, 63t
  barriers of, 63t
  crowdfunding, 60
  energy performance contracting, 60
  green bonds, 60
  guarantee funds, 61
  revolving funds, 61
  soft loans, 61
  third-party financing, 62
Installation contracting, 267
Institutional and regulatory barriers, 305

Intergovernmental Panel on Climate Change, 298
Internal financial market, 310
Internal revolving funds, 61
International Energy Agency (IEA), 65−66, 155, 216, 252−253, 294−295
International financing constraints, 310
International Standard ISO 7730, 85−86
Internet of things (IoT), 89−91, 100
Irish draft NECP, 48
Iron industry, 220−221
Italian draft NECP, 48

J
Jessica Fund, 67

K
Key actors identification, financial schemes for energy efficiency, 67−68
  descriptions and benefits, 70t
  key stakeholders' interactions, 68, 69f
Key stakeholders' interactions, 68, 69f
Knowledge transfer, financial schemes for energy efficiency, 68−72
  capacity building activities, 72
  peer-to-peer learning, 69−71, 71t

L
Large-scale CHP plants, 19−20
Latent heat storage (LHS) systems, 174, 195
Lead acid batteries, 171−173
Light-emitting diode bulbs, 151−152, 151t
Liquid thermal energy storage systems, 193−194
Lisbon objectives, 240
Lithium batteries, 173
Lithuania, financial schemes for energy efficiency, 66−67
Logistic regression models, 89−91
Logit multinomial model, 243−244
Low-carbon development pathway, 214−215
Low-energy buildings, 83−84, 88−89, 98−99
Low-temperature heat, 214, 224−225
Low thermal efficiency, 11

M
Machine energy (ME), 11

Machine learning techniques, 88−89
Market-based energy community, 101−103
Market size, 311
Mechanical chiller systems, 170
Metal hydride, 21
Mini split heat pumps (MSHPs), 200−201
Monocrystalline PV cells, 159

N
National Energy and Climate Plans (NECPs), 37, 47−49
National Energy Efficiency Action Plan (NEEAP), 229−230, 230f, 252−253, 259−262
Nearly zero-energy buildings (nZEBs), 83, 96−99, 97t
Neologism, 40
Nickel batteries, 171−173
Nickel−cadmium (NiCd) batteries, 171−173
Nickel−metal hydride (NiMH) batteries, 171−173
Nonconcentrating solar collectors, 161
Nonconcentrating technologies, 228−229
Nonconflict of interest, 265
Nonferrous metals, 221

O
Odds ratio, 244
Oil prices, 287−288, 288t
Organic Rankine cycle (ORC) systems, 162, 163f
  with recuperator, 164f
  regenerative system, 166f
  reheat system, 165f
Oslo Manual, 242−243

P
Paper-producing process, 221
PAREER Plan, 67
Paris Agreement, 56, 83, 131−132, 216, 223−224, 234−235, 251, 293−294
Passive solar water heating system, 185
Peer-to-peer learning, 57, 59, 69−71, 71t
Percentage of people dissatisfied, 84−85
Phase change materials (PCMs), 174, 195
Photovoltaic solar energy, 120, 121t
Piston engines, 158
Place-based energy community, 101−103
Plug loads, 147−149

Pluralistic market, 266
Policy makers, 296−297
Polycrystalline photovoltaic cells, 159
Polyurethane (PUR), 94
Power generation unit (PGU), 156−158
Power module, ground-source heat pump system, 120−121
  photovoltaic solar energy, 120, 121*t*
  wind energy, 120−121, 121*f*, 122*t*
Predicted mean vote (PMV), 84−85
Primary energy balance of building, 98−99, 98*f*
Primary energy consumption, 229
Procurement, 235−236
Producer responsibility, 26
Proposed solution, ground-source heat pump system, 111−115
  computerized system operation, 113−114, 114*f*
  decision-making in computerized system, 115*f*
  geothermal installation, 112*f*
  renewable module, 112−113, 113*f*
Prosumers, 40
Public Investment Bank (TPD), 66
Public procurement, 235−236

R
Rankine cycle, 170
Rebound effect, 89−91
Reciprocating engines, 158
RED II, 41−42, 44, 47
Renewable energy (RE), 5, 216, 224, 234, 287*t*
  cooperatives, 41
  resources, 281−282
  sources of, 19*t*
Renewable energy certificate system, 4−5
Renewable energy communities
  concept of, 41−43
  in draft National Energy and Climate Plans, 47−49
  efficiency gains, distributed generation of energy production, 39−41
  energy efficiency, increase of, 38−39
  in EU law, 44−47
Renewable Energy Directive (EU) 2018/2001, 41−42
Renewable Energy Expansion Act, 49

Renewable energy sources
  building energy, 183−191
    building-integrated photovoltaic thermal (BIPV/T) system, 188−191, 189*f*, 190*f*
    electrical/thermal loads analysis, 184
    local codes and requirements, consideration of, 184
    solar energy systems, 184−188
  industrial sector, 224−229, 225*f*
    bioenergy, 226−227, 227*f*
    solar heat, 227−229, 228*f*
Renewable energy technologies (RET), 4−5, 7, 12−25, 233−234
  combined heat and power, 19−20
  efficient bioenergy use, 14−19
  hydrogen production, 20−21
  hydropower generation, 21−22, 23*t*
  solar energy, 12−14
  wind energy, 22−25
Renewable module, 112−113, 113*f*
Revolving funds, 61
  advantages and barriers of, 63*t*
  external, 61
  internal, 61
Risk mitigation, 72

S
Scopus, 278
Seasonal energy efficiency ratio (SEER), 204
Self-consumption, 40, 48
Sensible cold storage systems, 192−195
Sensible heat storage (SHS) system, 173−174, 192−194
  liquid thermal energy storage systems, 193−194
  sensible cold storage mediums, 192−193
  sensible hot storage mediums, 192−193
  sensible solid heat storage system, 193
Sensible hot storage mediums, 192−193
Sensible liquid heat storage system, 193−194
Sensible solid heat storage system, 193
Sensitivity analysis, ground-source heat pump system, 126−128
  CO_2 emission factor, 127−128, 128*f*
  electricity price, 126, 127*f*
  electric rate, 126−127, 127*f*

Shared savings model, 308
Silicon, 159
Six-story secondary school building, 135–136
Small/micro wind turbines, 199
Small-to-medium-sized enterprises (SMEs), 264
Smart buildings, 83–84
  and home automation, 99–100
  to smart districts and cities, 100–103, 102f
Smart grid, 286t
Sodium sulfur (NaS) batteries, 173
Soft loans, 61
  advantages and barriers of, 63t
Solar absorption cooling systems, 170
Solar array power, 14
Solar collectors, 170
Solar cooling system, 170
Solar energy systems, 8–9, 11–14, 19t, 184–188
  solar PV systems, 184–185
  solar thermal systems, 184–185
  solar water heating (SWH), 184
    active systems, 185, 185f
    evacuated tube solar thermal collectors, 186–187, 187f
    flat-plate solar thermal collector, 186
    passive systems, 185
    solar thermal collectors, choice of, 187–188
Solar Graz, 101–103
Solar heat, 227–229, 228f
Solar mechanical cooling cycles, 170
Solar panels installation, 152, 152t
Solar photovoltaic (PV) systems, 13, 158–161, 184–185
Solar pumps, 13
Solar radiation data, 12–13
Solar thermal collectors, 185–188
  cost, 187
  installation, 188
  performance, 187–188, 188f
Solar thermal energy storage, 191–196
  latent heat storage systems, 195
  maturity of, 191f
  sensible cold storage system, 194–195
  sensible heat storage system, 192–194
  thermochemical storages, 195–196
  types of, 192–196
Solar thermal systems, 161–162, 184–185
Solar towers, 228
Solar water heating (SWH), 184
  active systems, 185, 185f
  evacuated tube solar thermal collectors, 186–187, 187f, 188f
  flat-plate solar thermal collector, 186
  passive systems, 185
  solar thermal collectors, choice of, 187–188
    cost, 187
    installation, 188
    performance, 187–188, 188f
Solid sensible thermal storage systems, 193
Solid-to-liquid phase change materials, 195
Spain
  financial schemes for energy efficiency, 67
  renewable energy communities, 48
Spanish economy, 239
Specific barriers, 299, 311
Standard effective temperature (SET), 84–86
Standard thermal comfort model, 86–87
State of California's Energy Design Resources Program, 135
Steel industry, 220–221
Strategic Plan on Energy and Climate, 67
Supply chain management, 26
Support schemes, 44–45
Surface water heat pump (SWHP) systems, 164–165, 166f
Sustainability, 133–134
  defined, 25
Sustainable development, 25–28, 26t, 214–215, 223–224, 229
  best practice tools, 26
  cleaner, leaner production processes, 26
  economic, social, and environment systems, 27t
  indicators for sustainable consumption and production, 28t
  innovation and ecodesign, 26
  issues, 25
  key variables defining facility sustainability, classification of, 27t
  management and measurement tools, 25
  openness and transparency, 26

performance assessment tools, 25
product stewardship, 26
resources and productivity, 27f
supply chain management, 26
Sustainable energy, 7−8, 12
Sustainable energy communities, 41
Sustainable energy communities network, 48
Sweating, 84−85
System operator (SO), 100−101
System Thermodynamics, 135

T

Tar-based solar thermal collector, 186
Technological Innovation Panel (PITEC)
    database, 242−243
Terminol, 193−194
Thematic index (TH), 279−281
Thermal comfort, 83−84, 89−91
    adaptive thermal comfort models, 87f
    ASHRAE Global Thermal Comfort
        Database II, 86−87
    ASHRAE psychrometric chart, limits in,
        86f
    classical thermal comfort models, 87f
    percentage of people dissatisfied, 84−85
    standard effective temperature (SET),
        84−85
Thermal discomfort, 83−84
Thermal energy storage (TES) systems,
    100−101, 161, 173
    latent heat storage, 174
    sensible heat storage, 173−174
    thermochemical heat storage systems, 174
    typical limitations for, 174t
Thermal insulation system, 94, 95t
    external cladding systems, 94
    thermohygrometric performance of, 95t
    ventilated facade, 95−96
Thermal module, ground-source heat pump
    system, 116−120
    geothermal energy, 116−118, 116t, 117f
    thermal solar energy, 118−120, 118t,
        119f, 120t
Thermal resistance, increase of, 182−183
Thermal solar energy, 118−120, 118t, 119f,
    120t
Thermal systems, 13
Thermochemical heat storage systems, 174
Thermochemical storages, 195−196

Thermohygrometric comfort, 84
Thermohygrometric performance of
    insulation materials, 95t
Thermoregulation, 84−85
Thin-film PV cells, 159
Third-party financing, 62
    advantages and barriers of, 63t
Time-use surveys (TUSs), 89−91
Tourism sector, energy efficiency in
    defined, 240−241
    discussion, 244−245
    methods and data, 242−244
    results, 244−245
    state of the arts, 240−242
TRACE 700, 135
Traditional geothermal system
    annual CO_2 emissions of, 125t
    annual expense of, 122−123, 124t
Training content and material development,
    72
Training courses, 72
Training, evaluation of, 72
Transaction costs, 56
Tree management, efficient bioenergy use,
    17−18, 18t
"Triple-A" energy efficiency investments, 57
TRNSYS, 135
Turkish bioenergy potential, 226−227
Two-stage variable capacity ASHPs (TS VC
    ASHPs), 188−189

U

University campus buildings, energy-saving
    strategies on
    building envelope of cafeteria, 140f
    centrographic approach, 132−133
    discussions, 139−152
    energy audit
        in cafeterias1 and 2, 139−141, 140f,
            141f, 143f
        in health center, 147−148, 147f, 148f
        in Mechanical Engineering building,
            141−144, 143f, 144f
        protocol, 133−134
        in students' halls of residence,
            148−150, 149f, 150f
        in university library, 145−146, 145f,
            146f
    energy conservation, 136

University campus buildings, energy-saving strategies on (*Continued*)
  energy modeling software, 134–136
  materials and methods, 136–139
    data analysis, instrumentation and procedure for, 137–138
    data collection, procedure for, 137
    daylighting control for cafeteria 1, 141*f*, 142*t*
    economic analysis, 138–139
    environmental impacts, assessment of, 139
    study location, 136–137
  qualitative recommendation analysis, 150–152
    light-emitting diode bulbs, lighting fixtures replacement with, 151–152, 151*t*
    solar panels installation, 152, 152*t*
  results, 139–152
  sustainability, 133–134
Upper middle income economy, 62
Upward modulation events, 99–100
Urban cool island (UCI) phenomenon, 92
Urban heat island (UHI) effect, 92
Urbanization, 91–93, 92*f*
User behavior, 88–91, 90*f*
User-interactive technologies, 83–84

V
Vasomotor thermoregulation, 84–85
Ventilated facades, 94–96
Ventilation, energy impact of, 30*f*
Ventilation systems, 170–171
Vertical-axis wind turbine (VAWT), 167, 168*t*, 196–197
Vertical closed loop systems, 164–165, 167*f*
Villes solaire, 49
Virtual net metering, 45, 48

W
Water-based solar thermal systems, 184–185
Weak economy, 62
Weak institutional development, 310
Weibull index, 120–121
Wind energy, 19*t*, 22–25, 196–199
  average wind farm capacity, 24–25, 25*f*
  building-augmented wind turbines, 198
  building-integrated wind turbine, 197–198
  building-mounted wind turbines, 198
  global prospects, 22–24, 24*f*
  ground-source heat pump system, 120–121
    velocity distribution, 121*f*
    wind generator, technical description of, 122*t*
  horizontal-axis wind turbines, 196–197
  optimizing building-integrated/mounted wind turbine devices, 198–199
  renewable cycle, 22, 24*f*
  small/micro wind turbines, 199
  turbines share, 22–24, 24*f*
  vertical-axis wind turbines, 196–197
  wind resource assessment, 197, 197*t*
  wind turbines, 196–197
Wind resource assessment, 197, 197*t*
Wind turbines, 167, 196–197
World hydro potential and development, 23*t*
World Summit (WS), 4
World Total Primary Energy Supply (TPES) from by fuel, 216*f*
Worldwide trends in energy market research
  affiliations, 281–282, 283*t*
  cluster analysis, keywords, 285–290
    China, 289*t*
    community detection, 285*f*
    district heating, 289*t*
    electricity market, 286*t*
    loss allocation/cost, 290, 290*t*
    oil prices, 288*t*
    renewable energy, 287*t*
    smart grid, 286*t*
  countries, 281–282
  data, 278
  geographical distribution of scientific production, 281*f*
  institutions in scientific production, 281–282, 282*f*
  journals metric analysis, 279–281
  keywords from worldwide publications, 282–284, 284*f*, 284*t*
  production and consumption, 277, 278*f*
  results, 278–290
  subjects from worldwide publications, 278–279

Z
Zero-energy school buildings, 135–136

Printed in the United States
By Bookmasters